CONTEMPORARY MATHEMATICS

827

Algebraic Structures in Knot Theory

AMS Western Sectional Meeting
Algebraic Structures in Knot Theory
May 6–7, 2023
California State University, Fresno, California

Carmen Caprau
J. Scott Carter
Neslihan Gügümcü
Sam Nelson
Editors

AMS
AMERICAN MATHEMATICAL SOCIETY

2020 *Mathematics Subject Classification.* Primary 57K12, 57K14, 57K31, 20F36, 20F38, 57K10, 57K45, 57K35, 57K99, 20C08.

Library of Congress Cataloging-in-Publication Data

Contemporary Mathematics ISSN: 0271-4132 (print); ISSN: 1098-3627 (online)
DOI: https://doi.org/10.1090/conm/827

∞ The paper used in this book is acid-free and falls within the guidelines
established to ensure permanence and durability.
Visit the AMS home page at `https://www.ams.org/`

10 9 8 7 6 5 4 3 2 1 30 29 28 27 26 25

Contents

Preface vii

On equivariant Khovanov homology
ROSTISLAV AKHMECHET and MELISSA ZHANG 1

Computing Khovanov homology via categorified Jones-Wenzl projectors
CHRISTINE RUEY SHAN LEE 31

A survey on skein modules via braids
IOANNIS DIAMANTIS 49

Topological symmetries of the Heawood family
BLAKE MELLOR and ROBIN WILSON 87

On the cohomology of two stranded braid varieties
TONIE SCROGGIN 121

Symmetry of three component links
KATE KEARNEY 157

The mixed Hilden braid group and the plat equivalence in handlebodies
PAOLO CAVICCHIOLI and SOFIA LAMBROPOULOU 163

Meridional rank, welded knots, and bridge trisections
JASON JOSEPH and PUTTIPONG PONGTANAPAISAN 189

On an invariant for colored classical and singular links
AUDREY BAUMHECKEL, CARMEN CAPRAU, and CONOR RIGHETTI 207

A study of Gram determinants in knot theory
DIONNE IBARRA and GABRIEL MONTOYA-VEGA 221

Preface

Knot theory in the early 21st century is a vast and deep subject area, encompassing not just the classical question of when two diagrams represent the same knotted circle in $\mathbb{R}^3$, but a wide range of generalization and questions in low-dimensional topology. Modern knot theory includes a variety of sub-disciplines generalizing the type of objects being knotted, the spaces in which they are knotted, the combinatorial rules imposed on the equivalence of diagrams, and the interpretations of these structures in terms of objects both physical and topological.

Knot invariants can be understood as tools we can use to probe the structure of knots and their generalizations, each with varying degrees of strength, effectiveness at separating knot types, and computability, with each carrying its own slice of information about the knotted objects under study.

Many of the most effective knot invariants arise from algebraic structures. Some of these structures are well-known from other areas of mathematics, such as groups, vector spaces, and homology theories; others arise more specifically from the combinatorics of knot theory, such as quandles, skein modules, and braid groups.

In this volume we collect some recent results on algebraic structures in knot theory. Several papers focus on Khovanov homology, a categorfication of the Jones polynomial. Another theme is understanding and exploiting symmetry in various forms. Several papers examine algebraic structures related to braids, and results on combinatorial structures such as colored links and welded knots are described. Finally, algebraic structures known as Gram determinants are examined in the context of knot theory.

Carmen Caprau
J. Scott Carter
Neslihan Gügümcü
Sam Nelson

Contemporary Mathematics
Volume **827**, 2025
https://doi.org/10.1090/conm/827/16530

On equivariant Khovanov homology

Rostislav Akhmechet and Melissa Zhang

ABSTRACT. We provide an exposition of equivariant Khovanov homology, with a focus on the $U(1) \times U(1)$-equivariant theory. We construct a reduced version, survey recent applications, and conclude by posing some questions.

1. Introduction

Khovanov's celebrated categorification of the Jones polynomial [**30**] has seen many generalizations since its introduction at the turn of the millennium. One of the most natural generalizations comes from perturbing the underlying TQFT by modifying the Frobenius algebra; see Section 2.2 for an exposition. Frobenius algebras are in bijection with *(1+1)-dimensional topological quantum field theories (TQFTs)*, i.e. functors that assign vector spaces or modules to collections of circles in the plane, and linear maps or module homomorphisms to smooth cobordisms between those collections of circles; see, for instance [**1**].

The most basic version of Khovanov homology uses the Frobenius algebra $\mathbb{F}[X]/(X^2)$ where $\mathbb{F} = \mathbb{C}$ or $\mathbb{Q}$. However, in applications, such as those that use Rasmussen's famous s-invariant [**44, 48**], we are forced to consider a perturbation such as Lee homology [**39**], which uses the Frobenius algebra $\mathbb{F}[X]/(X^2 - 1)$; the *Lee-Rasmussen spectral sequence* relates Khovanov homology built via $\mathbb{F}[X]/(X^2)$ to the relatively simple Lee homology. Rasmussen's s-invariant is sometimes interpreted using a Frobenius algebra which gives Lee's homology a "tower" structure, i.e. with an action of a variable [**35**, Proposition 8]. For example, one may work with Bar-Natan's perturbation $\mathbb{F}[H, X]/(X^2 - HX)$ (any field $\mathbb{F}$) or the Lee analogue $\mathbb{F}[T, X]/(X^2 - T)$ (any field $\mathbb{F}$ with characteristic not equal to 2); in these cases, the data of the Lee-Rasmussen spectral sequence is equivalent to the notion of H- or T-torsion, and this has been leveraged in many applications ([**6, 7, 18, 28**], to name a few).

In [**35**], Khovanov collected and studied the diagrammatics of various rank 2 Frobenius algebras that give rise to link homology. Khovanov homology built via the Frobenius algebra $A_E = \mathbb{Z}[E_1, E_2, X]/(X^2 - E_1 X + E_2)$ with ground ring $R_E = \mathbb{Z}[E_1, E_2]$ is universal in an appropriate sense [**35**, Proposition 5]. This theory is referred to as $U(2)$*-equivariant* Khovanov homology because R_E and A_E are the $U(2)$-equivariant cohomology of a point and $\mathbb{CP}^1$, respectively; it was denoted $\mathcal{F}_5$ in [**35**], and often in the literature the defining relation is written as $X^2 - HX - T$. This

2020 *Mathematics Subject Classification.* Primary 57K18.

theory specializes to usual Khovanov homology as well as to the graded or filtered Lee and Bar-Natan homologies. Equivariant link homology appears naturally in Robert-Wagner foam evaluation [**49**], and its generic deformations – analogues of Lee's theory – are crucial to establishing strict functoriality of exterior-colored Khovanov-Rozansky $\mathfrak{gl}_N$ homology [**24**].

In [**36**], Khovanov and Robert enlarge the ground ring to $R_\alpha = \mathbb{Z}[\alpha_1, \alpha_2]$, where E_1 and E_2 are elementary symmetric polynomials in α_1, α_2, so that the defining quadratic relation $X^2 - E_1 X + E_2$ in A_E splits as $(X - \alpha_1)(X - \alpha_2)$. Since R_α and $A_\alpha := R_\alpha[X]/((X - \alpha_1)(X - \alpha_2))$ are the $U(1) \times U(1)$-equivariant cohomology of a point and $\mathbb{CP}^1$, respectively, the resulting theory is called $U(1) \times U(1)$-*equivariant* Khovanov homology. From the point of view of link homology, moving from $U(2)$-equivariant to $U(1) \times U(1)$-equivariant Khovanov homology is a flat base change, and the homology in the $U(2)$ theory determines the homology in the $U(1) \times U(1)$ theory. However, in this article, we aim to illustrate that the latter provides additional flexibility that can both elucidate the structure of link homology and lead to new constructions in link homology. Namely, we discuss the unification of various Lee-type theories (Lee homology and Bar-Natan homology, for instance); functoriality of Khovanov homology; and the extension of equivariant homology to links in a thickened annulus. We hope for more applications to appear in the future.

We begin with an exposition of various flavors of equivariant Khovanov homology, including an overview of Bar-Natan's cobordism category, an introduction to Frobenius algebras, and a discussion of reduced homologies. Then, we survey recent developments and applications of equivariant link homology theories. We conclude with some open questions for future exploration.

Acknowledgments

We thank Mikhail Khovanov for many illuminating discussions on the subject. Thanks to Claudius Zibrowius for bringing to our attention the argument presented here as Theorem 3.16, which appears in [**28, 38, 43**]. Thanks to Lukas Lewark for enlightening conversations at Casa Matemática Oaxaca (BIRS-CMO). We thank Jen Hom, Gage Martin, and Taketo Sano for useful discussions. The first author was partially supported by NSF Grant DMS-2105467, NSF RTG Grant DMS-1839968, and the Jefferson Scholars Foundation during the writing of portions of this article. Part of this article is based on a preprint written by the authors while the second author was supported by the National Science Foundation under Grant No. DMS-1928930, while she was in residence at the Simons Laufer Mathematical Sciences Institute (previously known as MSRI) in Berkeley, California, during the Fall 2022 semester. Finally, we thank the referee, whose careful reading and thoughtful comments have helped us clarify several parts of the exposition.

2. Khovanov homology

This section gives an overview of the construction of Khovanov homology from a formal complex over the Bar-Natan cobordism category, introduced in [**11**]. Given a planar link diagram D of a link L, Bar-Natan defined a chain complex $[[D]]$ over an appropriate category and showed the chain homotopy type of $[[D]]$ is an invariant of L. Applying a suitable $(1+1)$-dimensional topological quantum field theory

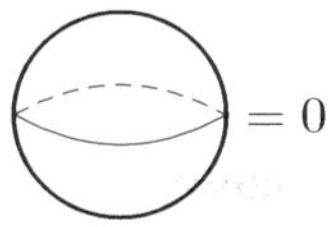

(A) Sphere

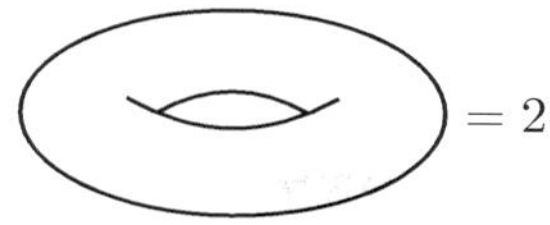

(B) Torus

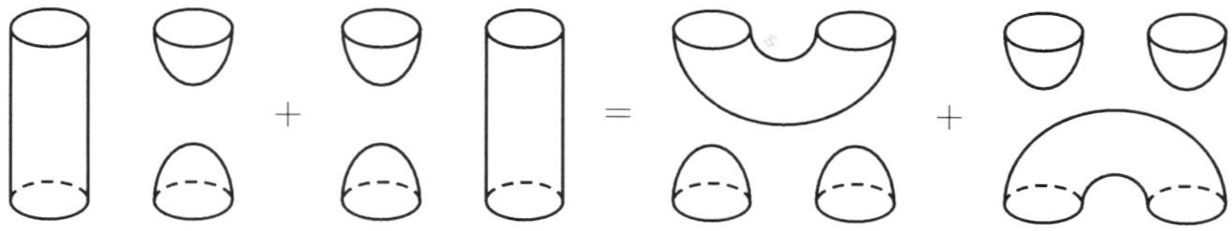

(C) Four tubes

FIGURE 1. Relations in $\mathcal{BN}$.

(TQFT) to $[[D]]$ yields homological invariants of oriented links. In the following sections, we describe these steps in detail.

2.1. The Bar-Natan category and formal chain complex. We start with a brief overview of the Bar-Natan category $\mathcal{BN}$ and the formal chain complex associated to oriented link diagrams, as introduced in [**11**]. Let $I := [0,1]$ denote the unit interval. Objects of $\mathcal{BN}$ are formal direct sums of formally graded collections of disjoint simple closed curves in the plane $\mathbb{R}^2$. Morphisms are matrices whose entries are formal $\mathbb{Z}$-linear combinations of cobordisms properly embedded in $\mathbb{R}^2 \times I$, modulo isotopy relative to the boundary, and subject to the local relations shown in Figure 1. The degree of a cobordism Σ is defined to be

$$\deg(\Sigma) = -\chi(\Sigma). \tag{2.1}$$

Note that deg is additive under composition of cobordisms and that the Bar-Natan relations are homogeneous, so deg descends to a grading on $\mathcal{BN}$.

REMARK 2.1. Throughout this article, we use the term "grading" to describe elements of a graded ring or graded module. We reserve the term "degree" for the change in grading associated to maps.

Let D be a diagram for an oriented link (L, o), where o is an orientation on the link $L \subset \mathbb{R}^3$. We describe the formal complex $[[D]]$ from [**11**]. First, form the so-called *cube of resolutions* as follows. Label the crossings of D by $1, \ldots, n$. There are two ways to smooth a crossing, called the *0-smoothing* and *1-smoothing*, shown in (2.2). For each sequence $u = (u_1, \ldots, u_n) \in \{0,1\}^n$, perform the u_i-smoothing at the i-th crossing. This yields a collection of disjoint simple closed curves in the plane, which we denote by D_u. Elements of $\{0,1\}^n$ are vertices of an n-dimensional cube, and we decorate the vertex u by the planar diagram D_u.

(2.2)

Let $u = (u_1, \ldots, u_n)$ and $v = (v_1, \ldots, v_n)$ be vertices which differ only in the i-th entry, in which $u_i = 0$ and $v_i = 1$. In other words, there is an edge from u to v in the n-dimensional cube; in this case, we say that v is an *immediate successor* of u. The two smoothings D_u and D_v are identical outside of a small disk around the i-th crossing. There is a cobordism from D_u to D_v consisting of the obvious saddle

(1-handle attachment) near the i-th crossing, and the identity (product cobordism) outside of the small disk. Denote this cobordism by $d_{u,v}$, and decorate each edge of the n-dimensional cube by the corresponding saddle cobordism. We now have a commutative cube-shaped diagram in the category $\mathcal{BN}$. There is a way to associate a sign $s_{u,v} \in \{\pm 1\}$ to each edge in the cube so that multiplying the edge map $d_{u,v}$ by $s_{u,v}$ results in an anti-commutative cube (see [**11**, Section 2.7] for details); any two such choices of edge sign assignments result in homotopy equivalent chain complexes.

For a vertex $u = (u_1, \ldots, u_n) \in \{0,1\}^n$, set $|u| = \sum_i u_i$. The chain complex $[[D]]$ is defined by setting

$$[[D]]^i = \bigoplus_{|u|=i+n_-} D_u\{n_- - n_+ - i\} \tag{2.3}$$

where n_- and n_+ are the number of negative and positive crossings in D, respectively, and the brackets $\{\cdot\}$ denote the upward formal grading shift in $\mathcal{BN}$. The differential d on the summand $D_u\{n_- - n_+ - i\}$ is the sum of all edge maps $s_{u,v}d_{u,v}$ where v is an immediate successor of u in the cube. Anti-commutativity of the cube implies that $d^2 = 0$.

THEOREM 2.2. ([**11**, Theorem 1]) *If diagrams D and D' are related by a Reidemeister move, then $[[D]]$ and $[[D']]$ are chain homotopy equivalent.*

Applying a functor from $\mathcal{BN}$ into an abelian category yields link homology, and Theorem 2.2 ensures that the homotopy class of the resulting chain complex is a link invariant. Constructing such a functor amounts to finding a $(1+1)$-dimensional TQFT which factors through the Bar-Natan relations in Figure 1; this is part of the content of Section 2.2.

REMARK 2.3. It is often convenient to consider a generalized Bar-Natan category. Let $\mathbb{D}^2 \subset \mathbb{R}^2$ denote the 2-disk. A *planar tangle* is a smoothly and properly embedded compact 1-manifold $T \subset \mathbb{D}^2$. Given a finite subset $B \subset \partial\mathbb{D}^2$, a planar tangle T is said to have *boundary* B if $\partial T = B$. If T_0, T_1 are planar tangles with boundary B, a *tangle cobordism* from T_0 to T_1 consists of a smoothly and properly embedded compact surface $\Sigma \subset \mathbb{D}^2 \times I$ such that $\partial\Sigma \cap (\mathbb{D}^2 \times \{i\}) = T_i$ for $i = 0, 1$ and $\partial\Sigma \cap (\partial\mathbb{D}^2 \times I) = B \times I$.

Let $\mathcal{BN}(B)$ denote the category whose objects are formal direct sums of formally graded planar tangles with boundary B, and morphisms are matrices whose entries are formal direct sums of tangle cobordisms modulo the local relations in Figure 1. The degree of a tangle cobordism Σ is defined to be

$$\deg(\Sigma) = -\chi(\Sigma) + \frac{|B|}{2}.$$

We will mostly be interested in the setting when B consists of 2 points (Section 3.6).

2.2. Frobenius algebras and Khovanov homology. We will now pass from Bar-Natan's formal complex to Khovanov homology.

DEFINITION 2.4. A *Frobenius algebra* consists of

- a commutative ring R;
- a commutative unital R-algebra A with unit and multiplication denoted $\iota : R \to A$ and $m : A \otimes_R A \to A$, respectively;

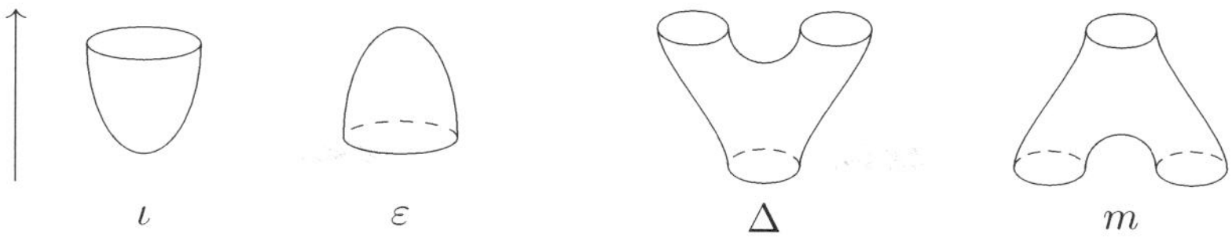

FIGURE 2. We read cobordisms from bottom to top. From left to right, this figure depicts the four elementary cobordisms: birth, death, split, and merge. These correspond to the unit, counit, comultiplication, and multiplication maps describing a $(1+1)$-dimensional TQFT, respectively.

- a map $\Delta : A \to A \otimes_R A$, called comultiplication, which is cocommutative, coassociative, and a map of (A, A)-bimodules; and
- a map $\varepsilon : A \to R$ called the counit (or the trace), satisfying $(\varepsilon \otimes \mathrm{id}) \circ \Delta = \mathrm{id}$.

We will sometimes refer to a Frobenius algebra as only the pair (R, A), or even simply by only A. Every Frobenius algebra gives rise to a $(1+1)$-dimensional TQFT, i.e. a monoidal functor from the category of (oriented) closed 1-manifolds and (oriented) cobordisms between them to the category of R-modules. Every collection of n circles is assigned $A^{\otimes n}$; every cobordism can be written as a composition of *elementary cobordisms*, and each elementary cobordism is assigned one of the structure maps (see Figure 2).

REMARK 2.5. Strictly speaking, in order to define a $(1+1)$-dimensional TQFT using a Frobenius algebra, all circles and cobordisms should be oriented. In our situation the embedding of circles and cobordisms in the plane and in the thickened plane, respectively, can be used to naturally orient all circles and cobordisms; see [**31**, Section 2.1].

Not every TQFT factors through the Bar-Natan relations; we now record several key Frobenius algebras which do yield link homology when applied to the Bar-Natan complex. All the examples considered in this article will be either graded (or filtered), with the structure maps being homogeneous (or filtered).

$$\begin{aligned} R_0 &= \mathbb{Z}, & A_0 &= \mathbb{Z}[X]/(X^2), \\ R_E &= \mathbb{Z}[E_1, E_2], & A_E &= \mathbb{Z}[E_1, E_2]/(X^2 - E_1 X + E_2), \\ R_\alpha &= \mathbb{Z}[\alpha_1, \alpha_2], & A_\alpha &= \mathbb{Z}[\alpha_1, \alpha_2]/((X - \alpha_1)(X - \alpha_2)). \end{aligned} \tag{2.4}$$

The pair (R_E, A_E) is sometimes referred to as $U(2)$-equivariant since R_E and A_E are the $U(2)$-equivariant cohomology of a point and $\mathbb{CP}^1$, respectively. For an analogous reason, (R_α, A_α) is referred to as $U(1) \times U(1)$-equivariant. In [**35**] the pair (R_E, A_E) is denoted $\mathcal{F}_5$ (up to renaming the variables), and the link homology obtained from it is sometimes called *universal* Khovanov homology. We note that (R_α, A_α) is an extension of (R_E, A_E), with the inclusion given by identifying E_1, E_2 with the elementary symmetric polynomials in α_1 and α_2:

$$E_1 \mapsto \alpha_1 + \alpha_2, \qquad E_2 \mapsto \alpha_1 \alpha_2. \tag{2.5}$$

For a symbol $\star \in \{0, E, \alpha\}$, $A_\star$ is a free $R_\star$-module with basis $\{1, X\}$. The trace $\varepsilon\colon A_\alpha \to R_\alpha$ is given by $\varepsilon(1) = 0$, $\varepsilon(X) = 1$. Comultiplication is

$$\begin{aligned}\Delta\colon A_\alpha &\to A_\alpha \otimes A_\alpha\\ 1 &\mapsto X \otimes 1 + 1 \otimes X - (\alpha_1 + \alpha_2) 1 \otimes 1\\ X &\mapsto X \otimes X - \alpha_1\alpha_2 1 \otimes 1.\end{aligned}$$

The Frobenius algebra (R_0, A_0) is also obtained from (R_E, A_E) and from (R_α, A_α) by specializing the equivariant parameters to zero: $E_1, E_2 \mapsto 0$ and $\alpha_1, \alpha_2 \mapsto 0$.

For $\star \in \{0, E, \alpha\}$, let $\mathcal{F}_\star$ denote the corresponding TQFT. It is straightforward to verify that $\mathcal{F}_\star$ factors through the Bar-Natan relations. The ground rings are all graded by setting R_0 to be concentrated in degree zero, $\deg(E_1) = 2$, $\deg(E_2) = 4$ in R_E, and $\deg(\alpha_1) = \deg(\alpha_2) = 2$ in R_α. Moreover, by defining the *quantum grading*

$$\mathrm{gr}_q(1) = -1 \qquad \mathrm{gr}_q(X) = 1,$$

we may view $\mathcal{F}_\star$ as a graded functor

$$\mathcal{F}_\star : \mathcal{BN} \to R_\star\text{-}\mathrm{gmod}$$

where $R_\star$-gmod is the category of $\mathbb{Z}$-graded $R_\star$-modules and homogeneous maps (of any degree) between them. A cobordism $S : Z_0 \to Z_1$ is assigned a map $\mathcal{F}_\star(S) : \mathcal{F}_\star(Z_0) \to \mathcal{F}_\star(Z_1)$ which is homogeneous of degree $\deg(S) = -\chi(S)$.

REMARK 2.6. In [**11**, Section 11.2], Bar-Natan also introduced a category where cobordisms are decorated with dots; in any of the TQFTs $\mathcal{F}_\star$ discussed above, a dot represents multiplication by X on the appropriate tensor factor. As originally defined, this category is not universal, in the sense that different Frobenius algebras require relations on the morphisms. Figure 3 shows relations if using (R_E, A_E). Relations for (R_0, A_0) and for (R_α, A_α) can be determined from the depicted ones by setting $E_1 = E_2 = 0$ and $E_1 = \alpha_1 + \alpha_2, E_2 = \alpha_1\alpha_2$, respectively. Note that in any of these cases, the relations in Figure 3 imply the undotted relations in Figure 1. Additional convenient diagrammatics are developed for dotted cobordisms in the equivariant settings in [**36**].

For an oriented link diagram D, we let

$$CKh_\star(D)$$

denote the chain complex of free graded $R_\star$-modules obtained by applying $\mathcal{F}_\star$ to the cube of resolutions. We denote by $Kh_\star(D)$ the homology of this complex.

As noted in [**36**], the extension $R_E \to R_\alpha$ in (2.5) is flat, since R_α is a free R_E-module (of rank 2, in fact). Then the natural map

$$Kh_E(D) \otimes_{R_E} R_\alpha \to Kh_\alpha(D)$$

is an isomorphism. In other words, $U(1)\times U(1)$-equivariant homology is determined by $U(2)$-equivariant homology. However, as we shall see, working in the larger $U(1) \times U(1)$-theory nevertheless adds flexibility that is useful for understanding and developing link homology.

We review a further extension $(R_\mathcal{D}, A_\mathcal{D})$ of the Frobenius pair (R_α, A_α), introduced in [**36**]. Let

$$\mathcal{D} = (\alpha_1 - \alpha_2)^2 \tag{2.6}$$

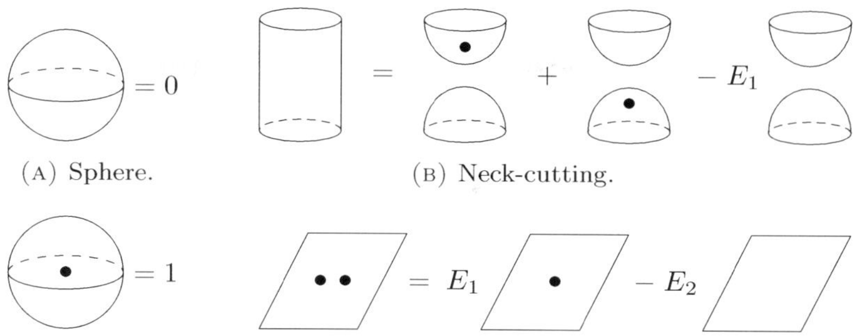

(A) Sphere. (B) Neck-cutting.

(C) Dotted sphere. (D) Two dots.

FIGURE 3. Relations in the equivariant dotted Bar-Natan category.

denote the discriminant of the quadratic polynomial $(X-\alpha_1)(X-\alpha_2) \in R_\alpha[X]$. Let

$$R_\mathcal{D} = R_\alpha[\mathcal{D}^{-1}]$$

denote the ring obtained by inverting $\mathcal{D}$ (equivalently, one may invert $\alpha_1 - \alpha_2$), and let

$$A_\mathcal{D} = A_\alpha \otimes_{R_\alpha} R_\mathcal{D}$$

be the extension of A_α to an $R_\mathcal{D}$-algebra. Let $\mathcal{F}_\mathcal{D}$ denote the corresponding TQFT. For a link $L \subset \mathbb{R}^3$ with diagram D, let

$$CKh_\mathcal{D}(D) := \mathcal{F}_\mathcal{D}([[D]])$$

be the resulting chain complex. It is an invariant of L up to chain homotopy equivalence, and we will denote its homology by $Kh_\mathcal{D}(L)$. Note that, since $R_\mathcal{D}$ contains invertible elements of both positive and negative gradings, the grading on $Kh_\mathcal{D}(L)$ contains little information.

The elements

$$e_1 = \frac{X-\alpha_1}{\alpha_2-\alpha_1}, \quad e_2 = -\frac{X-\alpha_2}{\alpha_2-\alpha_1} \in A_\mathcal{D}. \tag{2.7}$$

form a basis for $A_\mathcal{D}$ and satisfy

$$e_1 + e_2 = 1, \quad e_1^2 = e_1, \quad e_2^2 = e_2, \quad e_1 e_2 = 0,$$

so that the algebra $A_\mathcal{D}$ decomposes as a product, $A_\mathcal{D} = R_\mathcal{D} e_1 \times R_\mathcal{D} e_2$. With respect to the basis $\{e_1, e_2\}$, comultiplication in $A_\mathcal{D}$ is simply given by

$$\begin{aligned} \Delta(e_1) &= (\alpha_2 - \alpha_1) e_1 \otimes e_1, \\ \Delta(e_2) &= (\alpha_1 - \alpha_2) e_2 \otimes e_2. \end{aligned} \tag{2.8}$$

As noted in [**36**, Section 1.2], the TQFT $\mathcal{F}_\mathcal{D}$ is essentially the Lee deformation [**39**]; see also Section 2.3. By [**39**, Theorem 4.2], the Lee homology of an ℓ-component link is free (over $\mathbb{Q}$) of rank 2^ℓ. An alternate proof can be found in the final remark in [**58**]; see also [**12**]. The following proposition is readily proven using the same arguments.

PROPOSITION 2.7. *For a link $L \subset \mathbb{R}^3$ with ℓ components, the homology $Kh_\mathcal{D}(L)$ is a free $R_\mathcal{D}$-module of rank 2^ℓ.*

The above proposition unifies various flavors of link homology at generic specializations of the equivariant parameters. Precisely, suppose $\Bbbk$ is any commutative ring and $\varphi : R_\alpha \to \Bbbk$ is a ring homomorphism. As long as $\varphi(\alpha_1) - \varphi(\alpha_2)$ is a unit in $\Bbbk$, the homology obtained by extending scalars in CKh_α to $\Bbbk$ via φ will satisfy the same property as in Proposition 2.7; we refer to such a theory as *localized*. The filtered Lee and Bar-Natan theories, from which Rasmussen-type invariants are extracted, are obtained in this way. This is further discussed in Section 2.3.

Next, let

$$T(D) = \{\eta \in Kh_\alpha(D) \mid r \cdot \eta = 0 \text{ for some nonzero } r \in R_\alpha\}.$$

denote the R_α-torsion submodule of $Kh_\alpha(D)$. Proposition 2.7 implies the following; see also [**51**, Corollary 2.10].

COROLLARY 2.8. *The (non-localized) homology $Kh_\alpha(D)$ contains only $(\alpha_1 - \alpha_2)$-torsion. That is, every element of $T(D)$ is annihilated by a power of $\alpha_1 - \alpha_2$.*

PROOF. Exactness of localization at $\alpha_1 - \alpha_2$ yields an inclusion $T(D) \otimes_{R_\alpha} R_\mathcal{D} \hookrightarrow Kh_\alpha(D) \otimes_{R_\alpha} R_\mathcal{D} \cong Kh_\mathcal{D}(D)$. But $Kh_\mathcal{D}(D)$ is free (over the integral domain $R_\mathcal{D}$) by Proposition 2.7, while every element in $T(D) \otimes_{R_\alpha} R_\mathcal{D}$ is torsion. Therefore $T(D) \otimes_{R_\alpha} R_\mathcal{D} = 0$, which completes the proof. □

2.3. Canonical Lee classes. Lee's perturbation of Khovanov homology [**39**] was originally used to prove that the Khovanov homology of alternating knots is particularly simple, and is in fact determined by the Jones polynomial and the signature of the knot. Lee homology has since played an important role in applications of Khovanov homology: Rasmussen's invariant s [**48**] was defined by measuring the filtration level of Lee's *canonical classes*, which form a basis for Lee homology. A key feature of s is that it gives a lower bound on the slice genus of a knot. Within the framework we have discussed so far, Lee homology is the filtered version of Khovanov homology obtained from (R_α, A_α) by setting $\alpha_1 = \pm 1$, $\alpha_2 = \mp 1$ (or, equivalently, from (R_E, A_E) by setting $E_1 = 0, E_2 = -1$). Since $\mathcal{D} = 2$ for this Frobenius algebra, Lee's homology theory over a ring where 2 is invertible (such as $\mathbb{Q}$) is a localized theory, meaning the conclusion of Proposition 2.7 applies. Below, we discuss the generalization of Lee's canonical classes in localized $U(1) \times U(1)$-equivariant homology.

Let D be a link diagram. A *checkerboard coloring* for D is a choice of how to bicolor the regions of $\mathbb{R}^2 \setminus D$ such that the four regions abutting each crossing are alternately colored (traveling around the crossing). For us, we choose the checkerboard coloring that leaves the infinite region of $\mathbb{R}^2 \setminus D$ unshaded. Given an orientation o of D, the following algorithm is used to label the circles in a canonical Lee class, based on both the orientation and the checkerboard coloring:

ALGORITHM 2.9 (cf. [**52**] Algorithm 2.8). *Given an oriented link diagram (D, o), form the oriented resolution D_o. Let C denote a circle in D_o; the orientation on C determines a normal vector at each point on the circle. If these normal vectors all point toward a shaded region in the chosen checkerboard coloring of $\mathbb{R}^2 \setminus D$, label the circle C with $\boldsymbol{a}$; otherwise, label the circle with $\boldsymbol{b}$.*

We now define the canonical Lee classes.

DEFINITION 2.10 (cf. [**12, 39, 51, 58**]). The *canonical Lee class* $\mathfrak{s}_o \in Kh_\mathcal{D}(D)$ for (D, o) is the homology class of the cycle s_o in the summand $CKh_\mathcal{D}(D_o)$ of

$CKh_{\mathcal{D}}(D)$ where the circles in D_o are labeled according to Algorithm 2.9 by the rule $\boldsymbol{a} \mapsto e_1$, $\boldsymbol{b} \mapsto e_2$.

If the link represented by D has ℓ components, then the 2^ℓ homology classes

$$\{\mathfrak{s}_o \mid o \text{ is an orientation of } D\}$$

form a basis for $Kh_{\mathcal{D}}(D)$, and so we sometimes refer to these classes also as *Lee generators*.

For the opposite orientation $\bar{o}$ to o, the oriented resolution is the same: $D_o = D_{\bar{o}}$. Observe that the cycle $s_{\bar{o}}$ is obtained from the cycle s_o by swapping all the labels: $e_1 \leftrightarrow e_2$. We will return to the relationship between the homology classes $\mathfrak{s}_o$ and $\mathfrak{s}_{\bar{o}}$ later in Section 4.

There is an algebra involution $\mathbf{j} : A_\alpha \to A_\alpha$, defined in [**36**, Section 1.1], given by

$$\mathbf{j}(1) = 1, \quad \mathbf{j}(X) = \alpha_1 + \alpha_2 - X, \tag{2.9}$$

which interchanges $X - \alpha_1$ and $-(X - \alpha_2)$. Although $\mathbf{j}$ is an algebra involution, it does not preserve the Frobenius algebra structure, since $\Delta(\mathbf{j}(a)) = -(\mathbf{j}\otimes\mathbf{j})\Delta(a)$ and $\varepsilon(\mathbf{j}(a)) = -\varepsilon(a)$ for all $a \in A$. Consequently, applying $\mathbf{j}^{\otimes k(u)}$ at each vertex u of the cube of resolutions (where $k(u)$ is the number of planar circles in the resolution D_u) does not yield a chain map. However, one can make sign adjustments at each vertex to yield a chain map: for any vertex u, pick a path γ in the cube of resolutions from the all-0 resolution to u, and let $\#\Delta(u)$ denote the number of comultiplications (split maps) along the path γ. Note that $\#\Delta(u)$ does not depend on the choice of the path γ. Define the map at the vertex u to be $(-1)^{\#\Delta(u)}\mathbf{j}^{\otimes k(u)}$. See also the discussion of Schütz in [**55**, Section 2]. This results in an honest involution on the chain complex $CKh(D)$. We denote this involution by J.

Furthermore, $\mathbf{j}$ induces an involution on $A_{\mathcal{D}}$, which swaps e_1 and e_2, and by introducing signs as above we obtain an involution of the localized complex $CKh_{\mathcal{D}}(D)$ which agrees with the chain map induced by the chain map $J : CKh_\alpha(D) \to CKh_\alpha(D)$. By an abuse of notation, we will use $\mathbf{j}$ to denote the involutions on both A_α and $A_{\mathcal{D}}$, and use J to denote the involution on both chain complexes $CKh_\alpha(D)$ and $CKh_{\mathcal{D}}(D)$.

3. Based links and a reduced homology theory

In this section, we discuss the case of links with a basepoint. The data of the basepoint allows us to define a reduced version of $U(1)\times U(1)$-equivariant Khovanov homology. We discuss the self-duality of A_α and of A_E, mirror images, and $(1,1)$-tangles. A similar construction also appears in independent work of Sano-Sato [**53**].

3.1. The basepoint action. Let L be a *based* link, i.e. marked with a basepoint $p \in L$. Taking a generic projection we obtain a link diagram D of L with a basepoint p away from the crossings. Consider the ordinary Khovanov complex $CKh_0(D)$, built via the Frobenius algebra $A_0 = \mathbb{Z}[X]/(X^2)$. There is a *basepoint action* of A_0 on $CKh_0(D)$; see [**32**, Section 3]. The action is given my merging a small circle near p with the marked component at each resolution D_u. In particular, there is an endomorphism ξ_p of $CKh_0(D)$ given by the action of $X \in A_0$. The *reduced Khovanov complex* $\widetilde{CKh}_0(D,p)$ is defined to be

$$\widetilde{CKh}_0(D,p) = \mathrm{im}(\xi_p)\{-1\},$$

the (gr_q-shifted) image of the action of X via ξ_p. Equivalently, in a resolution D_u, let $Z_{u,p} \subset D_u$ denote the circle containing p. Then $\widetilde{CKh}_0(D,p)$ is freely generated by the Khovanov generators in each D_u which mark $Z_{u,p}$ by X, together with an overall downward gr_q-shift of 1.

REMARK 3.1. A careful study of the basepoint action on $U(1)\times U(1)$-equivariant homology is present in [**27**], where Gujral uses the $U(1)\times U(1)$ theory (referred to as α-homology therein) to give ribbon distance bounds, following work of Alishahi [**6**], Alishahi-Dowlin [**7**], and Sarkar [**54**].

3.2. The reduced equivariant chain complex. Let (L,o) be an oriented link, D a diagram for (L,o), and p a marked point on D. The equivariant chain complex $CKh_\alpha(D)$ has an action by A_α similar to the basepoint action above. In particular, let f_p denote the endomorphism of $CKh_\alpha(D)$ given by the action of $X-\alpha_1 \in A_\alpha$ on the tensor factor of $\mathcal{F}(D_u)$ corresponding to the marked circle $Z_{u,p}$, and identity on all other tensor factors. In other words, the action f_p merges a small circle labeled $X-\alpha_1$ into the marked circle.

DEFINITION 3.2. The *reduced complex* $\widetilde{CKh}_\alpha(D,p)$ of a based link diagram (D,p) is defined to be $\widetilde{CKh}_\alpha(D,p) = \mathrm{im}(f_p)\{1\}$. Its homology is denoted $\widetilde{Kh}_\alpha(D,p)$.

Both $\{1, X-\alpha_1\}$ and $\{1, X-\alpha_2\}$ form a basis for A_α. We have

(3.1)
$$(X-\alpha_1)(X-\alpha_1) = (\alpha_2-\alpha_1)(X-\alpha_1) \qquad \Delta(X-\alpha_1) = (X-\alpha_1)\otimes(X-\alpha_1)$$
$$(X-\alpha_2)(X-\alpha_2) = (\alpha_1-\alpha_2)(X-\alpha_2) \qquad \Delta(X-\alpha_2) = (X-\alpha_2)\otimes(X-\alpha_2).$$

It follows that the R_α-submodule of $CKh_\alpha(D,p)$ that is freely generated by the Khovanov generators for which the marked circle $Z_{u,p}$ in each resolution is labeled by $X-\alpha_1$ forms a subcomplex of $CKh_\alpha(D)$, which (after a gr_q-grading shift) is equal to $\widetilde{CKh}_\alpha(D,p)$.

Note that symmetry between α_1 and α_2 is broken by declaring the action of f_p to be multiplication by $X-\alpha_1$. Let f'_p be the endomorphism of $CKh_\alpha(D)$ given by multiplication by $X-\alpha_2$ near the basepoint, and let $\widetilde{CKh}'_\alpha(D,p) = \mathrm{im}(f'_p)\{-1\}$. Observe that J induces an isomorphism $\widetilde{CKh}_\alpha(D,p) \xrightarrow{\sim} \widetilde{CKh}'_\alpha(D,p)$.

PROPOSITION 3.3. *Let (L,p) be a based link with a based diagram (D,p). The chain homotopy type of $\widetilde{Kh}_\alpha(D,p)$ is an invariant of (L,p).*

PROOF. The argument is given in [**32**, Section 3] (see also [**10**, Lemma 2.2]), but we repeat it here. Any two based diagrams of (L,p) are related by a sequence of Reidemeister moves away from the marked points and a move which slides the basepoint past a crossing. This second basepoint sliding move can be realized as a sequence of Reidemeister moves away from basepoints, followed by the global move shown in Figure 4a, and finally followed by another sequence of Reidemeister moves away from basepoints. See Figure 4b for a schematic.

It is straightforward to see that the chain homotopy equivalences for these Reidemeister moves preserve the respective reduced complexes. For the global

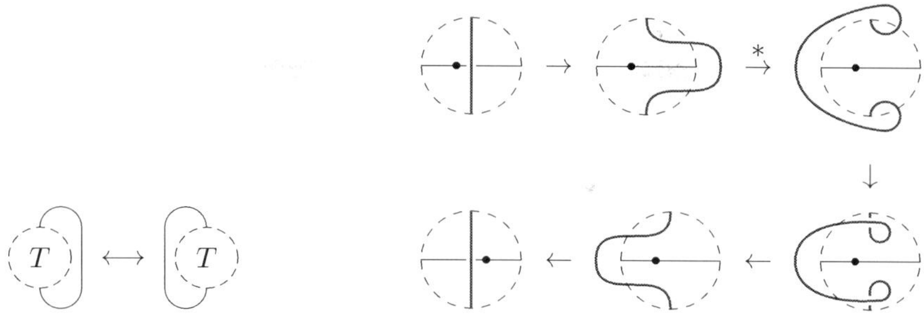

(A) The global move, where T is an oriented, based $(1,1)$-tangle.

(B) To move the basepoint past a crossing, when the basepoint is located on the under-strand, pull the the blue strand out of the page and above the rest of the diagram. The arrow marked $*$ is the global move. For the over-strand case, perform the same move, except with the blue strand pulled into the page, below the rest of the diagram.

FIGURE 4. A depiction of the chain homotopy equivalence assigned to moving the basepoint past a crossing.

move, use the obvious identification of circles in each resolution before and after the move; the associated isomorphism of chain complexes clearly preserves the respective reduced complexes. □

3.3. Based knots and connected cobordisms. We now restrict ourselves to the case of based knots and discuss maps on their reduced homology coming from connected cobordisms between them. We will view cobordisms in $\mathbb{R}^3 \times [0,1]$ as movies where each frame is a slice $\mathbb{R}^3 \times \{t\}$ at time t.

Let (K_i, p_i) for $i = 0,1$ be based knots in $\mathbb{R}^3$ with respective diagrams D_i. Consider an oriented, connected cobordism $F \subset \mathbb{R}^3 \times [0,1]$ from K_0 to K_1 (i.e. $K_i = F \cap (\mathbb{R}^3 \times \{i\})$ for $i = 0,1$). Choose an embedded arc Γ from $p_0 \in K_0$ to $p_1 \in K_1$; this is possible because F is connected. We call the pair (F,Γ) a *decorated cobordism* from K_0 to K_1. We claim that it is always possible to isotope F, fixing its boundary, such that the following conditions are met. Let $\pi : \mathbb{R}^3 \times [0,1] \twoheadrightarrow \mathbb{R}^2$ be the diagram projection map onto the first two coordinates of $\mathbb{R}^3$.

(DC-1) The path Γ is transverse to every slice $\mathbb{R}^3 \times \{t\}$ for $t \in [0,1]$. Let Γ_t denote the unique intersection point $\Gamma \cap (\mathbb{R}^3 \times \{t\})$.

(DC-2) The height function h_F on F given by projection to the time coordinate is Morse, the critical values (times when h_F has a critical point) are isolated, and there is only one critical point per critical value. Furthermore, we require that Γ be disjoint from the critical points of h_F.

(DC-3) For all but finitely many times t, the projection $\pi(F \cap (\mathbb{R}^3 \times \{t\}))$ is a link diagram (i.e. an immersion with isolated double points only).

(DC-4) For all but finitely many times t, the projection of the basepoint $\pi(\Gamma_t)$ is disjoint from the set of singular points of the projection under π of $F \cap (\mathbb{R}^3 \times \{t\})$ to the plane $\mathbb{R}^2$.

We briefly explain how to arrange this. Given any decorated connected cobordism (F,Γ) from K_0 to K_1, we can perform an ambient isotopy that transforms Γ into the "straight" path from p_0 to p_1 (that is, the path is given by the point p_0 crossed with the interval). This is possible because Γ is codimension 3 in $\mathbb{R}^3 \times [0,1]$. This achieves (DC-1). By a standard argument, (DC-2) and (DC-3) can be achieved by perturbations. Since transversality is an open condition, our new Γ may retain its transversality to the time slices, maintaining (DC-1). To arrange (DC-4), observe that there are three types of "diagrammatic" special points lying on the surface F:

(1) the Morse singular points, corresponding to the isolated critical points of (DC-2),
(2) the Reidemeister singular points, corresponding to the isolated critical points of (DC-3),
(3) and the diagram crossings, which comprise a 1-dimensional submanifold of F traced out by the double points under the projection π.

By perturbing, we may arrange that the 0-dimensional special points of (1) and (2) are disjoint from the crossing arcs in (3), as well as from each other. The union of all these special points is a 1-submanifold SP; since Γ is also 1-dimensional, we finish by slightly perturbing Γ in F so that it is transverse SP and the set of points where Γ intersects SP (points where the diagrammatic basepoint is passing through a crossing) is finite and disjoint from 0-dimensional components of SP, achieving (DC-4).

From now on, we will assume all decorated cobordisms (F, Γ) satisfy the above conditions; for clarity, we refer to them as *nice decorated cobordisms.* We say $t \in (0,1)$ is a *critical time* if at time t we see a special point from (DC-2), (DC-3), or an intersection of Γ with SP.

We may thus diagrammatically present (F, Γ) as a finite sequence of *elementary cobordisms* between based link diagrams. As noted in (DC-1), the intersection between Γ and each time slice $\mathbb{R}^3 \times \{t\}$ consists of a single point, which is viewed as the basepoint for the link (or singular link) $F \cap (\mathbb{R}^3 \times \{t\})$ at that time slice.

These consist of Morse moves away from the basepoint, Reidemeister moves disjoint from the basepoint, and moving the basepoint past a crossing. See also [**9**, Section 2.3].

REMARK 3.4. A nice cobordism may a priori contain a segment representing a Reidemeister move involving the basepoint. However, we may always replace this segment of the movie with an equivalent one that first moves the basepoint outside of the region affected by the Reidemeister isotopy and then later moves it back in.

DEFINITION 3.5. Given based link diagrams (D, p) and (D', p') which are related by one of the above elementary moves, we assign a chain map $\phi\colon \widetilde{CKh}_\alpha(D,p) \to \widetilde{CKh}_\alpha(D',p')$ as follows. For Morse moves and Reidemeister moves (both disjoint from the marked point), note that the usual chain map $CKh_\alpha(D) \to CKh_\alpha(D')$ preserves the respective reduced complexes, and we define ϕ to be the restriction of this usual map. For a move which slides the basepoint past a crossing, the chain map is realized as a sequence of Reidemeister moves away from basepoints, followed by the global move in Figure 4a, and finally followed by another sequence of Reidemeister moves (see Remark 3.6) away from basepoints. See Figure 4b for a schematic.

REMARK 3.6. There are various choices made, for example for the path Γ and Reidemeister moves before and after the global move in Definition 3.5. Any choice is sufficient for our purposes; we make no claims about the functoriality of these chain maps and only require their existence.

Note that the maps of reduced complexes assigned to Reidemeister moves and basepoint-moving are chain homotopy equivalences.

3.4. Localization and canonical Lee generators. Consider the reduced, localized chain complex

$$\widetilde{CKh}_{\mathcal{D}}(D,p) := \widetilde{CKh}_{\alpha}(D,p) \otimes_{R_\alpha} R_{\mathcal{D}}.$$

Recall the elements $e_1, e_2 \in A_{\mathcal{D}}$ from (2.7). The following definition is a reduced version of Lee's canonical generators.

Definition 3.7. Let (D,p) be a based link diagram. Given an orientation o of D, consider the checkerboard coloring of the oriented resolution D_o where the region to the left of p is shaded. For a circle C in D_o, label the circle e_1 if the region to the left of C is shaded, and otherwise label it e_2. Denote this element $\mathfrak{r}_o(D,p) \in \widetilde{CKh}_{\mathcal{D}}(D,p)$; note that the marked circle in D_o is labeled e_1.

The element $\mathfrak{r}_o(D,p)$ is equal to either $\mathfrak{s}_o$ or $\mathfrak{s}_{\overline{o}}$ (see Algorithm 2.9). Note also that $\mathfrak{r}_o(D,p) = \mathfrak{r}_{o'}(D,p)$, if and only if $o = o'$ or $\overline{o} = o'$, so if D has ℓ components then there are $2^{\ell-1}$ such generators. The following is immediate.

Lemma 3.8. *For a based link diagram (D,p) with ℓ components, $\widetilde{Kh}_{\mathcal{D}}(D,p)$ is a free $R_{\mathcal{D}}$-module of rank $2^{\ell-1}$, with basis given by $\{[\mathfrak{r}_o(D,p)]\}$ over all orientations o of D. Moreover, if (D,p') is obtained from (D,p) by moving the basepoint p past a single crossing, then $\mathbf{j}(\mathfrak{r}_o(D,p)) = \mathfrak{r}_o(D,p')$.*

Recall that the involution J swaps e_1 and e_2 in the localized theory (up to a sign). Below, we assume all states (pure tensors) are written with the component containing the basepoint first. For an element $y \in \widetilde{CKh}_{\mathcal{D}}(D,p)$, we write $e_1 \cdot y$ for the basepoint action that multiplies the based circle by e_1. Similarly, $e_2 \cdot$ indicates the action where the tensor factor corresponding to the based component is multiplied by e_2.

Remark 3.9. The basepoint action $e_1 \cdot$ does not quite agree with the action induced on the reduced complex by f_p, which is multiplication by only the numerator $X - \alpha_1$ of e_1. In other words, the two actions differ by the unit $\alpha_2 - \alpha_1$ in the localized theory.

The non-reduced localized complex splits as two copies of the reduced localized complex. Namely, we have a distinguished isomorphism

$$CKh_{\mathcal{D}}(D) \cong \widetilde{CKh}_{\mathcal{D}}(D,p) \oplus \widetilde{CKh}'_{\mathcal{D}}(D,p),$$

where the first summand is generated by generators where the marked component is labeled e_1, and the second summand $\widetilde{CKh}'_{\mathcal{D}}(D,p)$ is generated by those where the marked component is labeled e_2. It follows from the formulas in Equation (3.1) that both are indeed subcomplexes of $CKh_{\mathcal{D}}(D)$, and moreover the involution J gives an isomorphism $\widetilde{CKh}_{\mathcal{D}}(D,p) \cong \widetilde{CKh}'_{\mathcal{D}}(D,p)$. This splitting descends to a distinguished isomorphism

$$Kh_{\mathcal{D}}(K) \cong \widetilde{Kh}_{\mathcal{D}}(K,p) \oplus \widetilde{Kh}_{\mathcal{D}}(K,p) = R_{\mathcal{D}}\langle[\mathfrak{r}_o]\rangle \oplus R_{\mathcal{D}}\langle[\mathbf{j}(\mathfrak{r}_o)]\rangle.$$

for a knot K.

LEMMA 3.10. *Let (D,p) be a based link diagram, and suppose (D,p') is obtained from (D,p) by moving the basepoint past a single crossing. Let $\phi\colon \widetilde{Kh}_{\mathcal{D}}(D,p) \to \widetilde{Kh}_{\mathcal{D}}(D,p')$ denote the isomorphism from Definition 3.5. Then for any orientation o of D, we have $\phi([\mathfrak{r}_o(D,p)]) = \pm[\mathfrak{r}_o(D,p')]$.*

PROOF. Write ϕ as a composition

$$\widetilde{Kh}_{\mathcal{D}}(D,p) \xrightarrow{\phi_0} \widetilde{Kh}_{\mathcal{D}}(D_1,p_1) \xrightarrow{\phi_1} \widetilde{Kh}_{\mathcal{D}}(D_2,p_2) \xrightarrow{\phi_2} \widetilde{Kh}_{\mathcal{D}}(D,p'),$$

where ϕ_0 is induced by the first sequence of Reidemeister moves, ϕ_1 is the global move of Figure 4a, and ϕ_2 is another sequence of Reidemeister moves. Let o_1 and o_2 denote the orientations on D_1 and D_2. Without loss of generality, assume $\mathfrak{r}_o(D,p) = \mathfrak{s}_o(D)$. By [**51**, Proposition 2.13], $\phi_0([\mathfrak{s}_o(D)]) = \pm(\alpha_1-\alpha_2)^l[\mathfrak{s}_{o_1}(D_1)]$ for some l. We have $\phi_1([\mathfrak{s}_{o_1}(D_1)]) = [\mathfrak{s}_{\overline{o}_2}(D_2)]$ because the checkerboard coloring reverses. Applying [**51**, Proposition 2.13] again we obtain $\phi_2([\mathfrak{s}_{\overline{o}_2}(D_2)]) = \pm(\alpha_1-\alpha_2)^{l'}[\mathfrak{s}_{\overline{o}}(D)]$ for some $l' \in \mathbb{Z}$. By the last statement in Lemma 3.8, $\mathfrak{r}_o(D,p') = \mathbf{j}(\mathfrak{r}_o(D,p)) = \mathfrak{s}_{\overline{o}}$, so we have $\phi([\mathfrak{r}_o(D,p)]) = \pm(\alpha_1-\alpha_2)^{l+l'}[\mathfrak{r}_o(D,p')]$. Finally, $l+l'=0$ for quantum grading reasons. □

In fact, any decorated connected cobordism between knots gives an isomorphism on the localized reduced homology.

PROPOSITION 3.11. *Let (F,Γ) be a decorated connected cobordism from (K,o) to (K',o'). Pick based diagrams (D,p) and (D',p') for (K,p) and (K',p') respectively. There exists a chain map $\phi\colon \widetilde{CKh}_{\alpha}(D,p) \to \widetilde{CKh}_{\alpha}(D',p')$ of degree $-\chi(F)$ such that, after inverting $\mathcal{D}$, the induced map on homology is an isomorphism.*

PROOF. The argument is similar to [**48**, Proposition 4.1] and [**51**, Proposition 3.17]. Represent (F,Γ) as a movie of based link diagrams

$$(D,p) = (D_0,p_0), (D_1,p_1), \ldots, (D_k,p_k) = (D',p'),$$

where $(D_i,p_i) = F \cap (\mathbb{R}^3 \times \{i/k\})$, and (D_{i+1},p_{i+1}) is obtained from (D_i,p_i) by one of elementary moves as described in Definition 3.5. For each elementary move $(D_i,p_i) \to (D_{i+1},p_{i+1})$ consider the map

$$\phi_i : \widetilde{Kh}_{\alpha}(D_i,p_i) \to \widetilde{Kh}_{\alpha}(D_{i+1},p_{i+1})$$

as defined in Definition 3.5. Let $\phi = \phi_k \circ \cdots \circ \phi_0$ denote their composition and let $\psi_i = \phi_{i-1} \circ \cdots \circ \phi_0$ denote their composition until the i-th slice. We will show that $\phi([\mathfrak{r}_o(D,p)]) = \pm(\alpha_2-\alpha_1)^k \mathfrak{r}_{o'}(D',p')$ for some $k \in \mathbb{Z}$.

For $1 \le i \le k$, let $F_i = F \cap (\mathbb{R}^2 \times [0,i/k])$. An orientation of F_i is *permissible* if it induces the orientation o on D. Let O_i denote the set of permissible orientations on F_i. An orientation of F_i induces an orientation on D_i, and moreover since F is connected, distinct orientations on F_i induce distinct orientations on D_i. For $\alpha \in O_i$, we also write α to denote the induced orientation on D_i. The statement follows from three claims:

CLAIM 1. For any $\alpha \in O_i$,

$$\phi_i([\mathfrak{r}_\alpha(D_i,p_i)]) = \sum_{\beta \in \Omega_{i+1}} \pm(\alpha_2-\alpha_1)^{l_\beta}[\mathfrak{r}_\beta(D_{i+1},p_{i+1})],$$

where $\Omega_{i+1} \subset O_{i+1}$ is the set of orientations β that extend α. This set may be empty.

CLAIM 2. For any $y \in \widetilde{Kh}_\alpha(D_i, p_i)$ of the form $y = \sum_{\alpha \in \Omega_i} \pm(\alpha_2 - \alpha_1)^{l_\alpha}[\mathfrak{r}_\alpha(D_i, p_i)]$ where $\Omega_i \subset O_i$, we have that $\phi_i(y) = \sum_{\beta \in \Omega_{i+1}} \pm(\alpha_2 - \alpha_1)^{l_\beta}[\mathfrak{r}_\beta(D_{i+1}, p_{i+1})]$ for some $\Omega_{i+1} \subset O_{i+1}$.

CLAIM 3. For each $0 \leq i \leq k-1$, $\psi_i([\mathfrak{r}_o(D, p)]) = \sum_{\alpha \in \Omega_i} \pm(\alpha_2 - \alpha_1)^{l_\alpha}[\mathfrak{r}_\alpha(D_i, p_i)]$, where $\Omega_i \subset O_i$ is nonempty.

Before verifying the above claims, note that Claim 3 applied to $\psi_{k-1} = \phi$ implies the statement of the lemma, since there is only one permissible orientation on $F_1 = F$ (namely, the given orientation on F which induces o' on D').

To show Claim 1, suppose first that $(D_i, p_i) \to (D_{i+1}, p_{i+1})$ is a Morse or Reidemeister move away from the basepoint. Note that $\mathfrak{r}_\alpha(D_i, p_i)$ is equal to one of $\mathfrak{s}_\alpha(D_i)$ or $\mathfrak{s}_{\overline{\alpha}}(D_i)$. Without loss of generality, suppose $\mathfrak{r}_\alpha(D_i, p_i) = \mathfrak{s}_\alpha(D_i)$. By [**51**, Proposition 3.17],

$$\phi_i([\mathfrak{s}_\alpha(D_i)]) = \sum_{\beta \in \Omega_{i+1}} \pm(\alpha_2 - \alpha_1)^{l_\beta}[\mathfrak{s}_\beta(D_{i+1})].$$

Moreover, since in the checkerboard coloring used to define the reduced Lee generators, the unbounded region is colored the same in D_i and D_{i+1}, we also have $\mathfrak{s}_\beta(D_{i+1}) = \mathfrak{r}_\beta(D_{i+1}, p_{i+1})$, which verifies Claim 1 in this case. If the move $(D_i, p_i) \to (D_{i+1}, p_{i+1})$ slides the basepoint past a crossing, then Claim 1 follows from Lemma 3.10.

Claim 2 follows from Claim 1, since distinct permissible orientations on F_i extend to distinct permissible orientations on F_{i+1}.

Claim 3 follows from repeatedly applying Claim 2 and the fact that a generator $\mathfrak{r}_\alpha(D_i, p_i)$ is sent to zero by ϕ_i only if two oriented strands are incompatibly merged, which can never appear in the oriented cobordism F. □

3.5. Self-duality. Let S be a commutative ring and $\mathcal{G} = (B, \iota, \varepsilon, m, \Delta)$ a Frobenius algebra over S with underlying S-module B, trace $\varepsilon\colon B \to S$, unit $\iota\colon S \to B$, multiplication $m\colon B \otimes B \to B$, and comultiplication $\Delta\colon B \to B \otimes B$. Recall from [**35**] the *dual* Frobenius algebra $\mathcal{G}^* = (B^*, \varepsilon^*, \iota^*, \Delta^*, m^*)$ with underlying S-module the dual $B^* = \mathrm{Hom}_S(B, S)$, and structure maps given by dualizing the structure maps of $\mathcal{G}$. Following [**35**], $\mathcal{G}$ is said to be *self dual* if $\mathcal{G} \cong \mathcal{G}^*$ as Frobenius algebras; that is, there exists an S-linear isomorphism $\phi\colon B \to B^*$ which identifies $(\iota, \varepsilon, m, \Delta)$ with $(\varepsilon^*, \iota^*, \Delta^*, m^*)$.

Given a graded R-module M, let $M^* = \mathrm{Hom}_R(M, R)$ denote its dual and $M^*_{\mathrm{gr}} \subset M^*$ the graded dual,

$$M^*_{\mathrm{gr}} := \bigoplus_{i \in \mathbb{Z}} \mathrm{Hom}_i(M, R),$$

where $\mathrm{Hom}_i(M, R)$ is the abelian group of all R-linear maps $M \to R$ of degree i. Note that M^*_{gr} is a graded R-module and that $M^*_{\mathrm{gr}} = M^*$ when M is finitely generated. Moreover, if M is free with homogeneous basis $\{m_1, \ldots, m_k\}$, then the dual elements $\{m_1^*, \ldots, m_k^*\}$ form a homogeneous basis for $M^*_{\mathrm{gr}} = M^*$, with $\deg(m_i^*) = -\deg(m_i)$. Chain groups in Khovanov homology are finitely generated and free over the ground ring, so we will omit the subscript gr in the graded dual.

The following independently appeared in [**53**, Lemma 3.32].

PROPOSITION 3.12. *The Frobenius algebra $\mathcal{F}_\alpha$ corresponding to (R_α, A_α) is self dual via the degree zero isomorphism $\gamma\colon A_\alpha \to A_\alpha^*$: $\gamma(1) = X^*$, $\gamma(X) = 1^* + (\alpha_1+\alpha_2)X^*$.*

PROOF. We will define an auxiliary Frobenius algebra $\mathcal{F}'_\alpha$ and show that $\mathcal{F}_\alpha \cong \mathcal{F}'_\alpha \cong \mathcal{F}^*_\alpha$. Let $1^*, X^* \in A^*$ denote the dual vectors to 1 and X, so that $1^*(y) = \delta_{1,y}$, $X^*(y) = \delta_{X,y}$, for $y \in \{1, X\}$. Structure maps for $\mathcal{F}^*_\alpha$ are given below.

$$\begin{array}{ll} \iota^*\colon A_\alpha^* \to R_\alpha & m^*\colon A_\alpha^* \to A_\alpha^* \otimes A_\alpha^* \\ \quad 1^* \mapsto 1 & \quad 1^* \mapsto 1^*\otimes 1^* - \alpha_1\alpha_2 X^*\otimes X^* \\ \quad X^* \mapsto 0 & \quad X^* \mapsto 1^*\otimes X^* + X^*\otimes 1^* + (\alpha_1+\alpha_2)X^*\otimes X^* \end{array} \tag{3.2}$$

$$\begin{array}{ll} \varepsilon^*\colon R_\alpha \to A_\alpha^* & \Delta^*\colon A_\alpha^*\otimes A_\alpha^* \to A_\alpha^* \\ \quad 1 \mapsto X^* & \quad 1^*\otimes 1^* \mapsto -(\alpha_1+\alpha_2)1^* - \alpha_1\alpha_2 X^* \\ & \quad X^*\otimes 1^*, 1^*\otimes X^* \mapsto 1^* \\ & \quad X^*\otimes X^* \mapsto X^* \end{array} \tag{3.3}$$

We introduce an intermediate Frobenius algebra $\mathcal{F}'_\alpha$ over R_α:

$$A'_\alpha = R_\alpha[X]/((X+\alpha_1)(X+\alpha_2)),$$

with unit the natural inclusion $R_\alpha \hookrightarrow A'_\alpha$, trace $\varepsilon'(1) = 0, \varepsilon'(X) = 1$, and comultiplication

$$\Delta'(1) = (X+\alpha_1)\otimes 1 + 1\otimes(X+\alpha_2), \quad \Delta'(X) = X\otimes X - \alpha_1\alpha_2 1\otimes 1.$$

Note that $\mathcal{F}'_\alpha$ is just $\mathcal{F}_\alpha$ with α_1 and α_2 replaced with $-\alpha_1$ and $-\alpha_2$. From the above structure formulas, we see that[1] $\mathcal{F}^*_\alpha \cong \mathcal{F}'_\alpha$ via $1^* \mapsto X$, $X^* \mapsto 1$. Next, the R_α-algebra isomorphism $\varphi\colon A_\alpha \to A'_\alpha$ given by $\varphi(1) = 1$, $\varphi(X) = X + \alpha_1 + \alpha_2$ yields an isomorphism $\mathcal{F}'_\alpha \cong \mathcal{F}_\alpha$. The composition $A_\alpha \to A'_\alpha \to A^*_\alpha$ is precisely γ. □

REMARK 3.13. The proof of Proposition 3.12 also shows that (R_E, A_E) is self-dual via $1 \mapsto X^*$, $X \mapsto 1^* + E_1X^*$.

We are now ready to define the mirror complex. Given a link diagram D, denote by $\overline{D}$ its mirror image diagram; if D is based at p, let $\overline{p}$ be the image of p in $\overline{D}$. Let $CKh_\alpha(D)^!$ be the chain complex given by $\left(CKh_\alpha^{-i}(D)\right)^*$ in homological grading i and whose differential is obtained by dualizing the differential in $CKh_\alpha(D)$. Using well-known arguments [**30**, Section 7.3], Proposition 3.12 implies the following. We recall the argument since it will be used in Proposition 3.15.

COROLLARY 3.14. *There is an isomorphism $CKh_\alpha(D) \cong CKh_\alpha(\overline{D})^!$ of chain complexes of graded R-modules.*

PROOF. The isomorphism is defined on each vertex in the cube of resolutions. Let n denote the number of crossings of D, n_+ of which are positive and n_- of which are negative. Let $\overline{n}_+ = n_-$, $\overline{n}_- = n_+$ denote the number of positive and negative crossings in $\overline{D}$. For $u \in \{0,1\}^n$, let $\overline{u} \in \{0,1\}^n$ denote the n-tuple obtained by

[1] This was observed in [**35**, Example 2].

swapping all 0's to 1's and all 1's to 0's. We have $D_u = \overline{D}_{\overline{u}}$. The smoothing D_u contributes to the homological grading $|u| - n_-$ part of $CKh_\alpha(D)$, while $\overline{D}_{\overline{u}}$ contributes to the homological grading $|\overline{u}| - \overline{n}_- = -(|u| - n_-)$ part of $CKh_\alpha(\overline{D})$. Letting k_u denote the number of circles in D_u, the isomorphism $CKh_\alpha(D) \cong CKh_\alpha(\overline{D})^!$ is obtained by applying the isomorphism

$$\gamma_u \colon A_\alpha^{\otimes k_u} \to (A_\alpha^*)^{\otimes k_u}$$

induced by $A_\alpha \cong A_\alpha^*$ at each smoothing. Proposition 3.12 implies that the γ_u assemble into a chain map. Finally, one can check that γ_u is grading preserving after accounting for quantum grading shifts. □

PROPOSITION 3.15. *There is an isomorphism* $\widetilde{CKh}_\alpha(D,p) \cong \widetilde{CKh}_\alpha(\overline{D},\overline{p})^!$ *of chain complexes of graded R-modules.*

PROOF. Fix a resolution D_u of D, and order its circles $Z_1, \dots, Z_{k_u}$, with Z_1 the marked circle. Then D_u contributes the direct summand $R\alpha(X - \alpha_1) \otimes A_\alpha^{\otimes k_u - 1}$ to $\widetilde{CKh}_\alpha(D,p)$, while $\overline{D}_{\overline{u}}$ contributes the direct summand $(R_\alpha(X-\alpha_1))^* \otimes (A_\alpha^*)^{\otimes k_u - 1}$ to $\widetilde{CKh}_\alpha(\overline{D},\overline{p})^!$. Consider the R-linear isomorphism

$$\widetilde{\gamma}_u \colon R_\alpha(X-\alpha_1) \otimes A_\alpha^{\otimes k_u - 1} \to (R_\alpha(X-\alpha_1))^* \otimes (A_\alpha^*)^{\otimes k_u - 1}$$

given by $(X-\alpha_1) \otimes y_1 \otimes \cdots \otimes y_{k_u - 1} \mapsto (X-\alpha_1)^* \otimes \gamma(y_1) \otimes \cdots \otimes \gamma(y_{k_u-1})$. It is straightforward to verify that $\widetilde{\gamma}_u$ is grading preserving after the appropriate grading shifts are applied. We let $R_{\alpha_1} = (X - \alpha_1) \cdot A_\alpha$ be the free R-module with basis $X - \alpha_1$.

We check that $\widetilde{\gamma}_u$ commutes with each saddle map. When the saddle involves unmarked circles, commutativity follows from Proposition 3.12. It suffices then to check commutativity of the two diagrams in (3.4); the left (resp. right) diagram corresponds to a circle merging with (resp. splitting from) a marked circle.

$$\begin{array}{ccc} R_{\alpha_1} \otimes A_\alpha & \longrightarrow & R_{\alpha_1} \\ \downarrow{\scriptstyle \widetilde{\gamma}_u} & & \downarrow{\scriptstyle \widetilde{\gamma}_u} \\ (R_{\alpha_1})^* \otimes A_\alpha^* & \longrightarrow & (R_{\alpha_1})^* \end{array} \qquad \begin{array}{ccc} R_{\alpha_1} & \longrightarrow & R_{\alpha_1} \otimes A_\alpha \\ \downarrow{\scriptstyle \widetilde{\gamma}_u} & & \downarrow{\scriptstyle \widetilde{\gamma}_u} \\ (R_{\alpha_1})^* & \longrightarrow & (R_{\alpha_1})^* \otimes A_\alpha^* \end{array} \tag{3.4}$$

The bottom horizontal maps are obtained by dualizing comultiplication and multiplication in the reduced complex when at least one of the circles involved is marked:

$$\begin{array}{cc} (R_{\alpha_1})^* \otimes A^* \to (R_{\alpha_1})^* & (R_{\alpha_1})^* \to (R_{\alpha_1})^* \otimes A^* \\ (X-\alpha_1)^* \otimes 1^* \mapsto -\alpha_1(X-\alpha_1)^* & (X-\alpha_1)^* \mapsto (X-\alpha_1)^* \otimes (1^* + \alpha_2 X^*). \\ (X-\alpha_1)^* \otimes X^* \mapsto (X-\alpha_1)^* & \end{array}$$

Commutativity of the diagrams in (3.4) follows easily. □

3.6. Based links and (1,1)-tangles. We summarize some results regarding the classification of TQFTs which pass through Bar-Natan's cobordism category; these appear in works of Naot [**43**, Section 4], Kotelskiy-Watson-Zibrowius [**38**], and Iltgen-Lewark-Marino [**28**].

To start, consider a standard closed 3-ball $\mathbb{D}^3 = \{x \in \mathbb{R}^3 \mid |x| \leq 1\}$. An *oriented* $(1,1)$*-tangle* is a compact oriented 1-manifold T properly embedded in $\mathbb{D}^3$ such that the boundary of T lies in $\{(-1,0,0),(1,0,0)\}$ and such that the orientation of the unique interval component of T points from $(-1,0,0)$ to $(1,0,0)$. Two $(1,1)$-tangles are isotopic if there is an orientation preserving diffeomorphism of $\mathbb{D}^3$ which fixes the boundary point-wise and sends one tangle to the other in a manner that preserves the orientation on each component. A generic projection of an oriented $(1,1)$-tangle yields an oriented tangle diagram in $\mathbb{D}^2$, which are considered up to Reidemeister moves and ambient isotopy of $\mathbb{D}^2$ which fixes $\partial\mathbb{D}^2$ point-wise and preserves the orientation on each component. With some care, there is bijection between isotopy classes of based oriented links (L,p) in S^3 and isotopy classes of oriented $(1,1)$-tangles, and likewise for based oriented link diagrams and oriented $(1,1)$-tangle diagrams; we refer the reader to [**28**, Section 2.1] for a detailed discussion.

Consider the Frobenius algebra $A_H = \mathbb{Z}[H,X]/(X^2+HX)$ over $R_H = \mathbb{Z}[H]$. As usual A_H is a free R_H-module with basis $\{1, X\}$. We set H to be in grading 2, 1 to be in grading -1, and X to be in grading 1. The Frobenius algebra (R_H, A_H) is obtained from (R_E, A_E) by setting $E_1 = -H$, $E_2 = 0$, and is a graded version of Bar-Natan's theory. Let $\mathcal{F}_H$ denote the corresponding TQFT. A fact, implicit in [**43**] and explored further in [**28**], is that $\mathcal{F}_H$ is universal for *knots*. We aim to summarize this fact.

First, recall the Bar-Natan category with boundary $\mathcal{BN}(B)$ from Remark 2.3. Set $\mathcal{BN}_1 := \mathcal{BN}(B_\pm)$ where $B_\pm = \{(-1,0),(1,0)\}$ and fix an arc $\gamma \subset \mathbb{D}^2$ connecting the two points in $B_\pm$, considered as an object in $\mathcal{BN}_1$. Consider also the graded ring $\mathbb{Z}[H]$ with H in grading 2. As shown in [**28**, Appendix A], there is an isomorphism of graded rings

$$\mathcal{BN}_1(\gamma,\gamma) \cong \mathbb{Z}[H]$$

where $\mathcal{BN}_1(\gamma,\gamma)$ is the endomorphism algebra of γ in $\mathcal{BN}_1$. Explicitly, it is shown that $\mathcal{BN}_1(\gamma,\gamma)$ is freely generated by by the "curtain of genus[2] one:"

$\mathbb{H} =$ 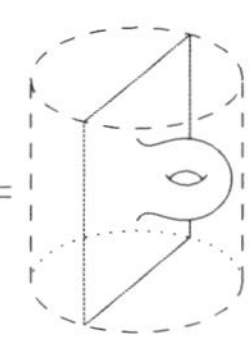

and the above isomorphism is then given by $\mathbb{H} \mapsto H$. Moreover, viewing $\mathcal{BN}_1(\gamma,\gamma)$ as a category with one object, the inclusion $\mathcal{BN}_1(\gamma,\gamma) \to \mathcal{BN}_1$ is an equivalence of categories [**28**, Proposition 2.8].

For simplicity, we will refer to a planar tangle with boundary $B_\pm$ as a planar $(1,1)$-tangle. Next, pick a rank 2 Frobenius algebra (R,A) which factors through the Bar-Natan relations in Figure 1, with corresponding TQFT $\mathcal{F}_A$. We view $\mathcal{F}_A$ as a functor $\mathcal{BN}_1 \to A\text{-mod}$ (if structure maps in A are appropriately graded or filtered, we may upgrade the target category correspondingly) as follows. For a planar tangle T with boundary $B_\pm$, write $T = T_0 \cup T_1 \cup \cdots \cup T_k$, where T_0 is the unique interval component of T. Set

$$\mathcal{F}_A(T) = \underline{A} \otimes_R A^{\otimes_R k}$$

where the underlined tensor factor of A corresponds to T_0, and multiplying with this underlined factor gives $\mathcal{F}_A(T)$ an A-module structure. Topologically, the action of A is obtained by merging a small circle near T_0 with T_0.

[2]We use H for "handle," as in [**43**], while [**28**] use G for "genus."

We may apply the above to (R_H, A_H) to obtain a functor $\mathcal{F}_H : \mathcal{BN}_1 \to A_H$-gmod (by a slight abuse of notation, we still denote this by $\mathcal{F}_H$). Following [**28**, Definition 2.14], consider the *reduced* functor $\widetilde{\mathcal{F}}_H : \mathcal{BN}_1 \to R_H$-gmod defined by

$$\widetilde{\mathcal{F}}_H(-) = (A_H/(X)) \otimes_{A_H} \mathcal{F}_H(-)\{1\}.$$

Note that $\widetilde{\mathcal{F}}_H$ takes values in R_H-mod rather than in A_H-mod. For a planar $(1,1)$-tangle T with $k+1$ components, $\widetilde{\mathcal{F}}_H(T)$ is a free graded R_H-module with basis $1 \otimes y_1 \otimes \cdots \otimes y_k$, where each $y_i \in \{1, X\}$. Due to the above grading shift, the tensor factor 1 corresponding to the interval component is in grading 0 rather than -1. If $T = T_0$ is a single arc, observe that $\widetilde{\mathcal{F}}_H(\mathbb{H})(1) = H \cdot 1$.

Now consider an oriented knot diagram D, pick a basepoint on K, and cut open along it to obtain an oriented $(1,1)$-tangle diagram D_T. Form the cube of resolutions of D_T over the category $\mathcal{BN}_1$, and consider the two complexes

$$CKh_A(D_T) \quad \text{and} \quad \widetilde{CKh}_H(D_T)$$

of A-modules and R_H-modules obtained by applying $\widetilde{\mathcal{F}}_H$ and $\mathcal{F}_A$ to the cube of resolutions, respectively. By viewing A as a $\mathbb{Z}[H]$-module where H acts via $\mathcal{F}_A(\mathbb{H})$, we form a chain complex of A-modules $\widetilde{CKh}_H(D_T) \otimes_{R_H} A\{-1\}$.

THEOREM 3.16 ([**28**, Theorem 2.17]). *There is an isomorphism of complexes*

$$CKh_A(D_T) \cong \widetilde{CKh}_H(D_T) \otimes_{R_H} A\{-1\}.$$

The above theorem says that $\widetilde{F}_H$ is universal for knots or, more generally, for based links.

4. Strict functoriality for Khovanov homology

After Khovanov introduced Kh_0 (refer to Section 2.2 for our notation), Khovanov [**34**], Jacobsson [**29**], and Bar-Natan [**11**] proved using various methods that the homology theory is *projectively functorial*, i.e. Khovanov homology is a TQFT that sends each cobordism between links to bigraded R-modules, *up to a sign*. By fixing signs for cobordisms, various authors were able to lift Khovanov homology to a strict functor [**13, 16, 19, 23, 24, 52, 57**].

Independently, Caprau [**19**] and Clark-Morrison-Walker [**23**] extended Khovanov's TQFT to achieve strict functoriality by making the following changes to Bar-Natan's construction of Khovanov homology. Smoothings are enhanced with *disorientations* or replaced by *bivalent webs*, which one can think of as planar circles with cusps that indicate a switch in orientation. Similarly, cobordisms between the new smoothings are enhanced with *seams*, which indicate the boundary between two faces whose orientations disagree. To keep track of these disorientations, both sets of authors work over a ground ring R containing a primitive fourth root of unity i, which enter into Bar-Natan-like relations between enhanced cobordisms, prescribing a sign for the cobordisms. For example, Caprau works over the ground ring $R = \mathbb{Z}[\frac{1}{2}, i, h, a]$, where h and a are indeterminates that can be specialized to recover Khovanov, Lee, or Bar-Natan homology. Once signs are chosen using the Lee-type homology (where $\mathcal{D}^{-1} \in R$, and whose canonical classes generate the homology), we may use these prescribed signs for cobordisms in other homology theories, including the original Kh.

After fixing the signs, all sets of authors construct explicit chain homotopy equivalences associated to Reidemeister moves, and check that their sign assignments are coherent under the Carter-Saito movie moves [**20**] which model isotopies of cobordisms between links.

4.1. Sano's sign adjustments and the $U(1) \times U(1)$-equivariant theory. More recently, Sano discovered yet another way to fix the sign indeterminacy in Khovanov homologies, not by modifying the objects and morphisms of Bar-Natan's TQFT, but rather by keeping track of the canonical Lee classes in localized $U(1) \times U(1)$-equivariant Khovanov homology.

We now sketch Sano's method for fixing the overall sign for cobordisms. Let (D, o) and (D', o') denote oriented diagrams representing (L, o) and (L', o'), respectively. Let Σ be an oriented cobordism from (L, o) to (L', o').

First, in Proposition 3.17 of [**51**], Sano extends [**48**, Proposition 4.1] to the equivariant theory Kh_α, proving that under any oriented cobordism $\Sigma : (D, o) \to (D', o')$ *without closed components*, the induced map $\phi_\Sigma : Kh_{\mathcal{D}}(D) \to Kh_{\mathcal{D}}(D')$ sends

$$\mathfrak{s}_o \mapsto \varepsilon c^j (\mathfrak{s}_{o'} + \text{ other terms}) \qquad \mathfrak{s}_{\bar{o}} \mapsto \varepsilon' (-c)^j (\mathfrak{s}_{\bar{o}'} + \text{ other terms}), \tag{4.1}$$

where c is a fixed square root of $\mathcal{D}$, $\varepsilon, \varepsilon' \in \{\pm 1\}$, and the integer j depends entirely on (1) the writhes of the diagrams D and D', (2) the number of Seifert circles in the oriented resolutions D_o and $D'_{o'}$, and (3) the Euler characteristic of Σ:

$$j = \frac{(\text{change in writhe}) - (\text{change in \# Seifert circles}) - \chi(\Sigma)}{2}.$$

REMARK 4.1. Let $-\Sigma$ denote the same surface as Σ but with the opposite orientation. Then $-\Sigma$ is an oriented cobordism from $-L$ to $-L'$, or more precisely, from $(L, \bar{o})$ to $(L', \bar{o}')$. If $(D, o), (D', o')$ are diagrams for $(L, o), (L', o')$ respectively, then $(D, \bar{o}), (D', \bar{o}')$ are diagrams for $(L, \bar{o}), (L', \bar{o}')$, respectively. The complexes for (D, o) and $(D, \bar{o})$ can be canonically identified; the same is true for (D', o') and $(D', \bar{o}')$. Under this identification, the roles of $\mathfrak{s}_o$ and $\mathfrak{s}_{\bar{o}}$ are reversed, as we are holding the choice of checkerboard coloring constant across this discussion.

Sano specializes to the theory with $(R, A) = (\mathbb{Z}, \mathbb{Z}[X]/(X^2 - X))$ to define his sign adjustment for the map $\phi_\Sigma : Kh_{\mathcal{D}}(D) \to Kh_{\mathcal{D}}(D')$. In this theory, $\mathcal{D} = 1$, which is indeed a unit in $\mathbb{Z}$, so this is a localized theory (we may fix c to be either ± 1).

The signs for cobordism maps are adjusted as follows. Given any movie decomposition $\Sigma = \Sigma_k \circ \cdots \circ \Sigma_1$ for Σ, where each Σ_i represents an elementary cobordism from an oriented link diagram (D_i, o_i) to an oriented link diagram (D_{i+1}, o_{i+1}), compute the induced map ϕ_{Σ_i} on $\mathfrak{s}_{o_i}$ to determine the sign ε_i in (4.1). Then, define $\varphi_{\Sigma_i} := \varepsilon_i \phi_{\Sigma_i}$ to be the sign-adjusted map for each elementary cobordism, and set $\varphi_\Sigma = \varphi_{\Sigma_k} \circ \cdots \circ \varphi_{\Sigma_1}$. This choice of signs is then also used for the more general $U(1) \times U(1)$ theory Kh_α, which in turn fixes the signs for all Khovanov homologies obtained from Kh_α. Again, let φ_Σ denote the sign-adjusted map in the equivariant theory.

REMARK 4.2. In light of the above paragraph, one may wonder why it is necessary to consider equivariant Kh_α in the first place, since the signs were determined by a specialized theory. The reason is two-fold. First, the main technical work in Sano's argument is his careful case study of how the canonical generators transform under Reidemeister and Morse moves, resulting in (4.1); with a more specialized theory, it is harder to see how the canonical generators transform. Second, the equivariant theory provides an umbrella theory pinning down the sign adjustments for all specializations.

To check strict functoriality, we need to check invariance of these adjusted cobordism maps under isotopy. In the diagrammatic category, such isotopies of surfaces in $\mathbb{R}^4$ are generated by Carter-Saito's movie moves [**20**].

Let Σ, Σ' denote two cobordisms (i.e. movies) from (D, o) to (D', o') that are related by one of Carter-Saito's moves. By inspection of these moves, we see that we may assume Σ, Σ' do not contain closed components. Projective functoriality ([**11**, Theorem 4]) ensures $\phi_\Sigma = \varepsilon\phi_{\Sigma'}$ for some $\varepsilon \in \{\pm 1\}$. However, since the signs for φ_Σ and $\varphi_{\Sigma'}$ are chosen so that the canonical class $\mathfrak{s}_o$ maps to $\mathfrak{s}_{o'}+$ other terms, we have $\varphi_\Sigma = \varphi_{\Sigma'}$, and therefore strict functoriality.

REMARK 4.3. Instead of writing the roots of the quadratic in the Frobenius algebra as α_1, α_2, Sano uses u, v, and denotes the canonical classes $\mathfrak{s}_o$ and $\mathfrak{s}_{\bar{o}}$ by $\alpha(D, o)$ and $\beta(D, o) = \alpha(D, \bar{o})$, respectively.

5. Equivariant annular link homology

Extending Khovanov homology to links in 3-manifolds besides 3-space is largely an open problem. The first such extension was introduced by Asaeda-Przytycki-Sikora [**8**]. A special case of their work gives homology of links in the thickened annulus, called *annular* Khovanov homology (or sometimes sutured annular Khovanov homology). In Section 5.1 we describe *equivariant* annular Khovanov homology via a filtration, following [**2**], which is analogous to the construction in [**50**]. We then give an overview of some of the ingredients used to define Khovanov-Rozansky [**37**] exterior-colored $\mathfrak{gl}_N$ homology, following the foam evaluation approach of Robert and Wagner [**49**]. This produces a theory that is naturally $U(N)$-equivariant. In Section 5.3 we review a generalization to equivariant annular $\mathfrak{gl}_N$ homology, introduced in [**5**] for $\mathfrak{sl}_2$ and $\mathfrak{sl}_3$ and in [**3**] for $\mathfrak{gl}_N$. In particular, throughout this section we stress that equivariant annular homology is not built in the $U(N)$ framework and instead requires passing to the larger $U(1)^N$ framework.

Let $\mathbb{A} = S^1 \times I$ denote the annulus, and embed $\mathbb{A}$ in the plane as $\mathbb{A} = \{x \in \mathbb{R}^2 \mid 1 \leq |x| \leq 2\}$. Identifying the interior of $\mathbb{A}$ with the punctured plane $\mathcal{P} := \mathbb{R}^2 \setminus \{(0,0)\}$, we represent the annulus in the plane by simply indicating a puncture using the symbol $\times$. By an *annular link* we mean an oriented link $L \subset \mathbb{A} \times I$. A generic projection of an annular link yields a diagram in $\mathbb{A}$, or, equivalently, in $\mathcal{P}$. There are two types of circles in $\mathbb{A}$, an *essential circle* which is nontrivial in $\pi_1(\mathbb{A})$, and a *trivial circle*, which is trivial in $\pi_1(\mathbb{A})$. For a collection of disjoint circles $Z \subset \mathbb{A}$, there is an ordering on the essential circles of Z, starting from innermost (closest to the puncture) to outermost.

5.1. A filtration approach. An equivariant version of annular $\mathfrak{sl}_2$ homology was introduced by the first author in [**4**] via a filtration analogous to the construction in [**26, 50**]. Subsequent work with Khovanov [**5**] generalized this construction to redefine the $\mathfrak{sl}_2$ theory and introduce equivariant annular $\mathfrak{sl}_3$ link homology using foam evaluation. A further extension to (exterior) colored $\mathfrak{gl}_N$ homology was introduced in [**3**]. A common feature of these constructions is they require working over the full polynomial ring rather than its subring of symmetric polynomials. In the

$\mathfrak{sl}_2$ setting, which will be discussed below, this means passing to the $U(1) \times U(1)$-equivariant theory.

We first demonstrate the incompatibility between annular Khovanov homology and the $U(2)$ theory; that is, the Frobenius algebra A_E. Let M be a free bigraded R_E-module with homogeneous basis m_- and m_+, viewed as the module assigned to a single essential circle. The first grading is called the *quantum* grading and the second grading is called the *annular* grading. The action of a homogeneous $x \in R_E$ on M is a map of bidegree $(\deg(x), 0)$, and we assume m_- and m_+ are in distinct annular gradings; both of these properties are expected from an annular link homology theory. Let $g : M \to M$ be the map assigned to the cobordism Σ in (5.1).

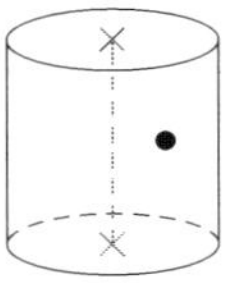

(5.1)

The defining relation in A_E implies that g should satisfy

$$g^2 = E_1 g - E_2 \operatorname{id}_M . \tag{5.2}$$

On the other hand, g is expected to have bidegree $(2, 0)$. For a homogeneous $m \in M$, this forces

$$g(m) = nE_1 m \tag{5.3}$$

for some $n \in \mathbb{Z}$, since the degree 2 part of R_E is spanned by E_1. However, applying (5.2) to m and using (5.3), we obtain $n^2E_1^2 = nE_1^2 - E_2$, which is impossible.

Passing to the larger Frobenius pair (R_α, A_α) introduces additional flexibility under which the above obstruction disappears, as follows. The degree 2 part of R_α is freely spanned by α_1 and α_2, so the map assigned to Σ sends a homogeneous m to $(n_1\alpha_1 + n_2\alpha_2)m$ for some $n_1, n_2 \in \mathbb{Z}$. Applying the defining relation (5.2) then yields

$$(n_1\alpha_1 + n_2\alpha_2)^2 = (\alpha_1 + \alpha_2)(n_1\alpha_1 + n_2\alpha_2) - \alpha_1\alpha_2,$$

which does have solutions, namely $(n_1, n_2) \in \{(1, 0), (0, 1)\}$.

We now define the equivariant annular TQFT from [**4**]. Consider the following elements of A_α:

$$v_- = 1 \qquad v_+ = X - \alpha_1 \qquad v'_- = 1 \qquad v'_+ = X - \alpha_2. \tag{5.4}$$

Note that $\{v_-, v_+\}$ and $\{v'_-, v'_+\}$ form a R_α-basis of A_α. Define the quantum grading gr_q and annular grading gr_a of these elements by setting

(5.5)

$$\operatorname{gr}_q(v_\pm) = \operatorname{gr}_q(v'_\pm) = 0 \qquad \operatorname{gr}_a(v_-) = \operatorname{gr}_a(v'_-) = -1 \qquad \operatorname{gr}_a(v_+) = \operatorname{gr}_a(v'_+) = +1.$$

Given a collection of circles $Z \subset \mathbb{A}$, e of which are essential and t of which are trivial, define the $(\operatorname{gr}_q, \operatorname{gr}_a)$ bigrading on $\mathcal{F}_\alpha(Z) = A_\alpha^{\otimes e} \otimes_{R_\alpha} A_\alpha^{\otimes t}$ by setting all tensor factors corresponding to trivial circles to be in annular grading zero, and by giving the tensor factor corresponding to the i-th essential circles, $1 \leq i \leq e$, the homogeneous basis $\{v_-, v_+\}$ if i is odd and $\{v'_-, v'_+\}$ if i is even.

REMARK 5.1. As elements of A_α, $v_\pm$ and $v'_\pm$ have quantum grading $\operatorname{gr}_q(v_\pm) = \operatorname{gr}_q(v'_\pm) = \pm 1$. However, in annular link homology it is more natural to consider the module assigned to an essential circle as concentrated in quantum grading

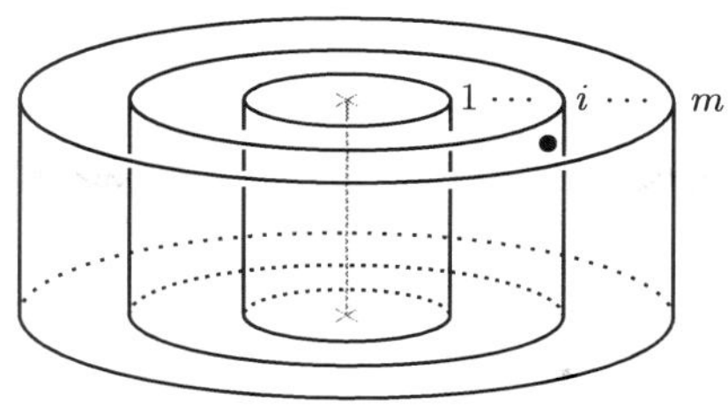

FIGURE 5. The product cobordism on $m > 0$ essential circles, with the i-th component dotted.

zero. This was considered in [**26**], where our gr_a is denoted j' (see [**26**, Remark 2]). This modified grading also appears naturally in Beliakova-Putyra-Wehrli's [**14**] identification of annular Khovanov homology with Hochschild homology of Stroppel and Chen-Khovanov bimodules over platform algebras [**17**, **22**, **56**], as well as in foam approaches to annular homology [**5**].

Given a cobordism $\Sigma \subset \mathbb{A} \times I$ from $Z_0 \subset \mathbb{A} \times \{0\}$ to $Z_1 \subset \mathbb{A} \times \{1\}$, we may view Σ as a cobordism in $\mathbb{R}^2 \times I$ and apply the TQFT $\mathcal{F}_\alpha$.

LEMMA 5.2 ([**4**, Corollary 4.3]). *With the above notation, $\mathcal{F}_\alpha(\Sigma) : \mathcal{F}_\alpha(Z_0) \to \mathcal{F}_\alpha(Z_1)$ splits as a sum $\mathcal{F}_\alpha(\Sigma) = \mathcal{F}_\alpha(\Sigma)_0 + \mathcal{F}_\alpha(\Sigma)_+$, where $\mathcal{F}_\alpha(\Sigma)_0$ preserves gr_a and $\mathcal{F}_\alpha(\Sigma)_+$ increases gr_a.*

For any annular cobordism Σ, we set $\mathcal{F}_{\mathbb{A}}(\Sigma)$ to be the gr_a-preserving part of $\mathcal{F}_\alpha(\Sigma)$:

$$\mathcal{F}_{\mathbb{A}}(\Sigma) := \mathcal{F}_\alpha(\Sigma)_0.$$

We set $\mathcal{BN}(\mathbb{A})$ to be the category whose objects are formal direct sums of formally bigraded collections of simple closed curves in $\mathbb{A}$, and whose morphisms are matrices whose entries are formal R_α-linear combinations of cobordisms properly embedded in $\mathbb{A} \times I$ modulo isotopy relative to the boundary and subject to the relations in Figure 1. Then $\mathcal{F}_{\mathbb{A}}$ constitutes a functor $\mathcal{F}_{\mathbb{A}} : \mathcal{BN}(\mathbb{A}) \to R_\alpha\text{-}\mathrm{ggmod}$, where R_α-ggmod is the category of $\mathbb{Z} \times \mathbb{Z}$-graded R_α-modules. An annular cobordism Σ with d dots induces a map $\mathcal{F}_{\mathbb{A}}(\Sigma)$ of bidegree $(-\chi(\Sigma) + 2d, 0)$.

Let us discuss the map assigned to the cobordism Σ in (5.1). In terms of the basis $\{v_-, v_+\}$ of A_α,

$$\mathcal{F}_\alpha(\Sigma)(v_-) = \alpha_1 v_- + \boxed{v_+}, \quad \mathcal{F}_\alpha(\Sigma)(v_+) = \alpha_2 v_+,$$

where the box indicates that gr_a has increased. Therefore $\mathcal{F}_{\mathbb{A}}(\Sigma)(v_-) = \alpha_1 v_-$, $\mathcal{F}_{\mathbb{A}}(\Sigma)(v_+) = \alpha_2 v_+$. More generally, letting Σ' denote the cobordism in Figure 5, we have that $\mathcal{F}_{\mathbb{A}}(\Sigma') : A_\alpha^{\otimes m} \to A_\alpha^{\otimes m}$ is the identity on all tensor factors except the i-th one, on which it is given by the left-hand side (resp. right-hand side) of (5.6) if i is odd (resp. even).

$$\begin{aligned} v_- &\mapsto \alpha_1 v_- \qquad & v'_- &\mapsto \alpha_2 v'_- \\ v_+ &\mapsto \alpha_2 v_+ \qquad & v'_+ &\mapsto \alpha_1 v'_+ \end{aligned} \tag{5.6}$$

Given an annular link L with diagram D, applying $\mathcal{F}_{\mathbb{A}}$ to the Bar-Natan complex $[[D]]$ yields a bigraded chain complex whose bigraded chain homotopy class is an invariant of L. By composing with the functor $R_\alpha\text{-}\mathrm{ggmod} \to \mathbb{Z}\text{-}\mathrm{ggmod}$ induced by the map $R_\alpha \to \mathbb{Z}$ where $\alpha_1, \alpha_2 \mapsto 0$, one obtains yet another functor

FIGURE 6. The flow condition near each trivalent vertex in a $\mathfrak{gl}_N$ web.

$\mathcal{BN}(\mathbb{A}) \to \mathbb{Z}$- ggmod. This is the functor defining Asaeda-Przytycki-Sikora's (non-equivariant) annular homology theory.

5.2. Webs, foams, and state spaces. Fix an integer $N \geq 1$. In this section we discuss Murakami-Ohtsuki-Yamada (MOY) webs [**42**]. These are certain trivalent graphs embedded in the plane which can be used to compute the exterior-colored Reshetikhin-Turaev $\mathfrak{sl}_N$ (or $\mathfrak{gl}_N$) link polynomial and which describe morphism spaces between tensor products of exterior powers of the defining $U_q(\mathfrak{sl}_N)$ representation.

DEFINITION 5.3. A $\mathfrak{gl}_N$ *web* (also called a *MOY web*) is a trivalent graph $\Gamma \subset \mathbb{R}^2$ embedded in the plane. It may contain closed loops with no vertices. Each edge and loop of Γ is oriented and carries a label in $\{0, \ldots, N\}$, called the *thickness* of the edge, such that at each vertex the condition shown in Figure 6 is satisfied. We will often simply write *web* instead of $\mathfrak{gl}_N$ web.

One approach to defining (exterior) colored Khovanov-Rozansky homology utilizes a cube of resolutions in which each resolution diagram is a web. Saddle cobordisms in the Khovanov complex are replaced by certain singular cobordisms called *foams*.

DEFINITION 5.4. A (closed) $\mathfrak{gl}_N$ *foam* F is a compact, piecewise-linear (PL), 2-dimensional CW complex with the following properties and data.

- Every point of F has neighborhood homeomorphic to either a disk, the letter Y times an interval, or the cone on the 1-skeleton of a tetrahedron; the latter two types of points, shown in Figure 7, are called *singular* points. The *singular graph of* F, denoted $s(F)$, consists of all singular points. A *singular vertex* is a vertex of $s(F)$, shown in Figure 7b. Edges and loops in $s(F)$ are called *bindings*.
- A *facet* of F is a connected component of $F \setminus s(F)$, and the set of facets is denoted $f(F)$. Each facet $f \in f(F)$ of F carries a label in $\{0, \ldots, N\}$ called its *thickness* and denoted $\mathrm{th}(f)$. Foams carry decorations consisting of a homogeneous symmetric polynomial P_f in $\mathrm{th}(f)$ variables for each facet f.
- The singular graph and facets of F must be oriented, such that three facets meeting along a binding must have thickness a, b, and $a + b$. The orientation of the binding must agree with the orientations of the thickness a and b facets and disagree with the orientation of thickness $a + b$ facet, a shown in Figure 7a.
- Finally, foams considered here will always be PL embedded in $\mathbb{R}^3$. We will often write *foam* instead of $\mathfrak{gl}_N$ foam.

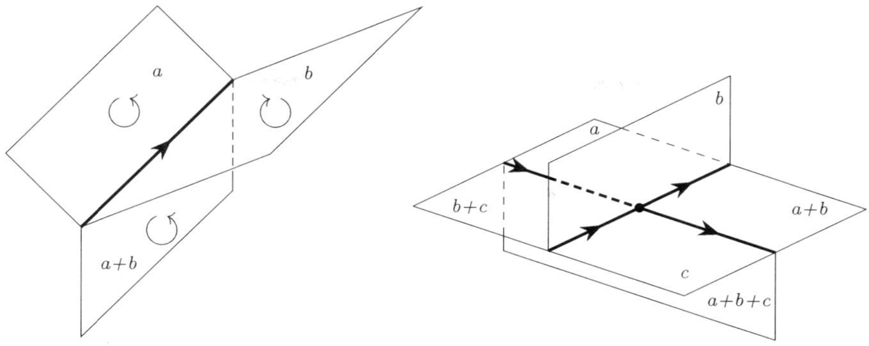

(A) A binding, where three facets meet. (B) A singular vertex, where six facets meet.

FIGURE 7. The thickness and orientation conventions at singular points.

REMARK 5.5. One may consider abstract foams (not embedded in $\mathbb{R}^3$). In this case a cyclic order of the three facets at each interval binding must be specified. In our situation all foams are embedded, and the cyclic order is determined by the left-hand rule.

To form state spaces, one also considers foams with boundary, defined below. These are viewed as cobordisms between webs.

DEFINITION 5.6. Let Γ_0, Γ_1 be $\mathfrak{gl}_N$ webs. A $\mathfrak{gl}_N$ *foam with boundary* (Γ_0, Γ_1) consists of an intersection of a $\mathfrak{gl}_N$ foam F with $\mathbb{R}^2 \times [0,1]$ such that, for some $\varepsilon > 0$, $F \cap (\mathbb{R}^2 \times [1-\varepsilon, 1]) = \Gamma_1 \times [1-\varepsilon, \varepsilon]$ and $F \cap (\mathbb{R}^2 \times [0, \varepsilon]) = (-\Gamma_0) \times [0, \varepsilon]$, where $-\Gamma_0$ denotes the embedded graph Γ_0 with orientation reversed.

Robert-Wagner [**49**] introduced a remarkable formula to evaluate closed $\mathfrak{gl}_N$ foams. Using the universal construction in the spirit of [**15**], they defined a graded R_N-module $\langle \Gamma \rangle_{\mathrm{RW}}$, called the *state space*, for any MOY web Γ. Their construction categorifies the MOY calculus, in the sense that $\langle \Gamma \rangle_{\mathrm{RW}}$ is a free graded R_N-module of graded rank equal to the MOY evaluation $\langle \Gamma \rangle_N$. A key step in defining $\mathfrak{gl}_N$ Khovanov-Rozansky homology is the functorial assignment of a module to each web.

We will use the following rings in the subsequent discussion:

- $R_N = \mathbb{Z}[x_1, \ldots, x_N]$
- $R_N^{sym} \subset R_N$ the subring of symmetric polynomials
- $R'_N = R_N[(x_i - x_j)^{-1} \mid 1 \leq i < j \leq N]$.

For a closed $\mathfrak{gl}_N$ foam F, the evaluation takes the form

$$\langle F \rangle_{\mathrm{RW}} = \sum_{c \in \mathrm{col}(F)} \langle F, c \rangle_{\mathrm{RW}} .$$

Where $\mathrm{col}(F)$ is the set of colorings, a certain assignment of a subset of $\{1, \ldots, N\}$ to each facet of F; see [**49**, Definition 2.4]. Each $\langle F, c \rangle_{\mathrm{RW}}$ is, in general, an element of R'_N. A delicate argument [**49**, Proposition 2.18] shows that no denominators appear in $\langle F \rangle_{\mathrm{RW}}$; that is, $\langle F \rangle_{\mathrm{RW}} \in R_N$. Moreover, $\langle F \rangle_{\mathrm{RW}}$ is also shown to be a *symmetric* polynomial. It follows that the state spaces $\langle \Gamma \rangle_{\mathrm{RW}}$ and the resulting link homology can be defined over the smaller ring R_N^{sym}. In this sense, the

Robert-Wagner construction is a priori of the $U(1)^N$-equivariant flavor, but can be restricted to a $U(N)$-equivariant theory.

Next, we review the functorial assignment of modules (state spaces) to each web via the *universal construction.* Let Γ be a $\mathfrak{gl}_N$ web and let $\mathrm{Fr}(\Gamma)$ be the free R_N^{sym}-module generated by all foams $F : \varnothing \to \Gamma$. For such a foam F, let $\overline{F}$ denote the foam $\Gamma \to \varnothing$ obtained by reflecting F through $\mathbb{R}^2 \times \{0\}$. Note that if $G : \varnothing \to \Gamma$ is another such foam, then the composition $\overline{F}G$ is a closed foam.

Define a bilinear pairing

$$\langle -, - \rangle_{\mathrm{RW}} : \mathrm{Fr}(\Gamma) \times \mathrm{Fr}(\Gamma) \to R_N^{sym}$$

by $\langle F, G \rangle_{\mathrm{RW}} = \langle \overline{F}G \rangle_{\mathrm{RW}}$, and extend bilinearly. Its kernel is defined to be

$$\ker(\langle -, - \rangle) = \{x \in \mathrm{Fr}(\Gamma) \mid \langle x, y \rangle = 0 \text{ for all } y \in \mathrm{Fr}(\Gamma)\}.$$

The *state space of* Γ is then

$$\langle \Gamma \rangle_{\mathrm{RW}} := \mathrm{Fr}(\Gamma)/\ker(\langle -, - \rangle).$$

The key feature of the above construction is that functoriality is obtained essentially for free, as follows. For a foam cobordism, $W : \Gamma_0 \to \Gamma_1$, consider the map $\mathrm{Fr}(\Gamma_0) \to \mathrm{Fr}(\Gamma_1)$ which sends a basis foam $F : \varnothing \to \Gamma_0$ to the composition $W \circ F : \varnothing \to \Gamma_1$. It is straightforward to verify this descends to a map $\langle W \rangle : \langle \Gamma_0 \rangle_{\mathrm{RW}} \to \langle \Gamma_1 \rangle_{\mathrm{RW}}$. This assignment is evidently functorial, in the sense that if $V : \Gamma_1 \to \Gamma_2$ is another foam cobordism then $\langle V \circ W \rangle = \langle V \rangle \circ \langle W \rangle$, and the identify foam $\Gamma \times I$ induces the identity map.

REMARK 5.7. The universal construction was formalized by Blanchet-Habegger-Masbaum-Vogel [**15**]. It was used in Khovanov's categorification of the $\mathfrak{sl}_3$ polynomial [**33**], and has been implemented extensively in the context of link homology. The TQFTs arising from the Frobenius algebras in (2.4) can also be obtained via the universal construction applied to a suitable evaluation of closed dotted surfaces.

5.3. Anchored foams and annular link homology. We review the construction of equivariant homology for links in the thickened annulus via foam evaluation, introduced in [**3**, **5**]. By an *annular* $\mathfrak{gl}_N$ *web* we mean a $\mathfrak{gl}_N$ web embedded in the punctured plane $\mathcal{P}$. Note that no surface in $\mathcal{P} \times I$ bounds a single noncontractible circle in $\mathcal{P} \times \{1\}$, an evident obstruction to the universal construction as sketched above. To address this, we let $\mathcal{L}$ denote a line in $\mathbb{R}^3$, fixed for convenience to be the z-axis, and we let foams generically intersect $\mathcal{L}$ as follows.

DEFINITION 5.8. An *anchored* $\mathfrak{gl}_N$ *foam* is a $\mathfrak{gl}_N$ foam F such that intersections of F with $\mathcal{L}$ occur transversely in the interior of its facets. A point in $F \cap \mathcal{L}$ is called an *anchor point*, and we set $\mathrm{an}(F) = F \cap \mathcal{L}$. For $p \in \mathrm{an}(F)$ lying on a facet f, its *thickness* $\mathrm{th}(p)$ is the thickness of f.

Anchored foams must also come equipped with a *label* of each anchor point p, which consists of a subset $\ell(p) \subset [N] = \{1, \ldots, N\}$ of cardinality equal to the thickness of the label on which p lies. Anchored foams with boundary are foam cobordisms in $\mathbb{R}^2 \times I$ whose boundary lies in $(\mathcal{P} \times \{0\}) \cup (\mathcal{P} \times \{1\})$, intersect the line segment $\{(0,0)\} \times I$ generically, and intersection points carry labels as above.

The *underlying foam* of F, denoted F^{un} is the $\mathfrak{gl}_N$ foam obtained by forgetting anchor points and their labels. A coloring of F means a coloring of the underlying foam.

Anchored foam evaluation $\langle - \rangle$ takes the form

$$\langle F, c\rangle = \langle F^{\mathrm{un}}, c\rangle_{\mathrm{RW}} \cdot \widetilde{Q}(F,c), \qquad \langle F\rangle = \sum_{c\in \mathrm{col}(F)} \langle F, c\rangle ,$$

where $\widetilde{Q}(F,c)$ is a product of contributions over the anchor points; see [**3**, Section 3.1] for details. One effect of introducing the factor $\widetilde{Q}(F,c)$ is that $\langle F,c\rangle = 0$ whenever there is an anchor point p whose label $\ell(p)$ is different from the color (according to c) of the facet on which p lies.

The evaluation $\langle F\rangle$ values in R_N [**3**, Proposition 3.8], but not, in general, in the subring R_N^{sym} of symmetric polynomials. State spaces of annular webs are built via the universal construction analogous to the procedure described in Section 5.2, but they must be defined over R_N rather than R_N^{sym}.

Recent work of Qi-Robert-Sussan-Wagner discovered symmetries of state spaces [**45**] and of equivariant (uncolored) Khovanov-Rozansky homology [**46**]. Their work in particular gives rise to an action of the Lie algebra $\mathfrak{sl}_2$ on these structures. As explained in [**46**, Section 6.1], foam evaluation and state spaces can be built over the ring of symmetric polynomials $\Bbbk[E_1,\ldots,E_N] = \Bbbk[x_1,\ldots,x_N]^{S_N}$, where $\Bbbk$ is a commutative ring. It is possible to specialize the equivariant parameters $E_1 = E_2 = \cdots = E_N = 0$ in order to obtain link homology over $\Bbbk$, but for the full $\mathfrak{sl}_2$ action to descend to this specialization it is required that $2 \in \Bbbk$ is invertible and that $\Bbbk$ has characteristic dividing N. Work in preparation of the first author [**2**] extends the Qi-Robert-Sussan-Wagner $\mathfrak{sl}_2$ action to anchored foams and annular link homology.

6. A few questions

While annular Khovanov homology is the most explored of Asaeda-Przytycki-Sikora's homologies [**8**] due to its relationship with braid closures and topological cabling procedures, it is natural to ask about other manifolds as well:

QUESTION 6.1. Can one define equivariant homology for links in I-bundles over surfaces, beyond the thickened annulus?

For example, Manolescu-Willis [**40**] (over characteristic $\neq 2$) and Chen [**21**] (over characteristic 2) recently defined Lee and Bar-Natan type deformations, respectively, for nullhomologous links in $\mathbb{RP}^3 = \mathbb{RP}^2 \times I \cup \{\text{point}\}$, building on work of [**8, 25, 41**]. This allows the authors to define analogous Rasmussen s-invariants, which in turn provide bounds on the genus of slice surfaces in $\mathbb{RP}^3 \times I$.

The next two questions require some context. Given an annular link L with diagram D, one can view D in the plane and consider the usual (non-equivariant) Khovanov complex $CKh_0(D)$. The puncture induces a filtration on this complex, and the annular chain complex of D can be taken to be the associated graded object [**50**]. Consequently, there is a spectral sequence from the annular Khovanov homology of L to the Khovanov homology of L, viewed as a link in $\mathbb{R}^3$. Moreover, Grigsby-Licata-Wehrli [**26**] established an $\mathfrak{sl}_2$ action on annular Khovanov homology (in fact, they establish a current algebra action). Queffelec-Rose generalized these results: they introduced non-equivariant annular Khovanov-Rozansky homology, which is shown to carry an $\mathfrak{sl}_N$ action as well as a spectral sequence to Khovanov-Rozansky homology [**47**]. In light of this, we pose the following:

QUESTION 6.2. Is there an analogous spectral sequence for equivariant annular $\mathfrak{gl}_N$ homology?

QUESTION 6.3. Can one extend the Grigsby-Licata-Wehrli $\mathfrak{sl}_2$ action, or more generally the current algebra action, to equivariant annular homology? Moreover, is there an analogous $\mathfrak{sl}_N$ or $\mathfrak{gl}_N$ action on equivariant annular $\mathfrak{gl}_N$ homology?

References

[1] Lowell Abrams, *Two-dimensional topological quantum field theories and Frobenius algebras*, J. Knot Theory Ramifications **5** (1996), no. 5, 569–587, DOI 10.1142/S0218216596000333. MR1414088

[2] Rostislav Akhmechet, *Symmetries of anchored foams and equivariant annular Khovanov-Rozansky homology*, in preparation.

[3] Rostislav Akhmechet, *Anchored $\mathfrak{gl}_N$ foams and annular Khovanov-Rozansky homology*, arXiv:2305.08944, 2023.

[4] Rostislav Akhmechet, *Equivariant annular Khovanov homology*, J. Knot Theory Ramifications **32** (2023), no. 2, Paper No. 2350002, 31, DOI 10.1142/S0218216523500025. MR4564614

[5] Rostislav Akhmechet and Mikhail Khovanov, *Anchored foams and annular homology*, Algebr. Geom. Topol. **23** (2023), no. 7, 3129–3204, DOI 10.2140/agt.2023.23.3129. MR4647674

[6] Akram Alishahi, *Unknotting number and Khovanov homology*, Pacific J. Math. **301** (2019), no. 1, 15–29, DOI 10.2140/pjm.2019.301.15. MR4007369

[7] Akram Alishahi and Nathan Dowlin, *The Lee spectral sequence, unknotting number, and the knight move conjecture*, Topology Appl. **254** (2019), 29–38, DOI 10.1016/j.topol.2018.11.020. MR3894208

[8] Marta M. Asaeda, Józef H. Przytycki, and Adam S. Sikora, *Categorification of the Kauffman bracket skein module of I-bundles over surfaces*, Algebr. Geom. Topol. **4** (2004), 1177–1210, DOI 10.2140/agt.2004.4.1177. MR2113902

[9] John A. Baldwin, Matthew Hedden, and Andrew Lobb, *On the functoriality of Khovanov-Floer theories*, Adv. Math. **345** (2019), 1162–1205, DOI 10.1016/j.aim.2019.01.026. MR3903915

[10] John A. Baldwin, Adam Simon Levine, and Sucharit Sarkar, *Khovanov homology and knot Floer homology for pointed links*, J. Knot Theory Ramifications **26** (2017), no. 2, 1740004, 49, DOI 10.1142/S0218216517400041. MR3604486

[11] Dror Bar-Natan, *Khovanov's homology for tangles and cobordisms*, Geom. Topol. **9** (2005), 1443–1499, DOI 10.2140/gt.2005.9.1443. MR2174270

[12] Dror Bar-Natan and Scott Morrison, *The Karoubi envelope and Lee's degeneration of Khovanov homology*, Algebr. Geom. Topol. **6** (2006), 1459–1469, DOI 10.2140/agt.2006.6.1459. MR2253455

[13] Anna Beliakova, Matthew Hogancamp, Krzysztof K. Putyra, and Stephan M. Wehrli, *On the functoriality of $\mathfrak{sl}_2$ tangle homology*, Algebr. Geom. Topol. **23** (2023), no. 3, 1303–1361, DOI 10.2140/agt.2023.23.1303. MR4598808

[14] Anna Beliakova, Krzysztof K. Putyra, and Stephan M. Wehrli, *Quantum link homology via trace functor I*, Invent. Math. **215** (2019), no. 2, 383–492, DOI 10.1007/s00222-018-0830-0. MR3910068

[15] C. Blanchet, N. Habegger, G. Masbaum, and P. Vogel, *Topological quantum field theories derived from the Kauffman bracket*, Topology **34** (1995), no. 4, 883–927, DOI 10.1016/0040-9383(94)00051-4. MR1362791

[16] Christian Blanchet, *An oriented model for Khovanov homology*, J. Knot Theory Ramifications **19** (2010), no. 2, 291–312, DOI 10.1142/S0218216510007863. MR2647055

[17] Jonathan Brundan and Catharina Stroppel, *Highest weight categories arising from Khovanov's diagram algebra I: cellularity* (English, with English and Russian summaries), Mosc. Math. J. **11** (2011), no. 4, 685–722, 821–822, DOI 10.17323/1609-4514-2011-11-4-685-722. MR2918294

[18] Carmen Caprau, Nicolle González, Christine Ruey Shan Lee, Adam M. Lowrance, Radmila Sazdanović, and Melissa Zhang, *On Khovanov homology and related invariants*, Research directions in symplectic and contact geometry and topology, Assoc. Women Math. Ser., vol. 27, Springer, Cham, ©2021, pp. 273–292, DOI 10.1007/978-3-030-80979-9_6. MR4417719

[19] Carmen Livia Caprau, sl(2) *tangle homology with a parameter and singular cobordisms*, Algebr. Geom. Topol. **8** (2008), no. 2, 729–756, DOI 10.2140/agt.2008.8.729. MR2443094

[20] J. Scott Carter and Masahico Saito, *Knotted surfaces and their diagrams*, Mathematical Surveys and Monographs, vol. 55, American Mathematical Society, Providence, RI, 1998, DOI 10.1090/surv/055. MR1487374

[21] Daren Chen, *Bar-Natan homology for null homologous links in* $\mathbb{RP}^3$, Pacific J. Math. **334** (2025), no. 2, 233–268, DOI 10.2140/pjm.2025.334.233. MR4862260

[22] Yanfeng Chen and Mikhail Khovanov, *An invariant of tangle cobordisms via subquotients of arc rings*, Fund. Math. **225** (2014), no. 1, 23–44, DOI 10.4064/fm225-1-2. MR3205563

[23] David Clark, Scott Morrison, and Kevin Walker, *Fixing the functoriality of Khovanov homology*, Geom. Topol. **13** (2009), no. 3, 1499–1582, DOI 10.2140/gt.2009.13.1499. MR2496052

[24] Michael Ehrig, Daniel Tubbenhauer, and Paul Wedrich, *Functoriality of colored link homologies*, Proc. Lond. Math. Soc. (3) **117** (2018), no. 5, 996–1040, DOI 10.1112/plms.12154. MR3877770

[25] Bovstjan Gabrovvsek, *The categorification of the Kauffman bracket Skein module of* $\mathbb{RP}^3$, Bull. Aust. Math. Soc. **88** (2013), no. 3, 407–422, DOI 10.1017/S0004972713000105. MR3189291

[26] J. Elisenda Grigsby, Anthony M. Licata, and Stephan M. Wehrli, *Annular Khovanov homology and knotted Schur-Weyl representations*, Compos. Math. **154** (2018), no. 3, 459–502, DOI 10.1112/S0010437X17007540. MR3731256

[27] Onkar Singh Gujral, *Ribbon distance bounds from Bar-Natan Homology and* α*-Homology*, `arXiv:2011.01190`, 2020.

[28] Damian Iltgen, Lukas Lewark, and Laura Marino, *Khovanov homology and rational unknotting*, `arXiv:2110.15107`, 2021.

[29] Magnus Jacobsson, *An invariant of link cobordisms from Khovanov homology*, Algebr. Geom. Topol. **4** (2004), 1211–1251, DOI 10.2140/agt.2004.4.1211. MR2113903

[30] Mikhail Khovanov, *A categorification of the Jones polynomial*, Duke Math. J. **101** (2000), no. 3, 359–426, DOI 10.1215/S0012-7094-00-10131-7. MR1740682

[31] Mikhail Khovanov, *A functor-valued invariant of tangles*, Algebr. Geom. Topol. **2** (2002), 665–741, DOI 10.2140/agt.2002.2.665. MR1928174

[32] Mikhail Khovanov, *Patterns in knot cohomology. I*, Experiment. Math. **12** (2003), no. 3, 365–374. MR2034399

[33] Mikhail Khovanov, *sl*(3) *link homology*, Algebr. Geom. Topol. **4** (2004), 1045–1081, DOI 10.2140/agt.2004.4.1045. MR2100691

[34] Mikhail Khovanov, *An invariant of tangle cobordisms*, Trans. Amer. Math. Soc. **358** (2006), no. 1, 315–327, DOI 10.1090/S0002-9947-05-03665-2. MR2171235

[35] Mikhail Khovanov, *Link homology and Frobenius extensions*, Fund. Math. **190** (2006), 179–190, DOI 10.4064/fm190-0-6. MR2232858

[36] Mikhail Khovanov and Louis-Hadrien Robert, *Link homology and Frobenius extensions II*, Fund. Math. **256** (2022), no. 1, 1–46, DOI 10.4064/fm912-6-2021. MR4361584

[37] Mikhail Khovanov and Lev Rozansky, *Matrix factorizations and link homology*, Fund. Math. **199** (2008), no. 1, 1–91, DOI 10.4064/fm199-1-1. MR2391017

[38] Artem Kotelskiy, Liam Watson, and Claudius Zibrowius, *Immersed curves in Khovanov homology*, `arXiv:1910.14584`, 2019.

[39] Eun Soo Lee, *An endomorphism of the Khovanov invariant*, Adv. Math. **197** (2005), no. 2, 554–586, DOI 10.1016/j.aim.2004.10.015. MR2173845

[40] Ciprian Manolescu and Michael Willis, *A Rasmussen invariant for links in* $\mathbb{RP}^3$, Trans. Amer. Math. Soc. Ser. B **12** (2025), 789–830, DOI 10.1090/btran/221. MR4904031

[41] Vassily Olegovich Manturov, *Khovanov homology for virtual knots with arbitrary coefficients*, J. Knot Theory Ramifications **16** (2007), no. 3, 345–377, DOI 10.1142/S0218216507005336. MR2320160

[42] Hitoshi Murakami, Tomotada Ohtsuki, and Shuji Yamada, *HOMFLY polynomial via an invariant of colored plane graphs*, Enseign. Math. (2) **44** (1998), no. 3-4, 325–360. MR1659228

[43] Gad Naot, *The universal Khovanov link homology theory*, Algebr. Geom. Topol. **6** (2006), 1863–1892, DOI 10.2140/agt.2006.6.1863. MR2263052

[44] Lisa Piccirillo, *The Conway knot is not slice*, Ann. of Math. (2) **191** (2020), no. 2, 581–591, DOI 10.4007/annals.2020.191.2.5. MR4076631

[45] You Qi, Louis-Hadrien Robert, Joshua Sussan, and Emmanuel Wagner, *Symmeties of $\mathfrak{gl}_n$ foams*, arXiv:2212.10106, 2022.

[46] You Qi, Louis-Hadrien Robert, Joshua Sussan, and Emmanuel Wagner, *Symmetries of equivariant Khovanov-Rozansky homology*, arXiv:2306.10729, 2023.

[47] Hoel Queffelec and David E. V. Rose, *Sutured annular Khovanov-Rozansky homology*, Trans. Amer. Math. Soc. **370** (2018), no. 2, 1285–1319, DOI 10.1090/tran/7117. MR3729501

[48] Jacob Rasmussen, *Khovanov homology and the slice genus*, Invent. Math. **182** (2010), no. 2, 419–447, DOI 10.1007/s00222-010-0275-6. MR2729272

[49] Louis-Hadrien Robert and Emmanuel Wagner, *A closed formula for the evaluation of foams*, Quantum Topol. **11** (2020), no. 3, 411–487, DOI 10.4171/qt/139. MR4164001

[50] Lawrence P. Roberts, *On knot Floer homology in double branched covers*, Geom. Topol. **17** (2013), no. 1, 413–467, DOI 10.2140/gt.2013.17.413. MR3035332

[51] Taketo Sano, *A description of Rasmussen's invariant from the divisibility of Lee's canonical class*, J. Knot Theory Ramifications **29** (2020), no. 6, 2050037, 39, DOI 10.1142/S0218216520500376. MR4125174

[52] Taketo Sano, *Fixing the functoriality of Khovanov homology: a simple approach*, J. Knot Theory Ramifications **30** (2021), no. 11, Paper No. 2150074, 12, DOI 10.1142/S0218216521500747. MR4376719

[53] Taketo Sano and Kouki Sato, *A family of slice-torus invariants from the divisibility of Lee classes*, Topology Appl. **357** (2024), Paper No. 109059, 37, DOI 10.1016/j.topol.2024.109059. MR4798100

[54] Sucharit Sarkar, *Ribbon distance and Khovanov homology*, Algebr. Geom. Topol. **20** (2020), no. 2, 1041–1058, DOI 10.2140/agt.2020.20.1041. MR4092319

[55] Dirk Schütz, *On an integral version of the Rasmussen invariant*, Michigan Math. J. **75** (2025), no. 1, 65–88, DOI 10.1307/mmj/20226211. MR4873797

[56] Catharina Stroppel, *Parabolic category $\mathcal{O}$, perverse sheaves on Grassmannians, Springer fibres and Khovanov homology*, Compos. Math. **145** (2009), no. 4, 954–992, DOI 10.1112/S0010437X09004035. MR2521250

[57] Pierre Vogel, *Functoriality of Khovanov homology*, J. Knot Theory Ramifications **29** (2020), no. 4, 2050020, 66, DOI 10.1142/S0218216520500200. MR4096813

[58] S. Wehrli, *A spanning tree model for Khovanov homology*, J. Knot Theory Ramifications **17** (2008), no. 12, 1561–1574, DOI 10.1142/S0218216508006762. MR2477595

Department of Mathematics, Columbia University, New York, New York 10027
Email address: akhmechet@math.columbia.edu

Department of Mathematics, UC Davis, One Shields Ave., Davis, California 95616
Email address: mlzhang@ucdavis.edu

Contemporary Mathematics
Volume **827**, 2025
https://doi.org/10.1090/conm/827/16531

Computing Khovanov homology via categorified Jones-Wenzl projectors

Christine Ruey Shan Lee

Abstract. Colored Khovanov homology is a categorification of the colored Jones polynomial, a generalization of the Jones polynomial that has been shown to be closely related to the geometry and the topology of the knot complement. We describe a computational approach for Khovanov homology based on understanding the categorification of the Jones-Wenzl projector used to define colored Khovanov homology.

1. Introduction

The colored Jones polynomial of a knot or a link in the 3-sphere is a generalization of the Jones polynomial which has been extensively studied and shown to be closely related to the geometry and topology of the associated knot complement, through conjectures such as the volume conjecture [**Kas97**, **MM01**], the AJ conjecture [**FGL02**, **Gar04**], and the slope conjecture [**Gar11**]. Similarly, the categorification of the colored Jones polynomial, called colored Khovanov homology, comes from the construction, due to Khovanov [**Kho00**], of a categorification of the Jones polynomial. As can be seen from the definitions (see Definitions 2.2 and 2.4), it is difficult to compute these quantum invariants which makes gathering evidence for these conjectures, or finding new relationships a persistent challenge.

The purpose of this article is to describe a conjectural approach for computing the colored Jones polynomial and Khovanov homology using the connection between the Khovanov homology of infinite torus braids, studied in [**GOR13**, **Hog19**], and the Jones-Wenzl projector, that can be used to define the colored Jones polynomial and its categorification, colored Khovanov homology.

Let $T(n,k)$ denote the (n,k)-torus link obtained by adding k positive full twists $\triangle_n^2 = (\sigma_1\sigma_2\cdots\sigma_n)^{n-1}$ on n strands, then taking the braid closure of the resulting torus braid $\beta(n,k) = (\sigma_1\sigma_2\cdots\sigma_n)^{k(n-1)}$. The Khovanov homology of the infinite torus link on n strands is the direct limit of the direct system formed by the Khovanov homology groups of $T(n,k)$ as $k \to \infty$. That this direct limit exists, which we denote by $Kh(T(n,\infty))$, is due to Stosic [**Sto07**], see Theorem 3. Extending to the case of braids, Rozansky [**Roz14**] showed the Khovanov homology of infinite torus braids $\beta(n,k)$, denoted by $Kh(\beta(n,\infty))$, gives a categorification of the Jones-Wenzl projector. The homology in the direct limit of both the torus link $T(n,k)$ and $\beta(n,k)$ as $k \to \infty$ are known as the *stable Khovanov homology of torus links* or

2020 *Mathematics Subject Classification.* Primary 57K14, 57K18.

of torus braids, respectively. Since the Jones-Wenzl projector can be used to define the colored Jones polynomial, see Definition 2.2, this results in a categorification of the colored Jones polynomial, which is called the colored Khovanov homology of a link. Working backwards, knowledge of the stable Khovanov homology of torus braids can be used to compute the colored Khovanov homology at low homological degrees and also the colored Jones polynomial through decategorification. The Gorsky-Oblomkov-Rasmussen conjecture (see Conjecture 4) comes into play as it predicts a simplified structure for $Kh(T(n,\infty))$, which we propose can greatly reduce the amount of inputs required to compute the colored Jones polynomial and colored Khovanov homology. More precisely, we conjecture that the chain complex of k positive full twists on n strands can be obtained from a truncation, see Definition 3.1, of a chain complex categorifying the Jones-Wenzl projector.

CONJECTURE 1. *Let $p(n,k)$ be an integer depending on k. The categorification complex $CKh(\beta(n,k))$ for a torus braid $\beta(n,k)$ is homotopy equivalent to a complex $\mathbf{B}^n$ truncated at $p(n,k)$ from that of a categorification complex $\mathbf{P}^n \simeq CKh(\beta(n,\infty))$ with bounded width of the n-th Jones-Wenzl projector.*

See Eq. (1) for our sense of width of a chain complex in this article. The conjecture is known to be true for torus braids on $n=2$ strands. While progress is ongoing for torus braids on $n \geq 3$ strands [**CGL**$^+$], our focus for this paper is to explain and explore the computational advantages afforded by the conjecture, already apparent in the $n=2$ case. It is our hope that this will help generate wider interest for the problem of computing colored Khovanov homology and related open problems.

Organization. In Section 2 we begin with the definition of the Jones polynomial in terms of the Kauffman bracket, leading to Khovanov homology and Bar-Natan's formulation in Section 2.2. Then we introduce the stable Khovanov homology of torus braids, define the colored Jones polynomial using the Jones-Wenzl projector, and discuss how the stable Khovanov homology of torus braids is connected to colored Khovanov homology through the identification of the stable Khovanov homology of torus braids with a categorification of the Jones-Wenzl projector in Sections 2.3–2.6.

In Section 3, we make precise our notion of truncating a chain complex and state Conjecture 1 in context. We discuss what is known for the conjecture, the $n=2$ case, some evidence from the decategorification, and how the conjecture can apply to the computation of the colored Jones polynomial and Khovanov homology.

To illustrate, in Section 4 we describe how to compute the Khovanov homology of a τ-tangle pretzel link $P(s_1, s_2, \ldots, s_\tau)$. An example for the chain complex is displayed for the 4-tangle pretzel knot $P(s_1=3, s_2=3, s_3=3, s_4=3)$ and an informal discussion of the reduction in computational complexity from this approach is given. A computer program that implements the algorithm for pretzel knots $P(s_1=s, s_2=s, \cdots, s_\tau=s)$ is available on GitHub and interested readers may email the author for the code. The author welcomes comments and suggestions for improving the program.

Brief accounts of related open problems and what are known about them are stated throughout as relevant, with the understanding that these accounts are by no means comprehensive. In particular, we do not discuss Khovanov thick/thinness and literature on the relationship to Floer-theoretic invariants or stable homotopy type.

Throughout the paper, we will stick with $\mathbb{Z}$ coefficients for the homology theory.

2. The colored Jones polynomial and colored Khovanov homology

2.1. The Jones polynomial. The Jones polynomial assigns a Laurent polynomial to a link in S^3 that is invariant under link isotopy. For a historical account of the discovery of the polynomial by Vaughan Jones, see the recent article by Kauffman [**Kau23**]. The discovery of the Jones polynomial has led to the construction of new link and 3-manifold invariants from the representation theory of quantum groups.

There are many ways to define the Jones polynomial. Here we present the definition using the Kauffman bracket, as it presents a natural context for discussing the categorification in the next section.

2.1.1. *The Kauffman bracket definition of the Jones polynomial.* Let A be an indeterminate complex number. Consider the $\mathbb{C}$-vector space generated by link diagrams in S^2 modulo the Kauffman bracket skein relations:

- $\langle \times \rangle = A\langle)\,(\rangle + A^{-1}\langle \asymp \rangle$
- $\langle D \sqcup \bigcirc \rangle = (-A^{-2} - A^2)\langle D \rangle$

The Kauffman bracket of the empty diagram is defined to be 1. Given an oriented link diagram D, its Kauffman bracket polynomial $\langle D \rangle$ is defined to be the Laurent polynomial in A obtained from reducing D (with orientations forgotten) to the empty diagram using the skein relations.

DEFINITION 2.1 ([**Lic97**]). Let D be an oriented diagram of a link $L \subset S^3$ and $w(D)$ be the writhe of the diagram D. The Jones polynomial of the link L is defined by

$$J_L(t) = \left((-A)^{-3w(D)}\langle D \rangle\right)_{t^{1/2}=A^{-2}}$$

A Kauffman state σ on a link diagram D is a choice of replacing every crossing $\times$ of the diagram by the 0-resolution $)\,($ or the 1-resolution $\asymp$.

$$\sigma : \text{Set of crossings of } D \to \{0, 1\}.$$

2.1.2. *Computational complexity of the Jones polynomial.* It is directly seen from Definition 2.1 that the computational complexity for the Jones polynomial in terms of data storage using the Kauffman bracket is exponential on the number of crossings of the diagram. This is because the number of disjoint circles for each Kauffman state needs to be stored, and there are 2^c Kauffman states on a link diagram with c crossings. In general, one cannot do better [**JVW90**] without restricting the class of knots (Quasi-alternating, adequate, alternating, torus knots, Montesinos links [**HMTY09**], etc.) or making assumptions on the diagram (for example, a braid presentation [**MS90**]).

2.2. Khovanov homology. In [**Kho00**], Khovanov introduced a homology theory categorifying the Jones polynomial. To a diagram of a link he associates a chain complex, whose homology groups are invariants of the link, and whose graded Euler characteristic recovers the Jones polynomial. The description here is close to that of Turner's excellent introduction [**Tur17**].

For a link L fix a diagram D. Denote by $|\sigma|$ the number of circles in the Kauffman state σ. Let V be the $\mathbb{Z}$-module generated by two generators denoted by v_- and v_+. We associate to each state a vector space V_σ that is the tensor product of $|\sigma|$-copies of V:

$$V_\sigma = \underbrace{V \otimes V \otimes \cdots \otimes V}_{|\sigma| \text{ copies}}.$$

We give a generator v of V_σ a homological grading $i(v)$ and a quantum grading $j(v)$ as follows. Let $r(\sigma)$ be the number of 1-resolutions chosen by the Kauffman state σ, and let $k(v)$ be the number of v_+'s minus the number of v_-'s in v. For ease of exposition we will incorporate the overall grading shift after evaluating the homology groups, so that generators from the Kauffman state that chooses the all 0-resolution everywhere is placed at homological grading 0. Define

$$i = r(\sigma) - c_- \qquad j(v) = i(v) + k(v) + c_+ - c_-.$$

Here c_+ is the number of positive crossings of the diagram D and c_- is the number of negative crossings. A Kauffman state σ is said to be an *immediate predecessor* of another Kauffman state σ' if these states assign the same resolution on every crossing in D except for one, and at this one crossing, σ chooses the 0-resolution and σ' chooses the 1-resolution. In this case, the state σ' is called an *immediate successor* of σ.

Changing the resolution from 0 to 1 at a crossing either merges a pair of circles in σ or splits a circle into two. We define a homomorphism $d_{\sigma,\sigma'} : V_\sigma \to V_{\sigma'}$ depending on these two cases.

$$d_{\sigma,\sigma'} = \begin{cases} m : V \otimes V \to V & \text{if a pair of circles corresponding to two factors } V \\ & \text{in } V_\sigma \text{ is merged into a circle in } \sigma' \\ \Delta : V \to V \otimes V & \text{if a circle in } \sigma \text{ corresponding to the factor } V \text{ in } V_\sigma \\ & \text{is split into two in } \sigma'. \end{cases}$$

Let the merge map $m : V \otimes V \to V$ be defined by

$$\begin{aligned} m(v_- \otimes v_-) &= 0 \\ m(v_- \otimes v_+) &= m(v_+ \otimes v_-) = v_- \\ m(v_+ \otimes v_+) &= v_+. \end{aligned}$$

Let the split map $\Delta : V \to V \otimes V$ be defined by

$$\begin{aligned} \Delta(v_-) &= v_- \otimes v_- \\ \Delta(v_+) &= v_- \otimes v_+ + v_+ \otimes v_-. \end{aligned}$$

The complex $CKh(D)$ is defined as a direct sum of V_σ over all Kauffman states σ of D with $i(\sigma)$ as the homological grading:

$$CKh(D) := \bigoplus_{i=i(\sigma)} V_\sigma.$$

The differential d applied to a generator v from V_σ is the sum of $d_{\sigma,\sigma'}$ over all immediate successors σ' of σ:

$$d(v) = \sum_{\sigma' \text{ an immediate successor of } \sigma} (-1)^{sgn(\sigma,\sigma')} d_{\sigma,\sigma'}(v).$$

Here sgn is a choice of sign on the differentials from σ to σ' so that d is a differential (i.e. $d \circ d = 0$.) There are various ways to define this sign. We will number the crossings of the link diagram $1, \ldots, c$ and define $sgn(\sigma, \sigma')$ to be the number of 1-resolutions chosen before the first crossing where the 0-resolution is chosen in σ and the 1-resolution is chosen in σ'.

Let $Kh_{i,j}$ denote the homology group of the chain complex CKh at homological grading i and quantum grading j. The unreduced Jones polynomial with specialization to the variable q (note $q = -A^{-2}$) is recovered as the graded Euler characteristic:

$$\widehat{J_L}(q) = \sum_{i,j} (-1)^i q^j rank(Kh_{i,j}(D)).$$

We will use the notation $Kh(D)$ to refer to the set of homology groups $\{Kh_{i,j}(D)\}$ of $(CKh(D), d)$. Next we describe what we need from Bar-Natan's formulation of Khovanov homology [**BN05**], which extends the theory to tangles.

2.2.1. *The Temperley-Lieb algebra.* The Temperley-Lieb algebra TL_n is the complex vector space over crossingless matchings of $2n$ points on two opposite sides of a rectangle with n points on each side, modulo planar isotopy. Multiplication in TL_n is performed by placing one rectangle on top of the other, joining the n points, and replacing a closed loop by $q + q^{-1}$, where $q = -A^{-2}$. See Figure 1 for an example of the multiplication in TL_3. We also allow a set of simple curves in the rectangle that do not intersect the boundary of the rectangle to be an element of TL_n.

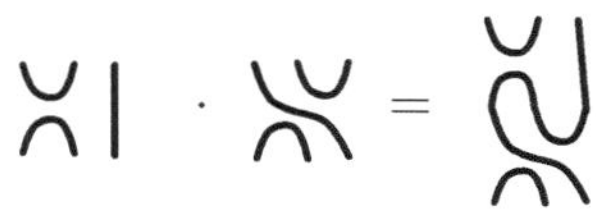

FIGURE 1. Multiplication in TL_3.

Let B be a finite set of points on the boundary of the rectangle. The category $Cob^3(B)_{/l}$ is an additive category[1] where

- **Objects** are crossingless matchings (or smoothings) of B in the rectangle considered up to planar isotopy, possibly with empty intersection with B;
- **Morphisms** are formal direct sums of isotopy classes of cobordisms in $S^2 \times [0,1]$ between the objects modulo the Bar-Natan skein relations shown below in Figure 2.

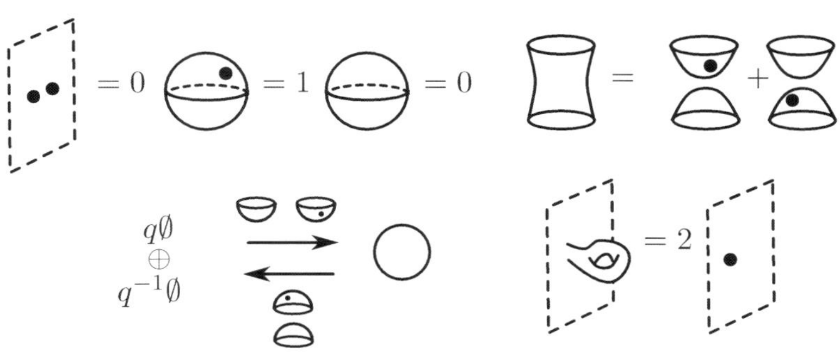

FIGURE 2. The Bar-Natan skein relations.

[1]We are simplifying from the exposition in [**BN05**]. Our $Cob^3(B)_{/l}$ is $Mat(Cob^3_{/l}(B))$ in [**BN05**].

By convention, we read cobordisms from bottom to top of the page. The first relation from the left of Figure 2 says that whenever a surface has more than one dot, it evaluates to zero. A handle on a surface evaluates to twice the surface with a dot, as on the bottom right of Figure 2.

For a tangle diagram T, Bar-Natan associates to it a formal chain complex $[T]$ in $Kom(Cob^3_{/l}(\partial T))$ whose homotopy type is an invariant of the isotopy class of T. If T is a link diagram, which means its intersection with B is empty as a special case of $Cob^3_{/l}$, then $[T]$ recovers the Khovanov homology of the link through a functor $\mathcal{F}$ Bar-Natan constructs from the (geometric) homotopy category of chain complexes $Kom_{/h}(\emptyset)$ to the algebraic category $Kom(\mathbb{Z}-mod)$ of chain complexes of $\mathbb{Z}$-modules. Interested readers may find the full technical details in [**BN05**].

Due to Bar-Natan's work, we may represent differentials of $[T]$ diagrammatically in terms of formal direct sums of cobordisms in $Cob^3(B)_{/l}$, and we will do so from now on for the rest of the paper.

2.2.2. *Complexity of computing Khovanov homology.* From the definition, computing Khovanov homology a priori involves first enumerating the set of Kauffman states. For a diagram with c crossings, there are 2^c Kauffman states. Then for each Kauffman state, the number of generators for a vector space associated to the Kauffman state is $2^{|\sigma|}$, where recall $|\sigma|$ is the number of circles in the Kauffman state.

Dominant algorithms employ various techniques to speed up the computation. We discuss the technique in the paper titled "Fast Kh Computation" by Bar-Natan [**BN07**] as it is most similar to the approach proposed by the article.

Lemma 2 (Gaussian elimination [**BN07**, Lemma 4.2]). *If $\phi : b_1 \to b_2$ is an isomorphism (in some additive category $\mathcal{C}$), then the four term complex segment in $Mat(\mathcal{C})$*

$$\cdots \longrightarrow [C] \xrightarrow{\begin{pmatrix}\alpha\\ \beta\end{pmatrix}} \begin{bmatrix} b_1 \\ D \end{bmatrix} \xrightarrow{\begin{pmatrix}\phi & \delta \\ \gamma & \epsilon\end{pmatrix}} \begin{bmatrix} b_2 \\ E \end{bmatrix} \xrightarrow{\begin{pmatrix}\mu & \nu\end{pmatrix}} [F] \longrightarrow \cdots$$

is isomorphic to the (direct sum) complex segment

$$\cdots \longrightarrow [C] \xrightarrow{(\beta)} [D] \xrightarrow{(\epsilon-\gamma\phi^{-1}\delta)} [E] \xrightarrow{(\nu)} [F] \longrightarrow \cdots$$

Using Lemma 2, one can pre-compute the Khovanov homology of tangles and assemble the result for a knot or a link containing those tangles. The spirit of the computational approach in this article is similar. We propose that the local computation of $CKh(T)$ when T is a torus braid $\beta(n,k)$ can be reduced to truncating the categorification of the Jones-Wenzl projector.

2.3. The Stable Khovanov homology of torus braids. Khovanov homology stabilizes under the action of adding more and more (all positive or all negative) full twists on a fixed number of strands. The simplest examples for this phenomenon are torus links $T(n,k)$ defined as the braid closure of torus braids $\beta(n,k) = (\triangle_n^2)^k$, which is k positive full twists $\triangle_n^2 = (\sigma_1\sigma_2\cdots\sigma_n)^{n-1}$ on n strands.

Consider the torus link $T(n,k)$ and their Khovanov homology $Kh(T(n,k))$. Stosic showed that the Khovanov homology groups stabilize as $k \to \infty$.

THEOREM 3 ([**Sto07**, partial statement of Theorem 6]). *Let n, k, i be integers such that $2 \leq n < k$, and $i < n + k - 3$. Then for every $j \in \mathbb{Z}$*

$$Kh_{i,j}(T(n, k-1)) = Kh_{i,j}(T(n,k)).$$

The proof uses the skein-exact sequence of Khovanov homology, which implies that at low homological degrees, the Khovanov homology groups of torus links $T(n,k)$ and $T(n, k+1)$ are isomorphic. Denote the set of stabilized homology groups by $Kh(T(n,\infty)) = \{Kh_{i,j}(T(n,\infty))\}$.

Understanding the structure of the stable Khovanov homology of torus links is an open problem. We discuss for example a conjecture by Gorsky-Oblomkov-Rasmussen.

CONJECTURE 4 ([**GOR13**, Conjecture 1.1]). *The unreduced stable Khovanov homology $Kh(T(n,\infty))$ is dual to the homology of the differential graded algebra generated by even variables $x_0, \ldots, x_{n-1}$ and odd variables $\xi_0, \ldots, \xi_{n-1}$ equipped with the differential d_2 defined by*

$$d_2(\xi_k) = \sum_{i=0}^{k} x_i x_{k-i} \text{ and } d_2(x_k) = 0.$$

Equivalently, this is the Koszul complex determined by the (nonregular) sequence $d_2(\xi_k)$ for $k = 0, \ldots n-1$.

The analogue of the conjecture for the categorification of the HOMFLY-PT polynomial has been resolved [**Hog18**] while Conjecture 4 remains open, though Hogancamp has also made significant progress [**Hog19**].

We now describe the connection between the stable Khovanov homology of torus links and the categorification of the colored Jones polynomial through the Jones-Wenzl projector.

2.4. The colored Jones polynomial. One can obtain the n-th colored Jones polynomial of a link from the representation theory of quantum groups by coloring the strands of a braid presentation of the link with n-dimensional irreducible representations of the quantum group $U_q(\mathfrak{sl}_2)$ (see [**Oht01**] for an explicit description of this setup). For the purpose of discussing colored Khovanov homology, we define the colored Jones polynomial from the Kauffman bracket of a certain cabling of a given diagram of the knot. We first describe the Jones-Wenzl projector, an element in the Temperley-Lieb algebra TL_n that is a key ingredient of this construction.

2.4.1. *The Jones-Wenzl projector.* We review the Jones-Wenzl projector as defined in [**Lic97**]. Let A be a complex number such that A^4 is not a kth root of unity for $k \leq n$ and let $e_i =$ 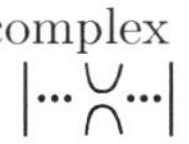 be a generator of the algebra TL_n. The

i $i+1$

n-th Jones-Wenzl projector is an element  in TL_n uniquely characterized by the following properties:

(1)  $\circ\, e_i = 0 = e_i \circ$  for $1 \leq i \leq n-1$.
(2) (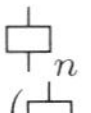 $- 1$) belongs to the algebra generated by $\{e_1, e_2, \ldots, e_{n-1}\}$.
(3)  $\circ$  $=$ 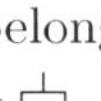, and
(4)  $= (-1)^n \frac{A^{2(n+1)} - A^{-2(n+1)}}{A^2 - A^{-2}}$.

Here $\bigcirc\!\!\!\!-_n$ denotes the sum of the closure of each term in $\boxed{}_n$ by n parallel strands. Let D be a link diagram. We define the n-cable of a link diagram decorated by the Jones-Wenzl projector, $\mathbf{D}^n$. First decompose each component D_j of the link diagram $D = D_1 \cup D_2 \cup \ldots \cup D_\ell$ into two (1, 1)-tangles $T_j \otimes T'_j$, where one is a simple arc denoted by T'_j. Let T^n denote the n-backboard cable of T. Then $\mathbf{D}^n$ is the tensor product of each (n, n)-tangle T^n_j in $T^n_1 \cup T^n_2 \cup \cdots \cup T^n_\ell$ with the n-th Jones-Wenzl projector (see Figure 3).

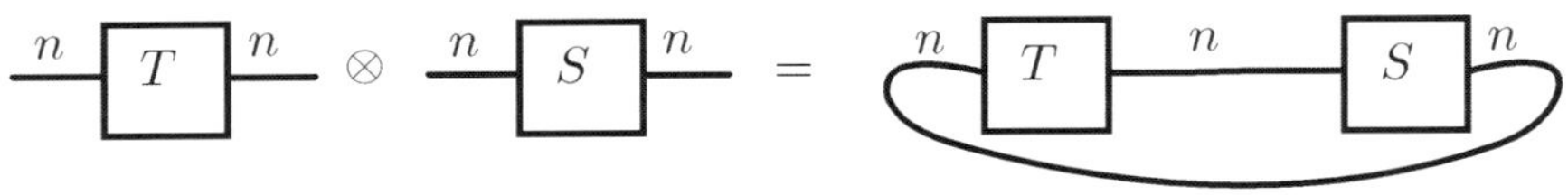

FIGURE 3. Tensor product of (n, n)-tangles.

DEFINITION 2.2. Let L be an oriented link in S^3 with diagram D. The unreduced n-th colored Jones polynomial is defined as follows.

$$\widehat{J^n_L}(t)/(-A^2 - A^{-2}) := ((-1)^n A)^{-w(D)(n^2+2n)} \langle \mathbf{D}^n \rangle \big|_{t^{1/2}=A^{-1/2}}$$

Here $w(D)$ is the writhe of the diagram D, and it is straightforward to see $\widehat{J^1_L}(t) = \widehat{J_L}(t)$ with the appropriate variable specialization to t from $q = -A^{-2}$.

2.4.2. *Computing the colored Jones polynomial.* Due to the cabling operation, the computational complexity of the n-th colored Jones polynomial is greater than that of the Jones polynomial, and closed formulas for the colored Jones polynomial are known for only a handful of families of knots. In general, the defining properties of the Jones-Wenzl projector can be used to simplify the diagram $\mathbf{D}^n$ on which the Kauffman bracket is evaluated to *spin networks.* For references of this method, see the original article [**MV94**] and [**Lic97**]. Briefly speaking, as many projectors as one desires can be inserted into $\mathbf{D}^n$ by the idempotent property (3) of the Jones-Wenzl projector. Then, crossings can be removed by applying fusion and uncrossing relations from properties (1) – (4) of the projector, resulting in sums over Kauffman brackets of crossingless diagrams decorated by projectors called *spin networks.* The Kauffman bracket of a spin network can be further evaluated using certain recurrence relations of the Jones-Wenzl projector, again derived from properties (1) – (4). Note Garoufalidis and Le also derived their own approach for computing the colored Jones in general [**GL05**] and this is what is implemented in KnotAtlas [**BNMea**].

2.4.3. *The Jones and colored Jones polynomials under twisting.* Champanerkar-Kofman showed the colored Jones polynomial exhibits interesting structural properties for links obtained by the operation of adding full (positive or negative) twists on a fixed number of strands in a link diagram [**CK13, CK05**], as another example of the principle of assembling local computations.

In [**CK05**], Champanerkar-Kofman decomposes copies of a full positive twist in terms of scalar multiples of minimal central orthogonal idempotents $\{p_i\}$ of TL_n. In their notation with the variable $t = A^{-4}$, they obtained ([**CK05**, Proof of Theorem 2.2])

$$\triangle_n^2 = \sum_{i=0}^{\lfloor n/2 \rfloor} t^{i(n-i+1)} p_i \qquad \text{and} \qquad (\triangle_n^2)^k = \sum_{i=0}^{\lfloor n/2 \rfloor} t^{ki(n-i+1)} p_i.$$

They used this to show the following.

THEOREM 5 ([**CK05**, Corollary 3.2]). *Suppose we twist n strands of L k times to obtain L_k such that the (unreduced) Jones polynomial for L_k changes (from L). If both N and n are fixed, then for k sufficiently large, the coefficient vector of $\widehat{J_{L_k}^N}(t)$ has fixed blocks separated by blocks of zeros which increase by constant lengths as k increases.*

The fixed blocks correspond to the evaluation of the rest of the link with the minimal central orthogonal idempotents in the Temperley-Lieb algebra. The larger the number k of full twists, the larger the power of the coefficient $t^{ki(n-i+1)}$, which contributes to the increasing separation of the constant blocks in Theorem 5.

2.5. Categorified n-th Jones-Wenzl Projectors. By a categorification of the colored Jones polynomial of a link L, we mean for each $n \geq 1$, a bigraded chain complex $CKh^n(L) = \{CKh^n_{i,j}(L)\}$ whose graded Euler characteristic recovers the unreduced n-th colored Jones polynomial.

$$\widehat{J^n_L}(q) = \sum_{i,j} (-1)^j q^j (rank(Kh^n_{i,j}(L))).$$

There are many constructions of this chain complex. For this paper we are using the construction by Cooper-Kruskal [**CK12**]. For each $n \geq 1$, they construct a chain complex in Kom^+, an extension of Kom that includes chain complexes with generators in (possibly) infinite positive homological degrees and bounded negative homological degrees.

The construction is built inductively on n starting with the categorification complex of the 2nd Jones-Wenzl projector shown in Figure 4. For the rest of the paper, we do not include quantum grading shifts in the pictorial description of a complex. The reader is referred to the reference [**CK12**] for the appropriate degree shifts.

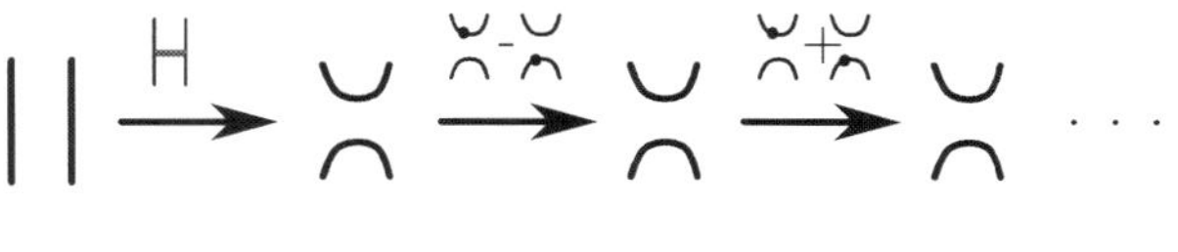

FIGURE 4. The chain complex for the 2nd Jones-Wenzl projector.

In Figure 4, the components of the differentials of the complexes with dots represent the identity cobordisms with dots, and red segments represent saddles between planar arcs.

In particular, Cooper-Krushkal constructs the complex for the n-th projector starting with a minimal set of generators built from the $(n-1)$-th projector. Additional generators and differentials are added to obtain a genuine chain complex with differentials squaring to 0, and so that the chain complex satisfies the homotopy version of the defining relations of the Jones-Wenzl projector. For details, the reader is referred to [**CK12**].

For the 3rd projector, Cooper-Krushkal describes the chain complex explicitly, which we will present here in Figure 5, again without grading shifts.

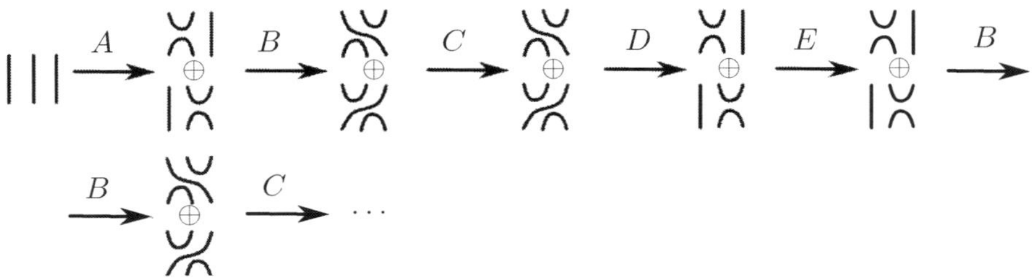

FIGURE 5. The complex for the 3rd projector. The maps and the generators cycle through B, C, D, E after the first 6 terms.

With the convention that the columns of a matrix give the images of the basis elements read from left to right and top to bottom, the maps are defined by

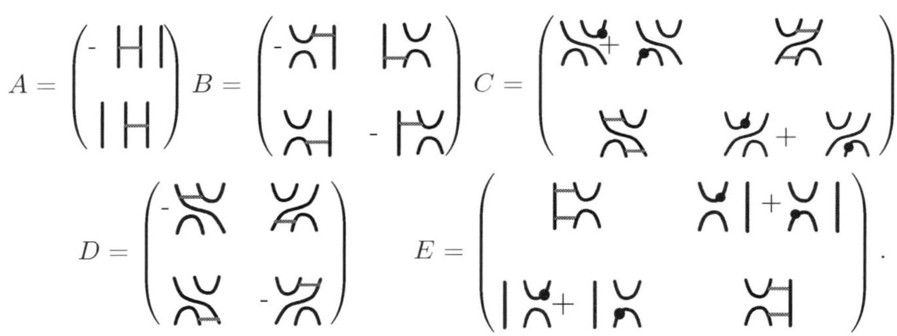

Take the braid closure of the Jones-Wenzl projector 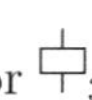, and consider the Khovanov chain complex of 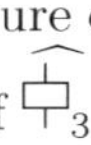 $=$ . We compute the 3-colored Khovanov homology of the unknot.

EXAMPLE 2.3. (The Khovanov homology of 3-colored unknot  $=$ ).

$$Kh_{0,*}(\widehat{\cdot_3}) = \mathbb{Z} \oplus \mathbb{Z}.$$

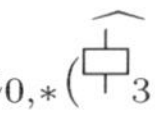

For $i \geq 1$,

$$Kh_{i,*}(\widehat{\cdot_3}) = \begin{cases} \mathbb{Z} & \text{if } i = 1 \mod 5 \\ \mathbb{Z} & \text{if } i = 2 \mod 5 \\ \mathbb{Z}/2\mathbb{Z} \oplus \mathbb{Z} & \text{if } i = 3 \mod 5 \\ \mathbb{Z} \oplus \mathbb{Z} & \text{if } i = 4 \mod 5 \\ \mathbb{Z} \oplus \mathbb{Z} & \text{if } i = 0 \mod 5 \end{cases}$$

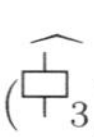

Working at the level of the categorification of torus braids $\beta(n,k)$ for $k \geq 1$, Rozansky defines a directed system (see Figure 6) on $\{CKh_{i,j}(\beta(n,k))\}_{k=1}^{\infty}$ and shows that the direct system has a unique limit[2] $Kh(\beta(n,\infty))$, called the *stable Khovanov homology of torus braids*. Decategorifying his result for $Kh(T(n,\infty))$ recovers Stosic's result.

$$CKh_{i,j}(\beta(n,k)) \xrightarrow{f_k} CKh_{i,j}(\beta(n,k+1)) \xrightarrow{\widetilde{f}_{k+1}} Kh_{i,j}(\beta(n,\infty)), \qquad \widetilde{f_k}: CKh_{i,j}(\beta(n,k)) \to Kh_{i,j}(\beta(n,\infty))$$

FIGURE 6. Directed system of $\{Kh_{i,j}(\beta(n,k))\}_{k=1}^{\infty}$.

Furthermore, Rozansky shows the chain complex $Kh(\beta(n,\infty))$ satisfies the axioms for a *universal Jones-Wenzl projector* [**CK12**, Section 3]. Therefore it gives a categorification of the Jones-Wenzl projector that is homotopy equivalent to Cooper-Krushkal's construction. It is through this identification of categorifications that the stable Khovanov homology of torus braids is connected to colored Khovanov homology, and therefore the colored Jones polynomial. We will denote this (universal) categorification of the n-th Jones-Wenzl projector by $\mathbf{P}^n$.

2.6. Colored Khovanov homology.

DEFINITION 2.4. Given a link diagram D, and its n-cable decorated by a Jones-Wenzl projector $\mathbf{D}^n$, the chain complex $CKh^n(D)$ for the categorification of the n-th colored Jones polynomial is a categorification of the Kauffman bracket of $\mathbf{D}^n$ constructed by taking the tensor product of the Khovanov chain complex for the tangle $\mathbf{T}^n$ and the chain complex $\mathbf{P}^n$ categorifying the n-th Jones-Wenzl projector. The homology groups $Kh^n(D)$ of the chain complex is called the *n-th colored Khovanov homology of the link*. In Figure 7 we show the 2nd colored Khovanov complex for the trefoil.

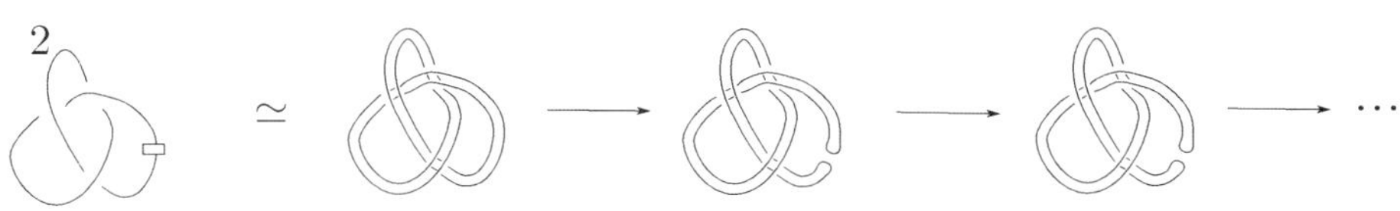

FIGURE 7. 2nd colored Khovanov homology of the trefoil where the differentials for the 2nd projector are shown.

Using Bar-Natan's formulation of Khovanov homology, we can generalize this definition to tangles and say the n-th colored Khovanov homology of the tangle T is the complex $CKh^n(T)$ formed by partially tensoring $\overline{\otimes}$ (decorating, see Figure 8) the categorification complex of T^n with that of the n-th Jones-Wenzl projector for each component of T.

[2]Note the definition differs from the usual definition of direct limit. Specifically, the universality properties have not been checked to hold for the limiting chain complex.

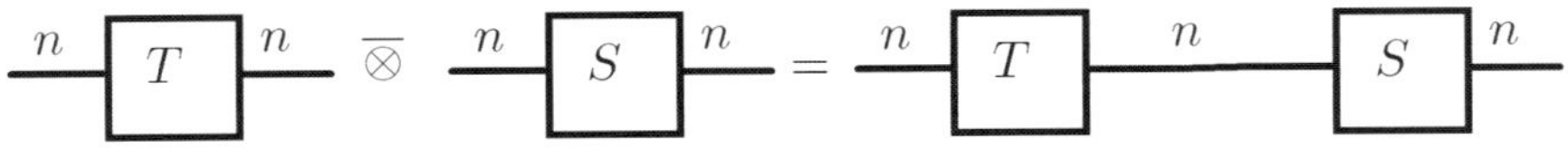

FIGURE 8. Partial tensoring of two (n, n)-tangles.

2.6.1. *Computations in colored Khovanov homology.* It follows from Sections 2.5 and 2.6 that the structure of the colored Khovanov homology of the unknot is an open problem that lies at the heart of the problem of computing colored Khovanov homology. Conjecture 4 presents the most concise (as far as the author knows) description of the chain complex for the Jones-Wenzl projector with a minimal set of generators and differentials. If true, one can apply the conjecture to compute colored Khovanov homology in the most efficient way possible. However, the difficulty lies in being able to show, whether in Rozansky's or Cooper-Kruskal's construction, that the differentials with these generators are as predicted by the statement of Conjecture 4. While Hogancamp has been able to give generators for the chain complex that satisfy the degree requirements of Conjecture 4 [**Hog19**], the exact forms of the differentials are not determined. In practice, one can simplify the generating set of the chain complex considerably using the homotopic version of the defining properties of the Jones-Wenzl projector and Gaussian elimination via Lemma 2, however, it remains difficult to pin down the exact form of the resulting differentials in the simplified complex on a large scale.

3. Truncating categorified Jones-Wenzl projectors

We distill conjectural implications on the structure of Khovanov homology from the previous sections. For a chain complex, we will consider the *width* on quantum grading. At a fixed homological degree i, let width_i be the number of nonzero generators at the homological grading i. The overall width of a chain complex C is

$$\text{width}_C = \sup_i \{\text{width}_i\}. \tag{1}$$

Note that this width is different from the Khovanov width that measures the largest horizontal distance between two lines spanning the support of Khovanov homology, see [**MO08**].

It follows from Stosic's and Rozansky's work that the Khovanov homology of torus braids injects into that of the stable Khovanov homology, and this injection can be used to determine the homology groups at low-homological degrees. Moreover, Cooper-Krushal shows that a truncated projector injects into the complex for the larger projector [**CK12**, Definition 7.14]. Based on these partial results and computational evidence, we ask a question in the opposite direction:

QUESTION 6. Can the categorification complex of the Jones-Wenzl projector be used to compute the Khovanov homology of torus braids?

We propose that the chain complex of the categorification of a torus braid $T(n,k)$ for any k may be obtained by restricting, or *truncating*, the chain complex for the categorification of the n-th Jones-Wenzl projector.

DEFINITION 3.1. Let (C, d) be a chain complex in Kom^+. A *truncation* at $k > 0$ of C is a chain complex (C', d') obtained from C such that for $0 \le i \le k$, $C_i = C'_i$ and d'_i is a component of d_i.

CONJECTURE 1. *Let $p(n,k)$ be an integer depending on k. The categorification complex $CKh(\beta(n,k))$ for a torus braid $\beta(n,k)$ is homotopy equivalent to a complex $\mathbf{B}^n$ truncated at $p(n,k)$ from that of a categorification complex $\mathbf{P}^n \simeq CKh(\beta(n,\infty))$ with bounded width of the n-th Jones-Wenzl projector.*

REMARK 3.2. Note that combining results by Hogancamp, Cooper-Krushkal, and Rozansky as discussed prior to Conjecture 1 implies that a truncated version of $CKh(\beta(n,k))$ is homotopy equivalent to a truncated version of $CKh(\beta(n,\infty))$. However, Conjecture 1 makes a stronger statement that *all* of the complex $CKh(\beta(n,k))$, including the terms with homological degrees towards the end of the nonzero portion of the complex may be obtained from a truncation of $CKh(\beta(n,\infty))$. Thus Conjecture 1 is a new prediction that, as far as the author knows, has not appeared in the literature.

A result by Islambouli-Willis [**IW18**] implies that we do not need to restrict ourselves to torus braids when constructing the chain complex categorifying the Jones-Wenzl projector.

THEOREM 7 ([**IW18**, Theorem 1.1]). *Let b be any complete semi-infinite positive braid, viewed as the limit of positive braid words*

$$b = \lim_{\ell \to \infty} \sigma_{j_1}\sigma_{j_2}\cdots\sigma_{j_\ell}.$$

Then the limiting Khovanov complex exists for b as $\ell \to \infty$ and is a categorification of the Jones-Wenzl projector.

A direct corollary is that the categorification of the n-cable of a twist region $T(2,k)$ with k half-twists decorated by a Jones-Wenzl projector, denoted by $\mathbf{T(2,k)^n}$, forms a directed system as $k \to \infty$, and has the categorification of the $2n$-th Jones-Wenzl projector as the direct limit. This fact was used to determine the direct limit of the direct system of colored Khovanov homologies of highly twisted links in an analogue to Thurston's hyperbolic Dehn surgery theorem in [**Lee**], and we use it to generalize Conjecture 1.

CONJECTURE 8. *Let β be any positive braid on n strands in which every braid generator σ_i, $1 \le i \le n$ occurs. Then the categorification complex $CKh(\beta^k)$ is homotopy-equivalent to a complex truncated from the categorification complex of bounded width of the n-th Jones-Wenzl projector.*

REMARK 3.3. We remark that β^k gives the braid word

$$\beta^k = \underbrace{\beta \cdot \beta \cdots \beta}_{k \text{ times}},$$

and the superscript k is not to be confused with the superscript n for the n-cable of a link or tangle diagram.

The computational advantages afforded by Conjectures 1 and 8 are that a categorification complex $\mathbf{P}^n$ for the Jones-Wenzl projector, while potentially infinite in positive homological degrees, is bounded in width and periodic in both its differentials and generators. Provided the truncation is understood meaning that one knows precisely how to obtain the differentials and generators for $CKh(\beta^k)$ from that of $\mathbf{P}^n$, we can replace the categorification complex for the n-cables of twist regions decorated by the Jones-Wenzl projector by much simpler chain complexes, which occur in the computation of the colored Khovanov homology of a link. This could significantly reduce the computational complexity of both the colored Jones polynomial and colored Khovanov homology of a highly twisted link by avoiding projectors.

For 2-braids, Conjecture 1 follows from a Proposition in [**Kho00**]. We give here the version in [**Man18**], rephrased in the context of the article.

LEMMA 9 ([**Man18**, Lemma 2.1]). *The (formal) Khovanov chain complex of the positive k-half twisted strands (viewed as a braid on two strands)* 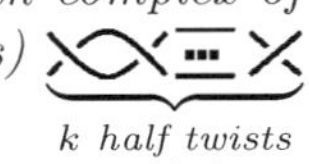

is homotopy equivalent to the dotted cobordism complex in Figure 9 *truncated from the complex of Figure* 4.

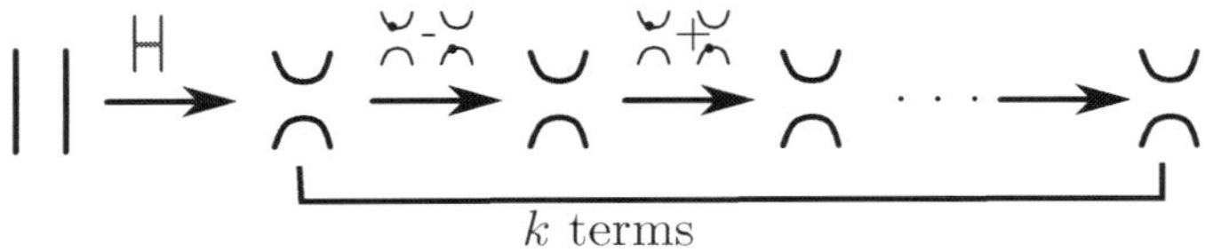

FIGURE 9. The truncated 2nd Jones-Wenzl projector.

The 3-braid case is done in [**CGL**$^+$]. The version of Lemma 9 for negative half twists is obtained by taking the dual of the complex in Kom^-, where the homological degrees of the complex are bounded from above.

4. Examples

4.1. The Khovanov homology of τ-tangle pretzel links. Due to the simplification of the categorification complex of half twists on two strands by Lemma 9, the Jones polynomial for pretzel links (and any link L in general) only depends on the planar weighted graph G whose edges correspond to maximal twist regions in the link diagram, such that if we replace the vertices of G by disks and the edges by twisted bands (specified by the weights on the edges) to obtain a surface F, then $\partial F = L$. Landvoy [**Lan98**] used this to give an algorithm and a computer program to compute the Jones polynomial of a pretzel link. We describe how to do this for the Khovanov homology of a pretzel link $P(s_1, s_2, \ldots, s_\tau)$. A computer program which currently does the case for $P(s_1 = s, s_2 = s, \ldots, s_\tau = s)$ can be shared via GitHub by contacting the author.

The algorithm is as follows: For each twist region with s_i crossings, $1 \leq i \leq \tau$, we define a generalized Kauffman state that assigns to each crossing in the twist region an integer from 0 to s_i, where an output of 0 represents the choice of the 0-resolution as before, the number 1 represents the choice of the 1-resolution, while the rest of the integers represent the repeated choice of the 1-resolution as we progress through truncated chain complex of Figure 9. Record the image of the Kauffman state as a vector.

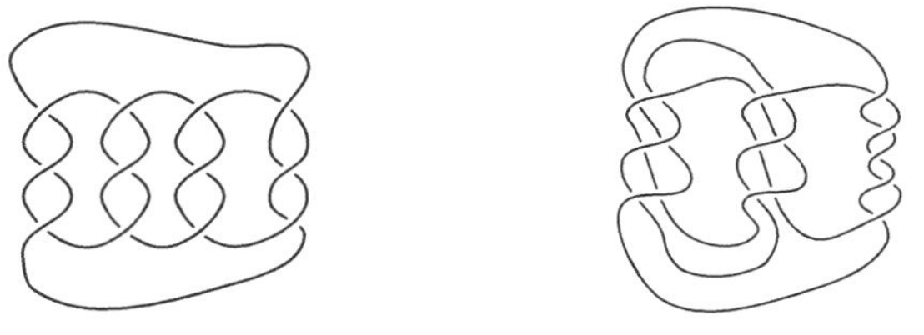

FIGURE 10. Left: $P(3,3,3,3)$-pretzel link (standard diagram). Right: Link diagram with full twists on 3 strands.

For each homological degree of the chain complex, build the list of Kauffman states and generators based on choosing the 0- or 1-resolution at a twist region. This list will completely determine the Kauffman states for the rest of the complex.

Starting with homological degree 0, evaluate the image of each generator in the list and modify the state to its successor by incrementing each entry of the state vector by 1, and applying the corresponding differential in the chain complex of Figure 9. If the entry in the new state vector after incrementing is ≥ 2, then the image will be identical, and a new Kauffman state is added to the list in the next homological grading. At this step, we also construct the matrix of the differential between homological degree 0 and homological degree 1. Repeat the process for each consecutive homological degree. Finally, find the homology groups using the matrices of differentials.

For the pretzel link $P(3,3,3,3)$, compared to the original Khovanov complex of 2^{12} states, the new complex without duplicate states, partially shown in Figure 11 has $(3+1)^4 = 2^8 < 2^{12}$ states, a significant improvement in computational efficiency. In general, given a positive τ-tangle pretzel link $P(s_1 = s, s_2 = s, \ldots, s_\tau = s)$, with $s \cdot \tau$ number of crossings, the number of states with the new algorithm is $(s+1)^\tau < 2^{s \cdot \tau}$. If applied to links containing full twists on $n \geq 3$ strands, such as the link shown in Figure 10 on the right, the efficiency gain is expected to be even greater. We will investigate the Khovanov homology of such links with twisting on a high number of strands $n \geq 3$ in a future project.

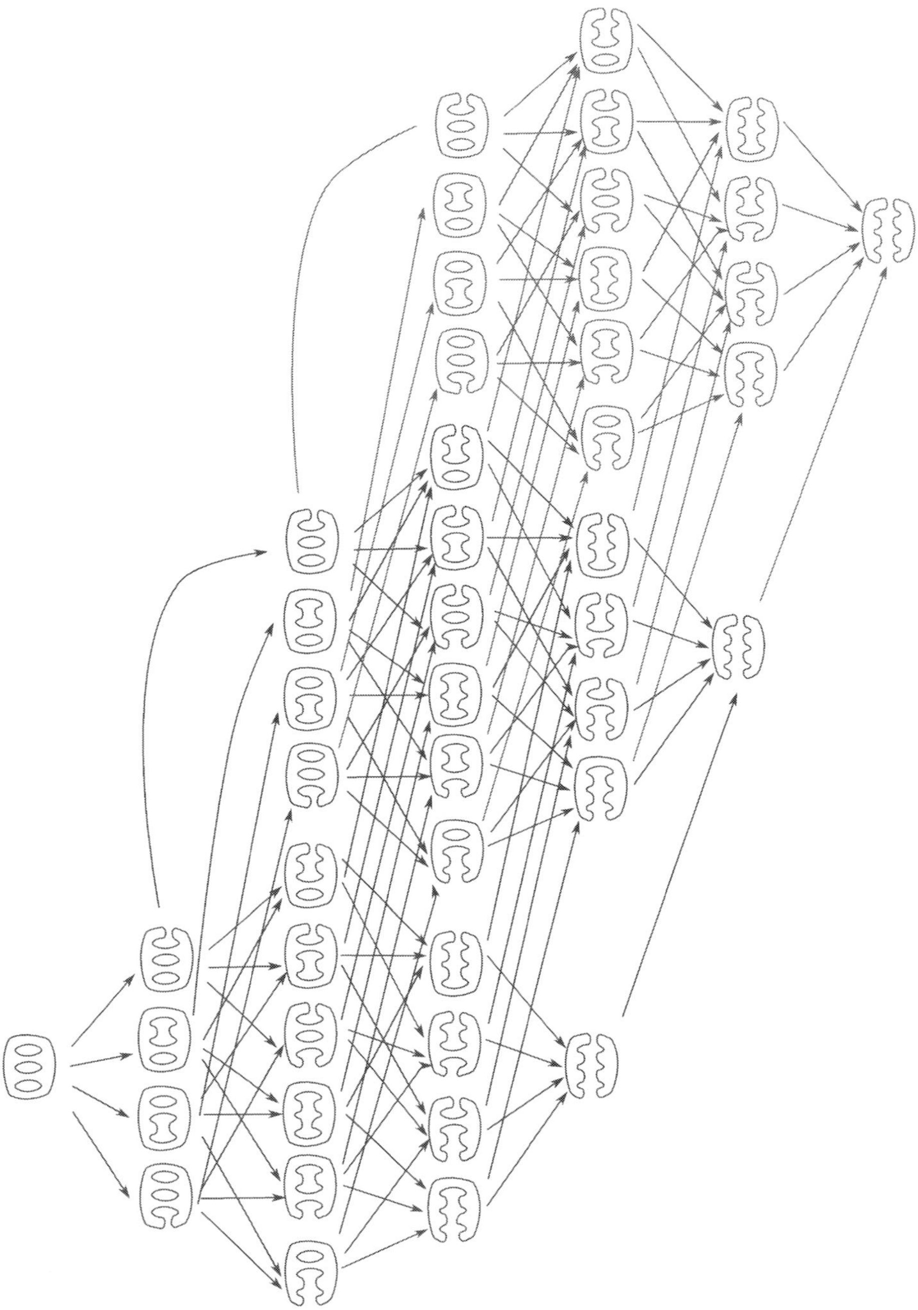

FIGURE 11. A portion of the complex for the pretzel knot $P(3,3,3,3)$. Different colored arrows represent differentials of the truncated 2nd Jones-Wenzl projector.

Acknowledgments

The author would like to thank the anonymous referee for their comments that have greatly improved the quality of the paper. The author would also like to acknowledge the partial support by NSF grant No. 2244923.

References

[BN05] Dror Bar-Natan, *Khovanov's homology for tangles and cobordisms*, Geom. Topol. **9** (2005), 1443–1499, DOI 10.2140/gt.2005.9.1443. MR2174270

[BN07] Dror Bar-Natan, *Fast Khovanov homology computations*, J. Knot Theory Ramifications **16** (2007), no. 3, 243–255, DOI 10.1142/S0218216507005294. MR2320156

[BNMea] Dror Bar-Natan, Scott Morrison, et al., *The Knot Atlas*, katlas.org.

[CGL+] Carmen Caprau, Nicolle Gonzalez, Christine Ruey Shan Lee, Radmila Sazdanovic, and Melissa Zhang, *A whittled complex for the Khovanov homology of torus links*, in preparation.

[CK05] Abhijit Champanerkar and Ilya Kofman, *On the Mahler measure of Jones polynomials under twisting*, Algebr. Geom. Topol. **5** (2005), 1–22, DOI 10.2140/agt.2005.5.1. MR2135542

[CK12] Benjamin Cooper and Vyacheslav Krushkal, *Categorification of the Jones-Wenzl projectors*, Quantum Topol. **3** (2012), no. 2, 139–180, DOI 10.4171/QT/27. MR2901969

[CK13] Abhijit Champanerkar and Ilya Kofman, *On the tail of Jones polynomials of closed braids with a full twist*, Proc. Amer. Math. Soc. **141** (2013), no. 7, 2557–2567, DOI 10.1090/S0002-9939-2013-11555-8. MR3043035

[FGL02] Charles Frohman, Răzvan Gelca, and Walter Lofaro, *The A-polynomial from the noncommutative viewpoint*, Trans. Amer. Math. Soc. **354** (2002), no. 2, 735–747, DOI 10.1090/S0002-9947-01-02889-6. MR1862565

[Gar04] Stavros Garoufalidis, *On the characteristic and deformation varieties of a knot*, Proceedings of the Casson Fest, Geom. Topol. Monogr., vol. 7, Geom. Topol. Publ., Coventry, 2004, pp. 291–309, DOI 10.2140/gtm.2004.7.291. MR2172488

[Gar11] Stavros Garoufalidis, *The Jones slopes of a knot*, Quantum Topol. **2** (2011), no. 1, 43–69, DOI 10.4171/QT/13. MR2763086

[GL05] Stavros Garoufalidis and Thang T. Q. Lê, *The colored Jones function is q-holonomic*, Geom. Topol. **9** (2005), 1253–1293, DOI 10.2140/gt.2005.9.1253. MR2174266

[GOR13] Eugene Gorsky, Alexei Oblomkov, and Jacob Rasmussen, *On stable Khovanov homology of torus knots*, Exp. Math. **22** (2013), no. 3, 265–281, DOI 10.1080/10586458.2013.798553. MR3171092

[HMTY09] Masao Hara, Masahiko Murakami, Seiichi Tani, and Makoto Yamamoto, *A fast algorithm for computing Jones polynomials of Montesinos links*, Sci. Math. Jpn. **69** (2009), no. 1, 1–26. MR2483781

[Hog18] Matthew Hogancamp, *Categorified Young symmetrizers and stable homology of torus links*, Geom. Topol. **22** (2018), no. 5, 2943–3002, DOI 10.2140/gt.2018.22.2943. MR3811775

[Hog19] Matthew Hogancamp, *A polynomial action on colored $\mathfrak{sl}_2$ link homology*, Quantum Topol. **10** (2019), no. 1, 1–75, DOI 10.4171/QT/122. MR3900776

[IW18] Gabriel Islambouli and Michael Willis, *The Khovanov homology of infinite braids*, Quantum Topol. **9** (2018), no. 3, 563–590, DOI 10.4171/QT/114. MR3827809

[JVW90] F. Jaeger, D. L. Vertigan, and D. J. A. Welsh, *On the computational complexity of the Jones and Tutte polynomials*, Math. Proc. Cambridge Philos. Soc. **108** (1990), no. 1, 35–53, DOI 10.1017/S0305004100068936. MR1049758

[Kas97] R. M. Kashaev, *The hyperbolic volume of knots from the quantum dilogarithm*, Lett. Math. Phys. **39** (1997), no. 3, 269–275, DOI 10.1023/A:1007364912784. MR1434238

[Kau23] Louis H. Kauffman, *The Jones polynomial, knots, diagrams, and categories*, Bull. Amer. Math. Soc. (N.S.) **60** (2023), no. 4, 507–537, DOI 10.1090/bull/1792. MR4642117

[Kho00] Mikhail Khovanov, *A categorification of the Jones polynomial*, Duke Math. J. **101** (2000), no. 3, 359–426, DOI 10.1215/S0012-7094-00-10131-7. MR1740682

[Lan98] Ryan A. Landvoy, *The Jones polynomial of pretzel knots and links*, Topology Appl. **83** (1998), no. 2, 135–147, DOI 10.1016/S0166-8641(97)00100-4. MR1606488

[Lee] Christine Ruey Shan Lee, *Stable Khovanov homology and volume*, Proc. Amer. Math. Soc. **153** (2025), no. 5, 2219–2234, DOI 10.1090/proc/17029. MR4881406

[Lic97] W. B. Raymond Lickorish, *An introduction to knot theory*, Graduate Texts in Mathematics, vol. 175, Springer-Verlag, New York, 1997, DOI 10.1007/978-1-4612-0691-0. MR1472978

[Man18] Andrew Manion, *The Khovanov homology of 3-strand pretzels, revisited*, New York J. Math. **24** (2018), 1076–1100. MR3890966

[MM01] Hitoshi Murakami and Jun Murakami, *The colored Jones polynomials and the simplicial volume of a knot*, Acta Math. **186** (2001), no. 1, 85–104, DOI 10.1007/BF02392716. MR1828373

[MO08] Ciprian Manolescu and Peter Ozsváth, *On the Khovanov and knot Floer homologies of quasi-alternating links*, Proceedings of Gökova Geometry-Topology Conference 2007, Gökova Geometry/Topology Conference (GGT), Gökova, 2008, pp. 60–81. MR2509750

[MS90] H. R. Morton and H. B. Short, *Calculating the 2-variable polynomial for knots presented as closed braids*, J. Algorithms **11** (1990), no. 1, 117–131, DOI 10.1016/0196-6774(90)90033-B. MR1041170

[MV94] G. Masbaum and P. Vogel, *3-valent graphs and the Kauffman bracket*, Pacific J. Math. **164** (1994), no. 2, 361–381. MR1272656

[Oht01] Tomotada Ohtsuki, *Quantum invariants*, Series on Knots and Everything, vol. 29, World Scientific Publishing Co., Inc., River Edge, NJ, 2002. A study of knots, 3-manifolds, and their sets. MR1881401

[Roz14] Lev Rozansky, *An infinite torus braid yields a categorified Jones-Wenzl projector*, Fund. Math. **225** (2014), no. 1, 305–326, DOI 10.4064/fm225-1-14. MR3205575

[Sto07] Marko Stošić, *Homological thickness and stability of torus knots*, Algebr. Geom. Topol. **7** (2007), 261–284, DOI 10.2140/agt.2007.7.261. MR2308944

[Tur17] Paul Turner, *Five lectures on Khovanov homology*, J. Knot Theory Ramifications **26** (2017), no. 3, 1741009, 41, DOI 10.1142/S0218216517410097. MR3627709

Texas State University, Department of Mathematics, San Marcos, Texas
Email address: `vne11@txstate.edu`

Contemporary Mathematics
Volume **827**, 2025
https://doi.org/10.1090/conm/827/16532

A survey on skein modules via braids

Ioannis Diamantis

Abstract. In this paper we present recent results on the computation of skein modules of 3-manifolds using braids and appropriate knot algebras. Skein modules generalize knot polynomials in S^3 to knot polynomials in arbitrary 3-manifolds and they have become extremely essential algebraic tools in the study of 3-manifolds. In this paper we present the braid approach to the HOMFLYPT and the Kauffman bracket skein modules of the Solid Torus ST and the lens spaces $L(p,1)$ and $S^1 \times S^2$.

1. Introduction

Skein modules were introduced independently by Przytycki [**34**] and Turaev [**36**] as generalizations of knot polynomials in S^3 to knot polynomials in arbitrary 3-manifolds. They are quotients of free modules over isotopy classes of links in 3-manifolds by properly chosen local (skein) relations. A skein module of a 3-manifold yields all possible isotopy invariants of knots which satisfy a particular skein relation. Skein modules based on the Kauffman bracket skein relation

$$L_+ - AL_0 - A^{-1}L_\infty$$

where L_∞ and L_0 are represented schematically by the illustrations in Figure 1, are called *Kauffman bracket skein modules* (KBSM).

The precise definition of KBSM is as follows:

Definition 1.1. Let M be an oriented 3-manifold and $\mathcal{L}_{\mathrm{fr}}$ be the set of isotopy classes of unoriented framed links in M. Let $R = \mathbb{Z}[A^{\pm 1}]$ be the Laurent polynomials in A and let $R\mathcal{L}_{\mathrm{fr}}$ be the free R-module generated by $\mathcal{L}_{\mathrm{fr}}$. Let $\mathcal{S}$ be the ideal generated by the skein expressions $L_+ - AL_0 - A^{-1}L_\infty$ and $L \bigsqcup \mathrm{O} - (-A^2 - A^{-2})L$. Note that blackboard framing is assumed and that $L \bigsqcup \mathrm{O}$ stands for the union of a link L and the trivially framed unknot in a ball disjoint from L.

2020 *Mathematics Subject Classification.* Primary 57K31, 57K14, 20F36, 20F38, 57K10, 57K12, 57K45, 57K35, 57K99, 20C08.

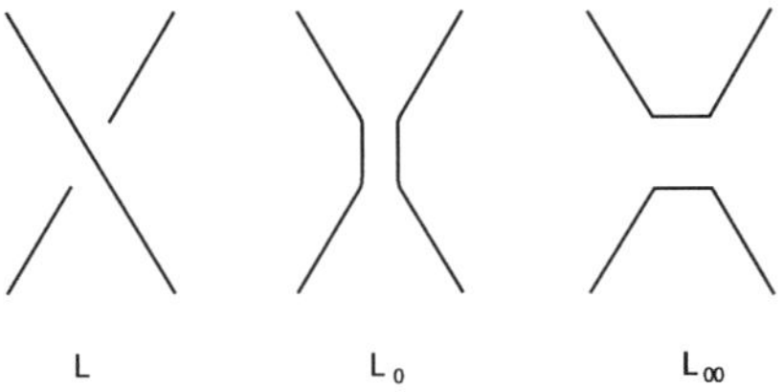

FIGURE 1. The links L, L_0 and L_∞ locally.

Then the *Kauffman bracket skein module* of M, KBSM(M), is defined to be:

$$\mathrm{KBSM}\,(M) = {}^{R\mathcal{L}_{\mathrm{fr}}}/_{S}.$$

For example, the Kauffman bracket skein module of S^3, KBSM(S^3), is freely generated by the unknot and it is equivalent to the Kauffman bracket for framed links in S^3 up to regular isotopy. It's ambient isotopy counterpart is the well-known Jones polynomial. Recall that the algebraic counterpart construction of the Kauffman bracket is the Temperley-Lieb algebra together with a unique Markov trace constructed by V.F.R. Jones (see § 2.3 or [**26**] and references therein). It is worth pointing out the importance of the Temperley-Lieb algebras. Through the pioneering work of V.F.R. Jones ([**26, 27**]), these algebras related knot theory to statistical mechanics, topological quantum field theories and the construction of quantum invariants for 3-manifolds (works of Witten, Reshetikhin-Turaev, Lickorish, etc).

Skein modules based on the HOMFLYPT skein relation

$$u^{-1}L_+ - uL_- - zL_0$$

where L_+, L_- and L_0 are represented schematically by the illustrations in Figure 2, are called *HOMFLYPT skein modules* (HOM).

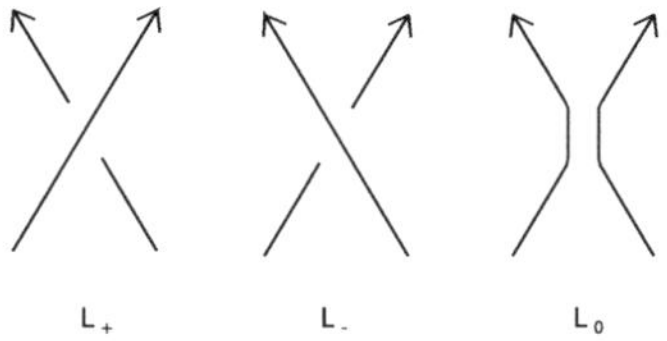

FIGURE 2. The links L_+, L_-, L_0 locally.

We have the following definition:

DEFINITION 1.2. Let M be an oriented 3-manifold, $R = \mathbb{Z}[u^{\pm 1}, z^{\pm 1}]$, $\mathcal{L}$ the set of all oriented links in M up to ambient isotopy in M and let S be the submodule of $R\mathcal{L}$ generated by the skein expressions $u^{-1}L_+ - uL_- - zL_0$, where L_+, L_- and L_0 comprise a Conway triple represented schematically by the illustrations in Figure 2.

For convenience we allow the empty knot, $\emptyset$, and add the relation $u^{-1}\emptyset - u\emptyset = zT_1$, where T_1 denotes the trivial knot. Then the *HOMFLYPT skein module* of M is defined to be:

$$\mathcal{S}(M) = \mathcal{S}\left(M; \mathbb{Z}\left[u^{\pm 1}, z^{\pm 1}\right], u^{-1}L_+ - uL_- - zL_0\right) = {}^{R\mathcal{L}}/_{S}.$$

For example, $\mathcal{S}(S^3)$ is freely generated by the unknot ([**18**, **35**]). For a survey on skein modules see [**25**]. It is worth mentioning that the HOMFLYPT skein module of a 3-manifold is very hard to compute.

Skein modules of 3-manifolds have become very important algebraic tools in the study of 3-manifolds, since their properties renders topological information about the 3-manifolds. For example, the existence of non-separating spheres and tori in a 3-manifold M is captured by the torsion part of KBSM(M).

The paper is organized as follows: In § 2 we recall Jones's construction of the Jones and the HOMFLYPT polynomials via a trace function on appropriate knot algebras. For the sake of completion, in § 2.1 we also present the Kauffman bracket polynomial following [**28**]. In § 3 we recall the setting and results from [**1**, **30**–**33**] and we present the universal invariants for knots and links in the Solid Torus ST, of the HOMFLYPT type (§ 3.2) and of the Kauffman bracket type (§ 3.3). In § 4 we generalize these universal invariants for knots and links in the lens spaces $L(p,1)$, taking into consideration the extra isotopy move for knots in $L(p,1)$. This is possible with the use of the new bases of the corresponding skein modules presented in § 4.2 for HOM(ST) and in § 4.3 for KBSM(ST). Finally, in § 5 we present recent results on KBSM($S^1 \times S^2$) via braids from [**3**].

2. Invariants of knots in S^3

2.1. Preliminaries. An oriented link on m-components is an embedding of m-copies of S^1 to S^3, in which each component is assigned an orientation. The main question of knot theory is *"How can we distinguish one link from another?"* In order to answer this question, we first define an equivalence relation on the set of links. We shall call two links L_1, L_2 in S^3 equivalent (or isotopic), denoted by $L_1 \sim L_2$, if there is an orientation-preserving piecewise linear homeomorphism $h : S^3 \to S^3$, such that $h(L_1) = L_2$. We may reduce the complexity of the theory by considering link projections on the plane, that we call *link diagrams*. More precisely, a link diagram is a projection of the link onto the plane, where we keep track of over and under crossings. Note that we only allow double points and no more than two points are allowed to be superposed. Since now a link diagram depends on the plane that we project the link on, we translate the notion of equivalence of links on the diagrammatic level as follows:

THEOREM 2.1 (Reidemeister [**29**]). *Two link diagrams correspond to isotopic links if and only if one can be obtained from the other by a finite sequence of Reidemeister moves and planar isotopies (or Delta moves), as illustrated in Figure* 3.

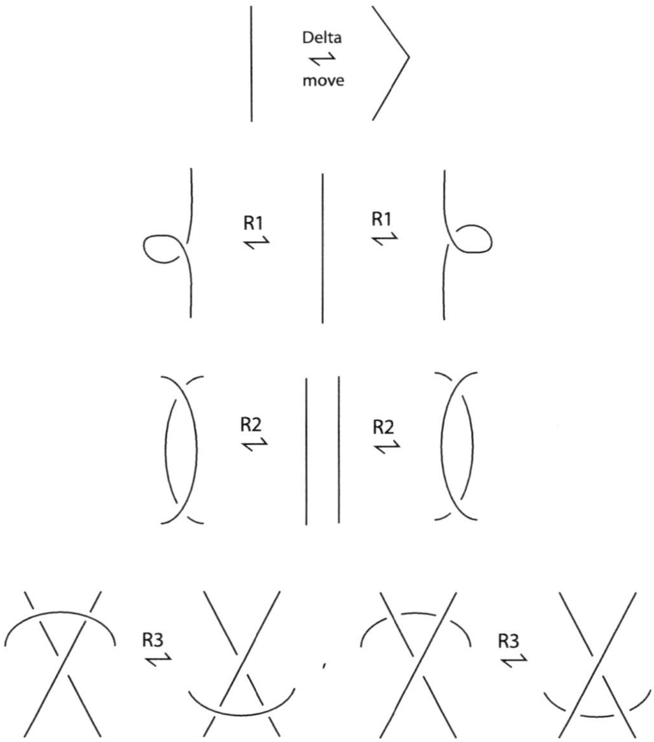

FIGURE 3. Isotopy moves.

A framed link on m-components is an embedding of a disjoint union of annuli in S^3 (see Figure 4). In the theory of framed links the Reidemeister move 1 is forbidden.

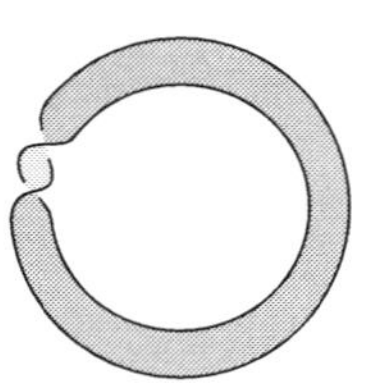

FIGURE 4. A framed knot.

If two knots are isotopic, using Reidemeister moves we can turn the diagram of one to the diagram of the other. But how can we tell if two knots are not isotopic? We do that by defining functions from the set of links, that we call *link invariants*, with the use of which we may determine whether two links are not equivalent. A very important invariant is the Kauffman bracket defined as follows:

DEFINITION 2.2. Let L be a framed link. The *Kauffman bracket polynomial* of L is defined by means of the following relations:

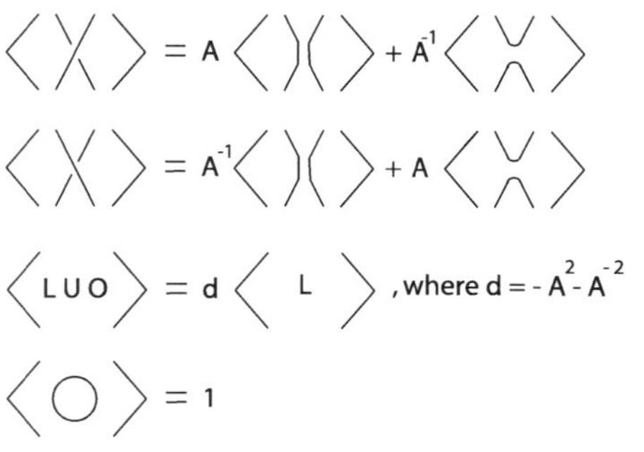

Simple computations show that $< L >$, which is a Laurent polynomial in $\mathbb{Z}\left[A^{\pm 1}\right]$, is an invariant of framed links. In order to obtain an invariant for links in S^3, we normalize the Kauffman bracket by considering the product of $< L >$ by the factor $\left(-A^3\right)^{-wr(L)}$, where $wr(L)$ is the writhe of the link L, defined as the number of positive crossings minus the number of negative crossings of L (see Figure 5).

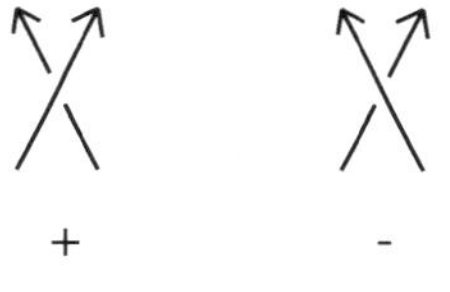

FIGURE 5. The sign of the crossings.

Then, the polynomial

$$P_L(A, A^{-1}) \;=\; (-A^{-3})^{w(L)} \; < L >,$$

is an invariant of links in S^3.

We now perform isotopy moves on a link diagram and we transform it to a very specific "knotted object" that we call a braid. A *braid* on n strands is defined as a set of pairwise non-intersecting descending polygonal lines (strands) joining the points $A_1, A_2, \ldots, A_n$ to the points $B_1, B_2, \ldots, B_n$ in any order, where $A_i = (i, 0, 0)$ and $B_i = (i, 0, 1)$ for $i = 1, 2, \ldots, n$. Two braids are called *isotopic* if and only if one can be transformed into the other by a finite sequence of *elementary deformations* (see Figure 6).

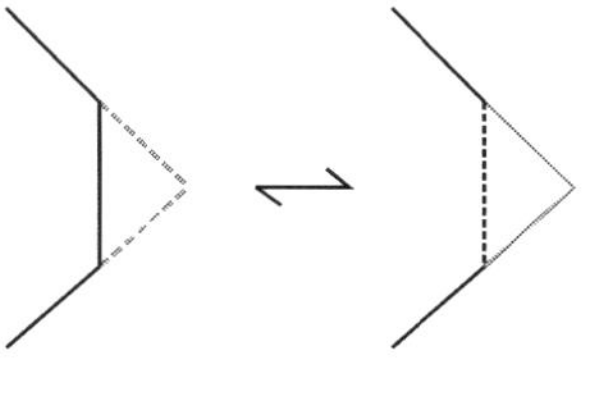

FIGURE 6. Elementary deformations.

The set of (equivalent classes of) braids on n strands has a natural group structure. We define the product of two braids a and b by joining the bottom points of the first braid to the top points of the second. The set of braids in n strands under this operation is called the braid group and is denoted by B_n. The unit element is the braid consisting of n parallel vertical strands and the inverse of a braid a, a^{-1}, is the mirror image of a in the plane. B_n has the following presentation:

$$B_n = \left\langle \quad \sigma_1, \ldots, \sigma_{n-1} \quad \left| \begin{array}{ll} \sigma_i\sigma_{i+1}\sigma_i = \sigma_{i+1}\sigma_i\sigma_{i+1}, & 1 \leq i \leq n-2 \\ \sigma_i\sigma_j = \sigma_j\sigma_i, \quad |i-j| > 1 & \end{array} \right. \right\rangle,$$

where the generators σ_i and σ_i^{-1} are illustrated in Figure 7.

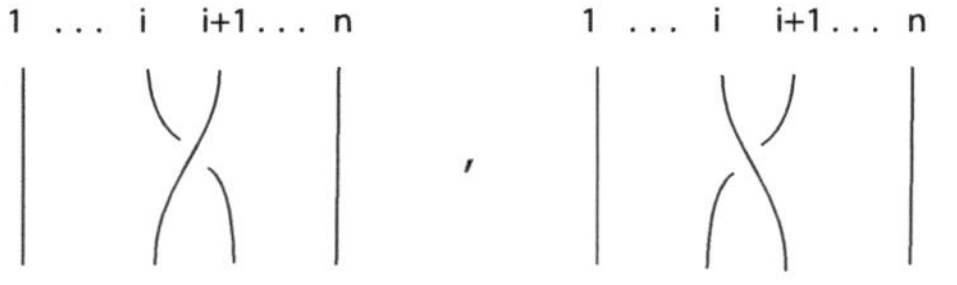

FIGURE 7. The generators σ_i and σ_i^{-1} of B_n.

From a topological point of view, closing a braid, that is, connecting corresponding ends in pairs, gives rise to an oriented link. The *closure of a braid* a is defined as the link $\widehat{a}$ obtained by joining the upper points of its strands to the lower ones (see Figure 8). The converse is also true. In particular, we have the following theorem by Alexander:

THEOREM 2.3 (Alexander). *Any oriented link is isotopic to the closure of a braid.*

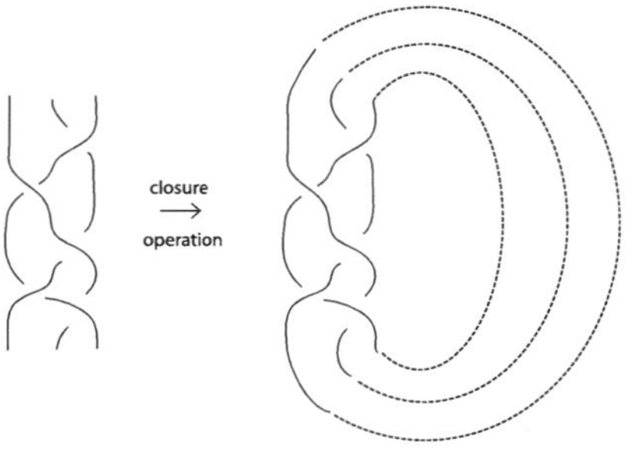

FIGURE 8. The closure of a braid.

Moreover, equivalence classes of braids are in one-to-one correspondence with the isotopy classes of oriented links. More precisely, we have the following theorem:

THEOREM 2.4 (Markov). *The closures of two braids are isotopic if and only if one braid can be taken to another by finite sequence of the following moves (for an illustration see Figure 9):*

$$\textit{Conjugation:} \quad a \quad \sim \quad bab^{-1}, \quad a, b \in B_n,$$

$$\textit{Stabilization:} \quad a \quad \sim \quad a\sigma_n^{-1}, \quad a \in B_n.$$

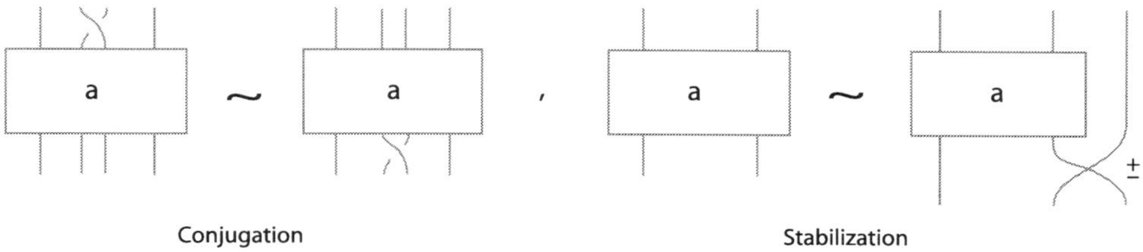

FIGURE 9. Braid equivalence.

REMARK 2.5. Jones noticed that instead of studying knots up to isotopy, we may study braids up to Markov's theorem. That is,

$$\frac{\text{Knots}}{\text{Reidemeister moves}} = \frac{\text{Braids}}{\text{Markov's theorem}}.$$

This idea lead to the discovery of the Jones polynomial in the 80's.

2.2. The HOMFLYPT polynomial. In this subsection we recall the construction of the HOMFLYPT polynomial via a unique Markov trace constructed on the Iwahori-Hecke algebras of type A. The Iwahori-Hecke algebra of type A, H_n, is a quotient of the braid group algebra $\mathbb{Z}[q^{\pm 1}]B_n$ by factoring out the quadratic relations

$$\sigma_i^2 = (q-1)\sigma_i + q.$$

These relations accurately reflect the skein relation of the HOMFLYPT polynomial. The algebra H_n has the following presentation:

$$\mathrm{H}_n(q) = \left\langle \quad g_1, g_2, \ldots, g_{n-1} \quad \left| \begin{array}{ll} g_i g_{i+1} g_i = g_{i+1} g_i g_{i+1}, & 1 \leq i \leq n-2 \\ g_i g_j = g_j g_i, \quad |i-j| > 1 & \\ g_i^2 = (q-1) g_i + q, \quad i = 1, 2, \ldots n-1 & \end{array} \right. \right\rangle,$$

that is

$$\mathrm{H}_n(q) = \frac{\mathbb{Z}\left[q^{\pm 1}\right] B_n}{\langle \sigma_i^2 - (q-1)\,\sigma_i - q \rangle}.$$

In [**26**] V.F.R. Jones gives the following linear basis for the Iwahori-Hecke algebra of type A, $\mathrm{H}_n(q)$:

$$S = \left\{ (g_{i_1} g_{i_1 - 1} \ldots g_{i_1 - k_1})(g_{i_2} g_{i_2 - 1} \ldots g_{i_2 - k_2}) \ldots (g_{i_p} g_{i_p - 1} \ldots g_{i_p - k_p}) \right\}, \quad \text{for } 1 \leq i_1 < \ldots < i_p \leq n-1\,. \tag{2.1}$$

The basis S yields directly an inductive basis for $\mathrm{H}_n(q)$, which is used in the construction of the Ocneanu trace, leading to the HOMFLYPT or 2-variable Jones polynomial. Recall that a linear function f from an algebra to some module is called a trace if it satisfies $f(xy) = f(yx)$ for all x, y in the algebra. We now have the following:

THEOREM 2.6 (Ocneanu). *There exists a unique linear Markov trace function:*

$$\mathrm{tr} : \bigcup_{n=1}^{\infty} \mathrm{H}_n(q) \to \mathbb{C},$$

determined by the rules:

$$\begin{array}{llcll} (1) & \mathrm{tr}(1) & = & 1 & \\ (2) & \mathrm{tr}(ab) & = & \mathrm{tr}(ba) & \textit{for } a, b \in \mathrm{H}_n(q) \\ (3) & \mathrm{tr}(ag_n) & = & z\mathrm{tr}(a) & \textit{for } a \in \mathrm{H}_n(q) \end{array}$$

THEOREM 2.7. *The function* $X : \mathcal{L} \to \mathbb{Z}[q^{\pm 1}, z, \lambda]$

$$X_L(q, \lambda) = \left[-\frac{1 - \lambda q}{\sqrt{\lambda}\,(1 - q)} \right]^{n-1} \left(\sqrt{\lambda}\right)^e \mathrm{tr}\left(\pi\left(\alpha\right)\right),$$

where $\alpha \in B_n$ *is a word in the* σ_i*'s,* e *is the exponent sum of the* σ_i*'s in* α*, and* π *the canonical map of* B_n *in* $\mathrm{H}_n(q)$*, such that* $\sigma_i \mapsto g_i$*, is an invariant of oriented links in* ST*.*

2.3. The Jones polynomial. We now present the definition of the Temperley-Lieb algebra as a quotient of the Iwahori-Hecke algebra of type A, given by Jones [**27**].

DEFINITION 2.8. The Temperley-Lieb algebra $TL_n(q)$ is defined as the quotient $\mathrm{H}_n(q)$ over the ideal I_n, where I_n is generated by the element $g_{1,2}$, where

$$g_{1,2} := 1 + q\,(g_1 + g_2) + q^2\,(g_1 g_2 + g_2 g_1) + q^3\, g_1 g_2 g_1.$$

REMARK 2.9. It is worth mentioning that the Temperley-Lieb algebra of type A was originally defined by generators $f_1, \ldots, f_{n-1}$ such that:

$$\begin{array}{lcll} f_i^2 & = & f_i & \\ f_i f_j f_i & = & \delta\, f_i & |i - j| = 1 \\ f_i f_j & = & f_j f_i & |i - j| > 1 \end{array}$$

where δ is an indeterminate. If we introduce the generators $g_i := (q + 1)f_i - 1$, where $\delta^{-1} = 2 + u + u^{-1}$, then the Temperley-Lieb algebra of type A can be defined as the quotient of H_n over $g_{1,2}$.

By annihilating the generator of the defining ideal of the quotient algebra TL_n, i.e. by solving $tr(g_{1,2}) = 0$, Jones redefined Ocneanu's trace on TL_n. In particular, setting $\lambda = q$ in the invariant X of Theorem 2.7, results in the Jones polynomial V. That is, $V(q) = X(q, q)$.

REMARK 2.10. It is important to recall that the Jones polynomial is equivalent to the Kauffman bracket defined in Definition 2.2 up to a change of variable. The Kauffman bracket skein module of a 3-manifold M is equivalent to constructing the Jones polynomial for knots and links in M via the Kauffman bracket.

3. Skein modules of the Solid Torus

In this section we present two universal invariants for knots and links in the Solid Torus ST. One is of the HOMFLYPT type, constructed in [**31**] via a trace function on the generalized Hecke algebra of type B, which recovers HOM(ST), and the other is of the Kauffman bracket type, constructed in [**1**], and which recovers KBSM(ST).

3.1. Topological and algebraic set up. We consider ST to be the complement of a solid torus in S^3. Representing the complement solid torus by the black simple curve on the left of Fig. 10, a knot has to avoid this black curve. Hence, an oriented link L in ST can be represented by an oriented *mixed link* in S^3, that is, a link in S^3 consisting of the unknotted fixed part $\widehat{I}$ representing the complementary solid torus in S^3, and the moving part L that links with $\widehat{I}$. A *mixed link diagram* is a diagram $\widehat{I} \cup \widetilde{L}$ of $\widehat{I} \cup L$ on the plane of $\widehat{I}$, where this plane is equipped with the top-to-bottom direction of I (for an illustration see Figure 10).

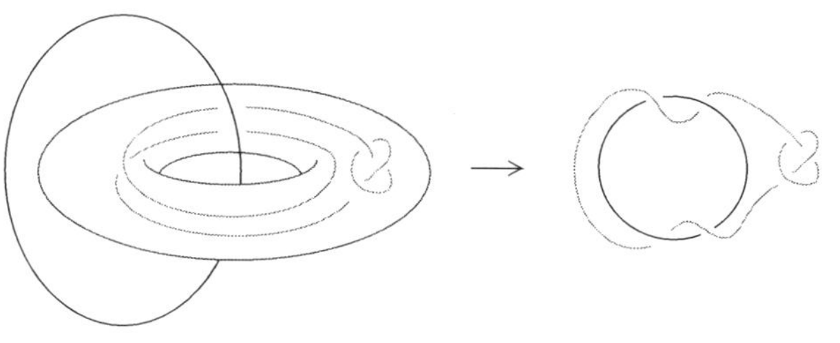

FIGURE 10. A link in ST as a mixed link in S^3.

Isotopy in ST is then translated to mixed link isotopy in S^3 as follows:

THEOREM 3.1 (Theorem 5.2 [**32**]). *Two links in ST are isotopic if and only if any two corresponding mixed link diagrams of theirs differ by planar isotopy and a finite sequence of the Reidemeister moves for the moving part of the mixed link and the mixed Reidemeister moves that involve both the fixed and the moving part of the mixed links and which are illustrated in Figure* 11.

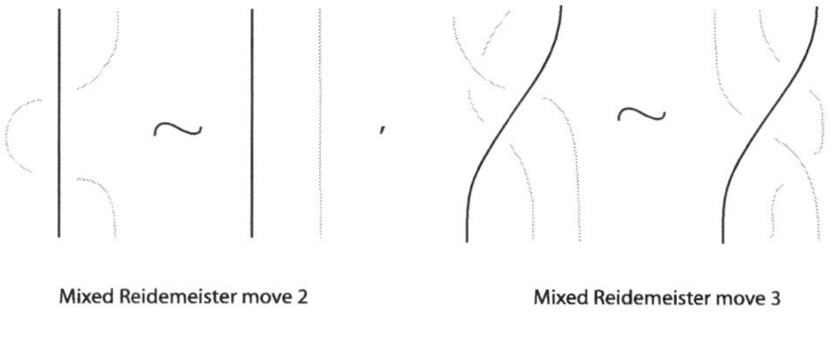

FIGURE 11. The mixed Reidemeister moves.

We now recall the analogue of the Alexander theorem for knots in ST (cf. Thm. 1 [**30**]):

THEOREM 3.2 (**The analogue of the Alexander theorem for ST**). *A mixed link diagram $\widehat{I} \cup \widetilde{L}$ of $\widehat{I} \cup L$ may be turned into a mixed braid $I \cup \beta$ with isotopic closure.*

By the Alexander theorem for knots in solid torus, a mixed link diagram $\widehat{I} \cup \widetilde{L}$ of $\widehat{I} \cup L$ may be turned into a *mixed braid* $I \cup \beta$ with isotopic closure (for an illustration see Figure 12). This is a braid in S^3 where, without loss of generality, its first strand represents $\widehat{I}$, the fixed part, and the other strands, β, represent the moving part L. The subbraid β shall be called the *moving part* of $I \cup \beta$.

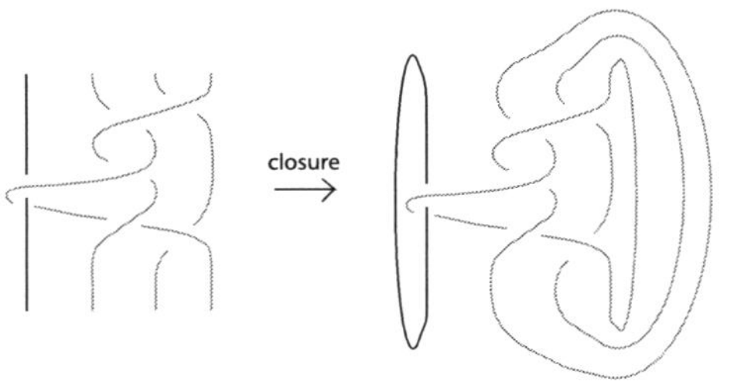

FIGURE 12. The closure of a mixed braid to a mixed link.

The sets of braids related to ST form groups, which are in fact the Artin braid groups type B, denoted $B_{1,n}$, with presentation:

$$B_{1,n} = \left\langle \quad t, \sigma_1, \ldots, \sigma_{n-1} \quad \left| \begin{array}{l} \sigma_1 t \sigma_1 t = t \sigma_1 t \sigma_1 \\ t\sigma_i = \sigma_i t, \quad i > 1 \\ \sigma_i \sigma_{i+1} \sigma_i = \sigma_{i+1} \sigma_i \sigma_{i+1}, \quad 1 \leq i \leq n-2 \\ \sigma_i \sigma_j = \sigma_j \sigma_i, \quad |i-j| > 1 \end{array} \right. \right\rangle,$$

where the generators σ_i and t are illustrated in Figure 13.

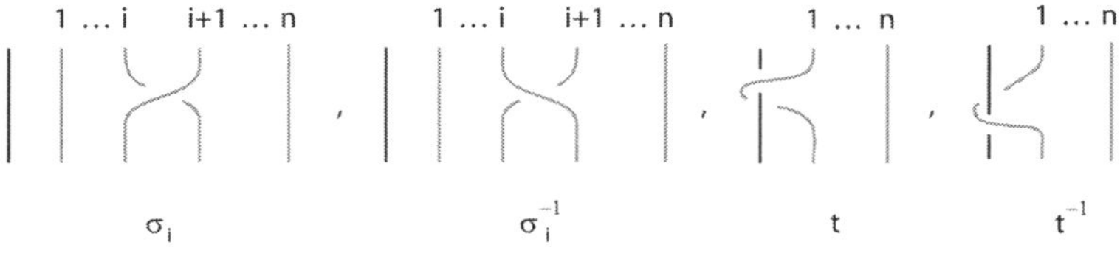

FIGURE 13. The generators of $B_{1,n}$.

Let now $\mathcal{L}$ denote the set of oriented knots and links in ST. Isotopy in ST is then translated on the level of mixed braids by means of the following theorem:

THEOREM 3.3 (Theorem 4, [**33**]). *Let L_1, L_2 be two oriented links in ST and let $I \cup \beta_1$, $I \cup \beta_2$ be two corresponding mixed braids in S^3. Then L_1 is isotopic to L_2 in ST if and only if $I \cup \beta_1$ is equivalent to $I \cup \beta_2$ in $\mathcal{B}$ by the following moves:*

(i)	*Conjugation :*	$\alpha \sim \beta^{-1}\, \alpha\, \beta,$	*if* $\alpha, \beta \in B_{1,n}$.
(ii)	*Stabilization moves :*	$\alpha \sim \alpha\, \sigma_n^{\pm 1} \in B_{1,n+1},$	*if* $\alpha \in B_{1,n}$.
(iii)	*Loop conjugation :*	$\alpha \sim t^{\pm 1}\, \alpha\, t^{\mp 1},$	*if* $\alpha \in B_{1,n}$.

For an illustration of the moves in Theorem 3.3, see Figure 22.

3.2. The HOMFLYPT skein module of ST. In [**22**] the HOMFLYPT skein module of the solid torus, HOM(ST), has been computed using diagrammatic methods. HOM(ST) is well-studied and understood by now. It forms a commutative algebra with multiplication induced by embedding two solid tori in one in a standard way. We now have the following theorem:

THEOREM 3.4 (Kidwell–Hoste [**22**], Turaev [**36**]). *The HOMFLYPT skein module of the solid torus, HOM(ST), is a free, infinitely generated $\mathbb{Z}[u^{\pm 1}, z^{\pm 1}]$-module isomorphic to the symmetric tensor algebra over conjugacy classes of nontrivial elements of $\pi_1(\mathrm{ST})$.*

The symmetric tensor algebra in this context is the commutative algebra generated by the conjugacy classes of the fundamental group of the solid torus, with coefficients in $\mathbb{Z}[u^{\pm 1}, z^{\pm 1}]$. A basic element of HOM(ST) in the context of [**22**], is illustrated in Figure 14. Note that in the diagrammatic setting of [**22**], ST is considered as Annulus $\times$ Interval.

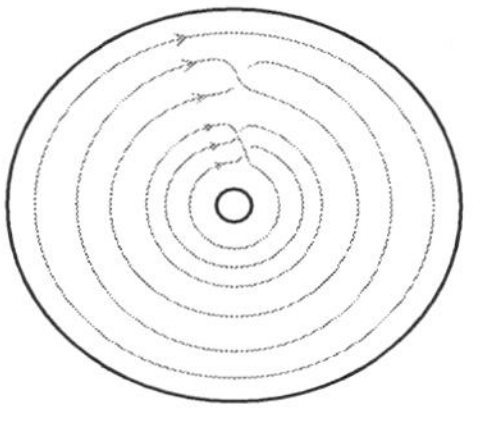

FIGURE 14. A basic element of HOM(ST).

Let now $\mathcal{B}^+$ denote the sub-algebra of HOM(ST), freely generated by elements that are clockwise oriented and let $\mathcal{B}^-$ denote the sub-algebra, freely generated by elements with counter-clockwise orientation. Let also $\mathcal{B}_k^+$ denote the sub-module generated by elements in $\mathcal{B}^+$ whose winding number is equal to $k \in \mathbb{N}$ and $\mathcal{B}_{-k}^-$ denote the sub-module generated by elements in $\mathcal{B}^-$ whose winding number is equal to k. As a linear space, $\mathcal{B}^+$ is graded by

$$\mathcal{B}^+ \cong \bigoplus_{k \geq 0} \mathcal{B}_k^+$$

and similarly, $\mathcal{B}^-$ is graded by

$$\mathcal{B}^- \cong \bigoplus_{k \geq 0} \mathcal{B}_{-k}^-.$$

Finally, it is worth mentioning the following module decomposition:

$$\mathrm{HOM(ST)} = \bigoplus_{\lambda, \mu \geq 0} \mathcal{B}_{-\lambda}^- \otimes \mathcal{B}_\mu^+.$$

We now recover HOM(ST) using the braid approach following [**31**]. We consider the *generalized Hecke algebra* of type B, $\mathrm{H}_{1,n}$ as the quotient of $\mathbb{C}\left[q^{\pm 1}\right] B_{1,n}$ over the quadratic relations $\sigma_i^2 = (q-1)\sigma_i + q$. Namely:

$$\mathrm{H}_{1,n}(q) = \frac{\mathbb{C}\left[q^{\pm 1}\right] B_{1,n}}{\langle \sigma_i^2 - (q-1)\,\sigma_i - q \rangle}.$$

This algebra is infinite dimensional, since the "looping" generator t satisfies no polynomial relation. Note also that in [**31**] the algebra $\mathrm{H}_{1,n}(q)$ is denoted as $\mathrm{H}_n(q,\infty)$. From [**31**] we have:

THEOREM 3.5 ([**31**, Proposition 1 & Theorem 1]). *The following sets form linear bases for* $\mathrm{H}_{1,n}(q)$*:*

$$\begin{array}{llll} (i) & \Sigma_n & = & \{t_{i_1}^{k_1} \ldots t_{i_r}^{k_r} \cdot \sigma\}, \ \textit{where}\ 0 \leq i_1 < \ldots < i_r \leq n-1, \\ (ii) & \Sigma_n' & = & \{{t'_{i_1}}^{k_1} \ldots {t'_{i_r}}^{k_r} \cdot \sigma\}, \ \textit{where}\ 0 \leq i_1 < \ldots < i_r \leq n-1, \end{array}$$

where $k_1, \ldots, k_r \in \mathbb{Z}$ *and* σ *a basic element in* $\mathrm{H}_n(q)$ *(recall Eq. (2.1)) and the "looping generators"* t_i *and* t_i' *are defined as follows (for an illustration see Figure 15):*

$$(3.1)\quad t_0' = t_0 := t, \quad t_i' = \sigma_i \ldots \sigma_1 t \sigma_1^{-1} \ldots \sigma_i^{-1} \quad \textit{and} \quad t_i = \sigma_i \ldots \sigma_1 t \sigma_1 \ldots \sigma_i$$

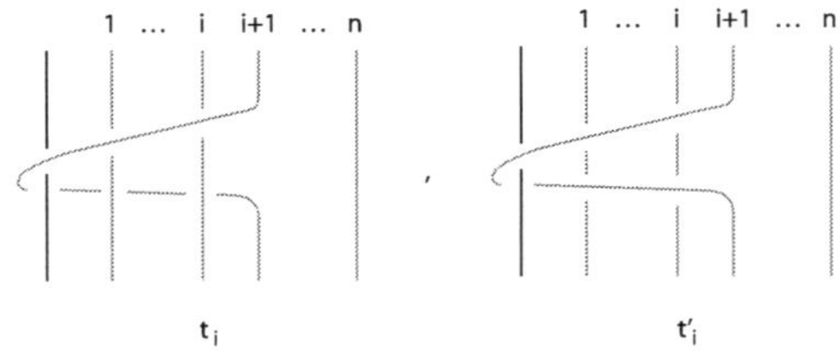

FIGURE 15. The looping generators.

We shall denote

$$\Sigma := \bigcup_n \Sigma_n \quad \text{and similarly} \quad \Sigma' := \bigcup_n \Sigma'_n. \tag{3.2}$$

In [**31**] the bases Σ'_n are used for constructing a Markov trace on $\bigcup_{n=1}^{\infty} \mathrm{H}_{1,n}(q)$.

THEOREM 3.6 ([**31**, Theorem 6]). *Given z, s_k, with $k \in \mathbb{Z}$ specified elements in $R = \mathbb{Z}\left[q^{\pm 1}\right]$, there exists a unique linear Markov trace function*

$$\mathrm{tr} : \bigcup_{n=1}^{\infty} \mathrm{H}_{1,n}(q) \to R\left(z, s_k\right), k \in \mathbb{Z}$$

determined by the rules:

$$\begin{array}{llcll} (1) & \mathrm{tr}(ab) & = & \mathrm{tr}(ba) & \textit{for } a, b \in \mathrm{H}_{1,n}(q) \\ (2) & \mathrm{tr}(1) & = & 1 & \textit{for all } \mathrm{H}_{1,n}(q) \\ (3) & \mathrm{tr}(ag_n) & = & z\mathrm{tr}(a) & \textit{for } a \in \mathrm{H}_{1,n}(q) \\ (4) & \mathrm{tr}(a{t'_n}^k) & = & s_k\mathrm{tr}(a) & \textit{for } a \in \mathrm{H}_{1,n}(q),\ k \in \mathbb{Z}. \end{array}$$

Note that the use of the looping elements t'_i enable the trace tr to be defined by just extending by rule (4) the three rules of the Ocneanu trace on the algebras $\mathrm{H}_n(q)$ [**26**]. Using tr a universal HOMFLYPT-type invariant for oriented links in ST is constructed. Namely, let $\mathcal{L}$ denote the set of oriented links in ST. Then:

THEOREM 3.7 ([**31**, Definition 1]). *The function $X : \mathcal{L} \to R(z, s_k)$*

$$X_{\widehat{\alpha}} = \Delta^{n-1}\mathrm{tr}\left(\pi\left(\alpha\right)\right),$$

where $\Delta := -\frac{1-\lambda q}{\sqrt{\lambda}(1-q)}\left(\sqrt{\lambda}\right)^e$, $\lambda := \frac{z+1-q}{qz}$, $\alpha \in B_{1,n}$ is a word in the σ_i's and t'_i's, $\widehat{\alpha}$ is the closure of α, e is the exponent sum of the σ_i's in α, and π the canonical map of $B_{1,n}$ on $\mathrm{H}_{1,n}(q)$, such that $t \mapsto t$ and $\sigma_i \mapsto g_i$, is an invariant of oriented links in ST.

REMARK 3.8. As shown in [**11, 31**] the invariant X recovers the HOMFLYPT skein module of ST. For a survey on the HOMFLYPT skein module of the lens spaces $L(p, 1)$ via braids, the reader is referred to [**12, 15**].

In the braid setting of [**31**], the elements of HOM(ST) correspond bijectively to the elements of the following set Λ':

$$\Lambda' = \{t^{k_0}{t'_1}^{k_1}\ldots{t'_n}^{k_n},\ k_i \in \mathbb{Z}\setminus\{0\},\ k_i \geq k_{i+1}\ \forall i,\ n \in \mathbb{N}\}. \tag{3.3}$$

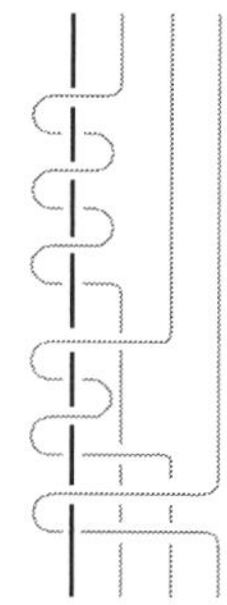

FIGURE 16. The basic element $t^3 {t'_1}^2 t'_2$ of Λ'.

So, we have that Λ' is a basis of HOM(ST) in terms of braids (for an illustration see Figure 16). Note that Λ' is a subset of $\bigcup_n \mathrm{H}_{1,n}$ and, in particular, Λ' is a subset of Σ'. Note also that in contrast to elements in Σ', the elements in Λ' have no gaps in the indices, the exponents are ordered and there are no 'braiding tails'.

REMARK 3.9. It is worth mentioning that the invariant X satisfies the following skein relation:

$$\frac{1}{\sqrt{q}\sqrt{\lambda}} X_{L_+} - \sqrt{q}\sqrt{\lambda} X_{L_-} = \left(\sqrt{q} - \frac{1}{\sqrt{q}}\right) X_{L_o}.$$

3.3. The Kauffman bracket skein module of ST. In this subsection we present results on the Kauffman bracket skein module of ST, KBSM(ST). A basis for KBSM(ST) is presented in [**22**] using diagrammatic methods (see also [**36**]). More precisely:

THEOREM 3.10 ([**22**]). *The Kauffman bracket skein module of ST, KBSM(ST), is freely generated by an infinite set of generators* $\{x^n\}_{n=0}^{\infty}$*, where* x^n *denotes* n *parallel copies of a longitude of ST and* x^0 *is the affine unknot (see Figure 17).*

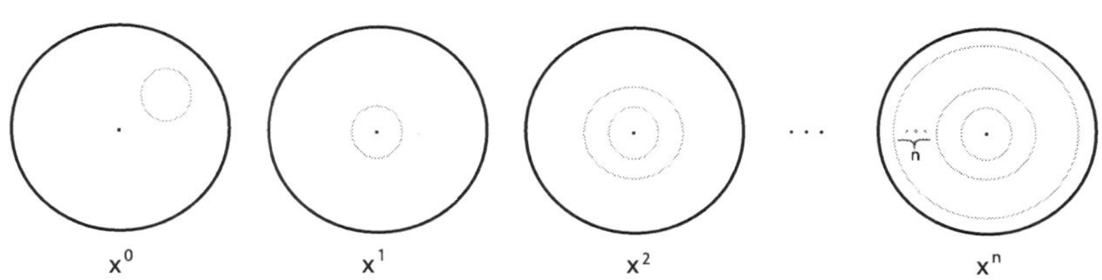

FIGURE 17. The standard basis of KBSM(ST).

In the braid setting, the elements of the standard basis of KBSM(ST) correspond bijectively to the elements of the following set (for an illustration see Figure 18):

$$B'_{\mathrm{ST}} = \{tt'_1 \dots t'_n,\ n \in \mathbb{N}\}. \tag{3.4}$$

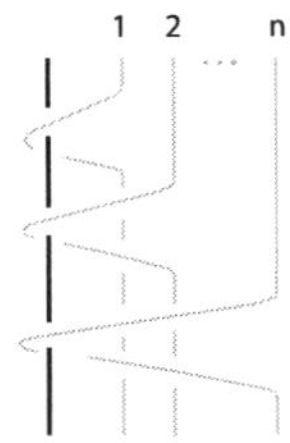

FIGURE 18. The element $tt'_1 \dots t'_{n-1}$ of B'_{ST}.

REMARK 3.11. Note now that the basis B'_{ST} is not "natural", since on the level of braids, elements in B'_{ST} have crossings. This was our main motivation for establishing a different and more suitable basis for KBSM(ST). We present this new basis in § 4.3. As we will see, this new basis is appropriate in extending the (universal) bracket polynomial, that recovers KBSM(ST), to an invariant of knots and links in the lens spaces $L(p, q)$.

We now compute the Kauffman bracket skein module of ST via braids. Our starting point is the unique Markov trace constructed by Lambropoulou in [**31**] on the generalized Hecke algebra of type B, $\mathrm{H}_{1,n}$, and the universal invariant for knots and links in ST of the HOMFLYPT type (recall Theorems 3.6 and 3.7).

Following the work of V.F.R. Jones, we define the *generalized Temperley-Lieb algebra of type B* as a quotient of $\mathrm{H}_{1,n}$ over the ideal generated by $\sigma_{i,i+1}$, $i \in \mathbb{N}\backslash\{0\}$, where

$$\sigma_{i,i+1} \quad := \quad 1 + u\ (\sigma_i + \sigma_{i+1}) + u^2\ (\sigma_i\sigma_{i+1} + \sigma_{i+1}\sigma_i) + u^3\ \sigma_i\sigma_{i+1}\sigma_i. \tag{3.5}$$

More precisely, we have the following definition:

DEFINITION 3.12. The *generalized Temperley-Lieb algebra of type B*, $TL_{1,n}$, is defined as the quotient of the generalized Hecke algebra of type B, $H_{1,n}(q)$, over the ideal generated by the elements $\sigma_{i,i+1}$, $i \in \mathbb{N}\backslash\{0\}$ defined in Eq. (3.5).

REMARK 3.13. Note that in [**17**] a different presentation for $\mathrm{H}_{1,n}$ is used, that involves the parameters u, v and the quadratic relations

$$\sigma_i^2 = (u - u^{-1})\sigma_i + 1. \tag{3.6}$$

One can switch from one presentation to the other by a taking $\sigma_i = u\sigma_i$, $t = vt$ and $q = u^2$. We will adapt this setting from now on when working on $TL_{1,n}$.

The necessary and sufficient conditions so that the Markov trace defined in $\mathrm{H}_{1,n}$ factors through to $TL_{1,n}$ are presented in [**1**]:

THEOREM 3.14. *The trace defined in* $\mathrm{H}_{1,n}$ *factors through to* $TL_{1,n}$ *if and only if the trace parameters take the following values:*

$$z = -\frac{1}{u(1+u^2)}. \tag{3.7}$$

REMARK 3.15. Note now that for $z = -\frac{1}{u(1+u^2)}$, one deduces $\lambda = u^4$.

We are now in position to present the universal invariant for knots and links in ST of the Kauffman bracket type [**1**]:

THEOREM 3.16. *The following invariant is the universal invariant of the Kauffman bracket type, for knots and links in ST:*

$$V_{\widehat{\alpha}}(u,v) \;:=\; \left(-\frac{1+u^2}{u}\right)^{n-1} (u)^{2e}\,\mathrm{tr}\left(\overline{\pi}\left(\alpha\right)\right),$$

where $\alpha \in B_{1,n}$ is a word in the σ_i's and t_i''s, $\widehat{\alpha}$ is the closure of α, e is the exponent sum of the σ_i's in α, $\overline{\pi}$ the canonical map of $B_{1,n}$ to $\mathrm{TL}_{1,n}$, such that $t \mapsto t$ and $\sigma_i \mapsto g_i$.

REMARK 3.17. The invariant V recovers the Kauffman bracket skein module of ST since it gives distinct values to distinct elements of the basis of KBSM(ST), B_{ST}'. Indeed, $tr(tt_1'\ldots t_n') = s_1^n$.

REMARK 3.18. It is worth emphasizing on the difference between the basis of the HOMFLYPT skein module of ST (recall Eq. (3.3)) and that of the Kauffman bracket skein module of ST (Eq. (3.4)). For the Kauffman bracket there are no crossings, and this is due the skein relation in the definition of the Kauffman bracket. On the other hand, the skein relation that characterizes the HOMFLYPT polynomial does not annihilate the crossings, but changes its "handiness" instead.

REMARK 3.19.

i. In [**7**] we introduce and study *knotoids* on the surface of the torus. Knotoids are open knotted curves in oriented surfaces Σ, that is, generic immersions of the unit interval $[0,1]$ into Σ. In [**7**] we extend the notion of skein modules for knotoids and in particular, we show that the Kauffman bracket skein module of the Solid Torus is freely generated by elements of the form illustrated in Figure 19.

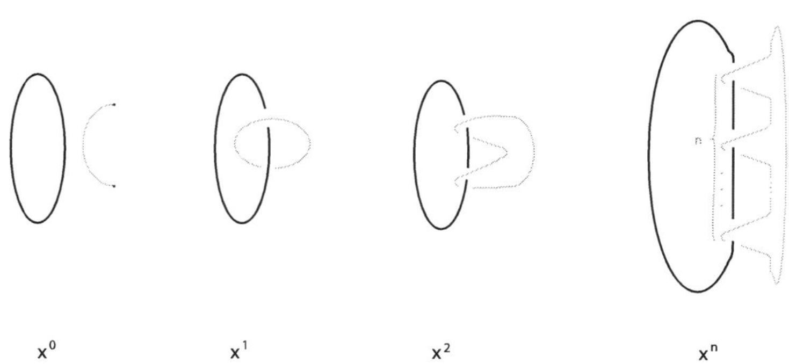

FIGURE 19. A basis for KBSM(ST) for knotoids.

ii. Finally, it is worth mentioning that in [**6, 8**] we construct the Kauffman bracket for pseudo knots in the Solid Torus and in the handlebody of genus g. Our intention is to generalize these invariants for other 3-manifolds, i.e., compute the corresponding skein module.

4. Skein modules of the lens spaces $L(p,1)$

4.1. Topological and algebraic set up. In this subsection we recall results from [**10**, **32**, **33**]. We consider the lens spaces $L(p,1)$ obtained from S^3 by integral surgery along the unknot with coefficient p. Namely, we start from S^3 and we drill out a tubular neighborhood of the unknot, that is, a solid torus ST_1. What is left is another solid torus ST_2. We then glue these two solid tori via a homeomorphism h on their boundaries, such that a meridian on $\partial\mathrm{ST}_1$ is mapped to a $(p,1)$-curve on $\partial\mathrm{ST}_2$ (for an illustration see Figure 20).

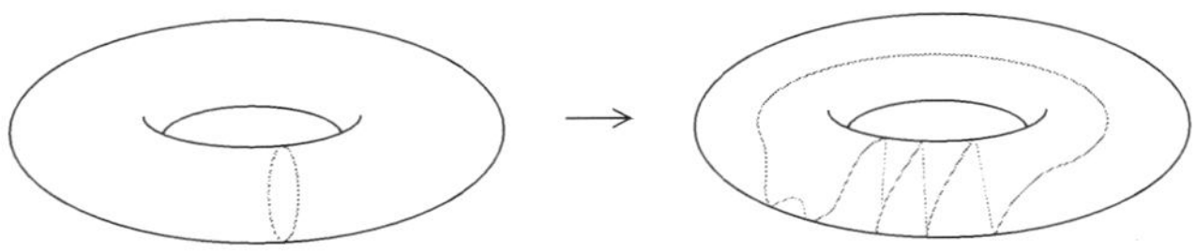

FIGURE 20. The homeomorphism h.

REMARK 4.1. Note that a meridian bounds a disc in ST, and since this meridian is glued to a $(p,1)$-torus knot, it follows that this $(p,1)$-torus knot bounds a disc in $L(p,1)$.

Isotopy in $L(p,1)$ can be viewed as isotopy in ST together with the *band moves* in S^3, which reflect the surgery description of the manifold, see Figure 21 (see [**32**]).

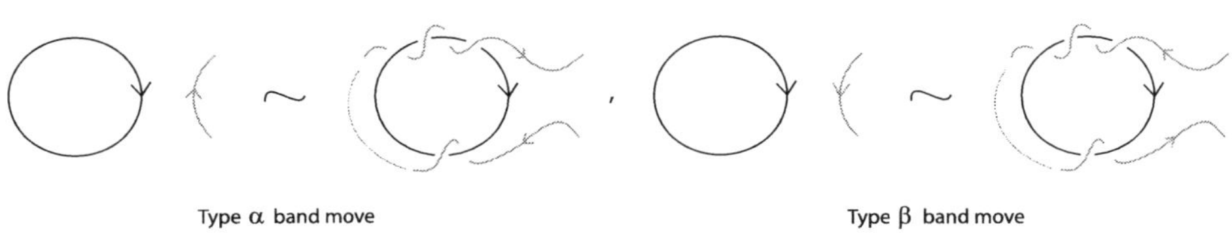

FIGURE 21. The two types of band moves.

In [**10**] it is shown that it suffices to consider only one type of band move in order to describe isotopy for knots and links in $L(p,1)$, and thus, isotopy between oriented links in $L(p,1)$ is reflected in S^3 by means of the following theorem (cf. [**32**, Theorem 5.8], [**10**, Theorem 6]):

THEOREM 4.2. *Two oriented links in $L(p,1)$ are isotopic if and only if two corresponding mixed link diagrams of theirs differ by isotopy in ST together with a finite sequence of the type α (or β) band moves.*

DEFINITION 4.3. We define a *braid band move* to be a move between mixed braids, which is a α-type band move between their closures. It starts with a little band oriented downward, which, before sliding along a surgery strand, gets one twist *positive* or *negative* (see Figure 22).

REMARK 4.4. For now we shall only consider the α-type band moves, since the result of an α-type band move remain braided, while the result of a β-band move does not. Note also that there are two different types of braid band moves; the positive and the negative braid band move, depending on the kind of the twist the

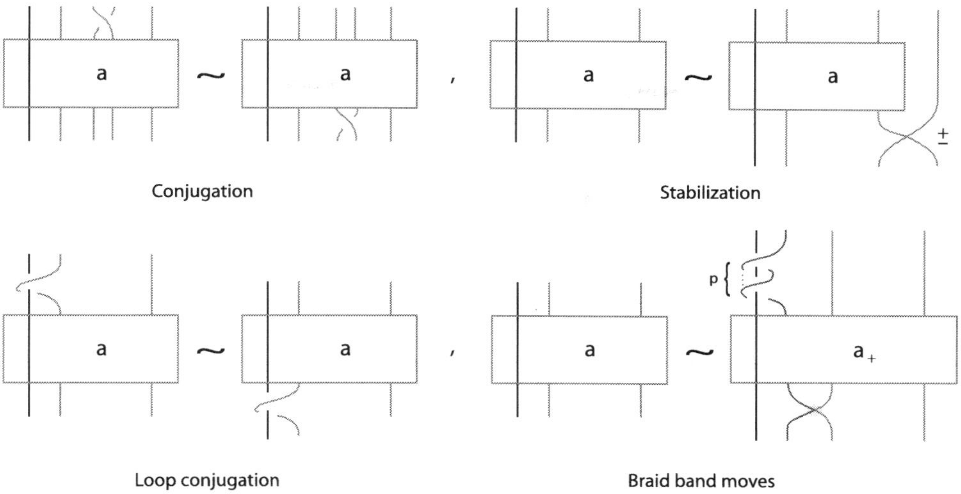

FIGURE 22. Mixed braid equivalence.

moving strand gets before sliding along the fixed strand. We denote a braid band move by bbm and, specifically, the result of a positive or negative braid band move performed on a mixed braid β by $bbm_{\pm}(\beta)$.

Isotopy in $L(p,1)$ is then translated on the level of mixed braids by means of the following theorem:

THEOREM 4.5 (Theorem 5, [**33**]). *Let L_1, L_2 be two oriented links in $L(p,1)$ and let $I \cup \beta_1$, $I \cup \beta_2$ be two corresponding mixed braids in S^3. Then L_1 is isotopic to L_2 in $L(p,1)$ if and only if $I \cup \beta_1$ is equivalent to $I \cup \beta_2$ in $\mathcal{B}$ by the following moves:*

(i)	*Conjugation :*	$\alpha \sim \beta^{-1}\,\alpha\,\beta,$	*if* $\alpha, \beta \in B_{1,n}.$
(ii)	*Stabilization moves :*	$\alpha \sim \alpha\,\sigma_n^{\pm 1} \in B_{1,n+1},$	*if* $\alpha \in B_{1,n}.$
(iii)	*Loop conjugation :*	$\alpha \sim t^{\pm 1}\,\alpha\,t^{\mp 1},$	*if* $\alpha \in B_{1,n}.$
(iv)	*Braid band moves :*	$\alpha \sim t^p\,\alpha_+\,\sigma_1^{\pm 1},$	$a_+ \in B_{1,n+1},$

where α_+ is the word α with all indices shifted by +1 and β_+ is the word β with all indices shifted by q. Note that moves (i), (ii) and (iii) correspond to link isotopy in ST (recall Theorem 3.3*).*

As noted before, our aim is to extend the universal invariants X and V for oriented (respectively framed) knots and links in ST, to invariants for knots and links in $L(p,1)$. Namely, we need to study the effect of bbm's to elements in a basis of the corresponding skein module of ST. Note now that, as shown in Figure 23, bbm's are not naturally described by elements in the bases Λ' for the HOM(ST),

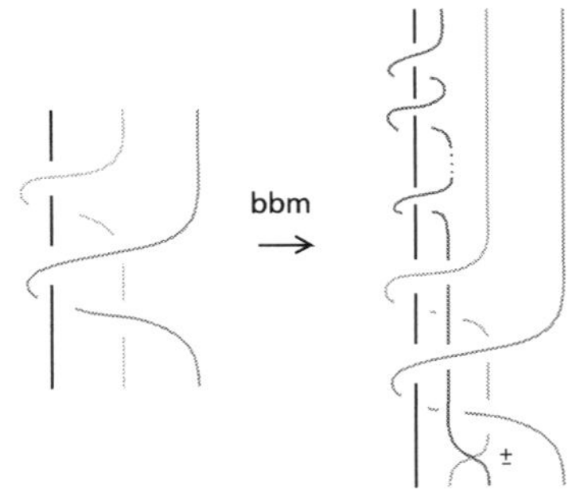

FIGURE 23. Braid band move on the element tt'_1.

and by elements in B'_{ST} for the case of the KBSM(ST). For this reason we first present more appropriate bases for both HOM(ST) and KBSM(ST).

4.2. A new basis for HOM(ST). A different basis, Λ, for HOM(ST) is presented in [**11**]. This basis is important for computing HOM($L(p,1)$), since, as noted before, it naturally describes the braid band moves, which extend the link isotopy in ST to link isotopy in $L(p,1)$. This was in fact our motivation for establishing this new basis. We have the following theorem:

THEOREM 4.6 ([**11**, Theorem 2]). *The following set forms a basis for HOM(ST):*

$$\Lambda = \{t^{k_0} t_1^{k_1} \dots t_n^{k_n},\ k_i \in \mathbb{Z} \setminus \{0\},\ k_i \geq k_{i+1}\ \forall i,\ n \in \mathbb{N}\}. \tag{4.1}$$

For an illustration of elements in the basis Λ see Figure 24. Note that comparing the set Λ with the set $\Sigma = \bigcup_n \Sigma_n$, we observe that in Λ there are no gaps in the indices of the t_i's and the exponents are in decreasing order. Also, there are no 'braiding tails' in the words in Λ.

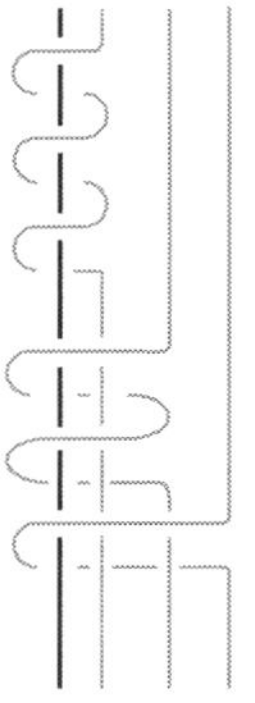

FIGURE 24. The element $t^3 t_1^2 t_2$ in the new basis of HOM(ST).

Our method for proving Theorem 4.6 is the following: We first define total orderings in the sets Λ' and Λ, and we relate the two ordered sets via a lower triangular infinite matrix with invertible elements on the diagonal. More precisely, we start with elements in the basic set Λ' and we convert them into sums of elements in Σ. These elements consist of two parts: arbitrary monomials in the t_i's followed by 'braiding tails' in the bases of the algebras $\mathrm{H}_n(q)$. Then, these elements are converted into elements in the set Λ by managing the gaps in the indices, by ordering the exponents of the t_i's and by eliminating the 'braiding tails'. These procedures are not independent since if for example we manage the gaps in the indices of the looping generators t_i's, 'braiding tails' may occur and also the exponents of the t_i's may alter. Similarly, when the 'braiding tails' are eliminated, gaps in the indices of the t_i's might occur. This is a long procedure that eventually stops and only elements in the set Λ remain. Using the change of basis matrix we finally show that the set Λ is linearly independent and hence, the set Λ forms a basis for HOM(ST).

In order to define the ordering relation, we introduce the notion of the *index* of a word w, denoted $ind(w)$. Note that for an element $t'_i \in \Lambda'$ (resp. $t_i \in \Lambda$), the index is defined to be i.

DEFINITION 4.7. Let $w \in \Lambda$ (respectively in Λ'). Then, $ind(w)$ is defined to be the highest index of the t_i's (resp. of the t'_i's) in w. If $w \in \Sigma$ (or in Σ'), then $ind(w)$ is defined as above by ignoring possible gaps in the indices of the looping generators and by ignoring the braiding parts in the algebras $\mathrm{H}_n(q)$. Finally, we define the index of a monomial in $\mathrm{H}_n(q)$ to be equal to 0.

EXAMPLE 4.8.

i.	$ind(t'^{k_0}{t'_1}^{k_1}\dots {t'_m}^{k_m})$	$=$ m	$=$	$ind(t^{k_0}{t_1}^{k_1}\dots {t_m}^{k_m})$
ii.	$ind(t^{k_0}t_3^{k_3}\cdot\sigma)$	$=$ 3		$,\ \sigma \in \mathrm{H}_n(q)$
iii.	$ind(\sigma)$	$=$ 0		$,\ \sigma \in \mathrm{H}_n(q)$

We are now in position to state the ordering relation.

DEFINITION 4.9 ([**11**, Definition 2]). Let $w = {t'_{i_1}}^{k_1}\dots {t'_{i_\mu}}^{k_\mu}\cdot\beta_1$ and $u = {t'_{j_1}}^{\lambda_1}\dots {t'_{j_\nu}}^{\lambda_\nu}\cdot\beta_2$ in Σ', where $k_c, \lambda_s \in \mathbb{Z}$ for all c, s and $\beta_1, \beta_2 \in H_n(q)$. Then, we define the following ordering in Σ':

(a) If $\sum_{i=0}^{\mu} k_i < \sum_{i=0}^{\nu}\lambda_i$, then $w < u$.
(b) If $\sum_{i=0}^{\mu} k_i = \sum_{i=0}^{\nu}\lambda_i$, then:
  (i) if $ind(w) < ind(u)$, then $w < u$,
  (ii) if $ind(w) = ind(u)$, then:
    (α) if $i_1 = j_1, \dots, i_{s-1} = j_{s-1}, i_s < j_s$, then $w > u$,
    (β) if $i_c = j_c$ for all c and $k_\mu = \lambda_\mu, k_{\mu-1} = \lambda_{\mu-1}, \dots, k_{i+1} = \lambda_{i+1}, |k_i| < |\lambda_i|$, then $w < u$,
    (γ) if $i_c = j_c$ for all c and $k_\mu = \lambda_\mu, k_{\mu-1} = \lambda_{\mu-1}, \dots, k_{i+1} = \lambda_{i+1}, |k_i| = |\lambda_i|$ and $k_i > \lambda_i$, then $w < u$,
    (δ) if $i_c = j_c\ \forall c$ and $k_i = \lambda_i,\ \forall i$, then $w = u$.

The ordering in the set Σ is defined as in Σ', where t'_i's are replaced by t_i's.

EXAMPLE 4.10.

i.	$t^3{t'_1}^4 < t^4{t'_1}^5$	since $3+7\ <\ 4+5$	, (Def.4.9a)
ii.	$t{t'_1}^2{t'_2}^3 < t^2{t'_1}^2t'_2t'_3$	since $\mathrm{ind}(t{t'_1}^2{t'_2}^3) < \mathrm{ind}(t^2{t'_1}^2t'_2t'_3)$	, (Def.4.9b(i))
iii.	$t^2{t'_1}^3{t'_3}^4{t'_4}^8 < t^8t'_1t'_2{t'_4}^7$	since $3\ =\ i_3\ >\ j_3\ =\ 2$	, (Def.4.9b(ii)(α))
iv.	$t^2{t'_1}^4t'_3{t'_4}^3 < t^{14}{t'_1}^{-8}t'_3{t'_4}^3$	since $\lvert 4\rvert\ <\ \lvert -8\rvert$	, (Def.4.9b(ii)(β))
v.	$t{t'_1}^4t'_3{t'_4}^3 < t^{10}{t'_1}^{-4}t'_3{t'_4}^3$	since $-4\ <\ 4$	, (Def.4.9b(ii)(γ))

REMARK 4.11. We shall denote a monomial of the form $t_i^{k_i} \dots t_{i+m}^{k_{i+m}}$ by $\tau_{i,i+m}^{k_{i,i+m}}$, and similarly ${\tau'}_{i,i+m}^{k_{i,i+m}} := {t'}_i^{k_i} \dots {t'}_{i+m}^{k_{i+m}}$, for $m \in \mathbb{N}$, $k_j \neq 0$ for all j.

In order to relate the two basic sets Λ' and Λ via an infinite matrix, we need to define the *subsets of level k*, $\Lambda_{(k)}$ and $\Lambda'_{(k)}$, of Λ and Λ' respectively ([**11**, Definition 3]), to be the sets:

$$\Lambda_{(k)} := \{t_0^{k_0} t_1^{k_1} \dots t_m^{k_m} \,|\, \textstyle\sum_{i=0}^{m} k_i = k,\ k_i \in \mathbb{Z} \setminus \{0\},\ k_i \leq k_{i+1}\ \forall i\}$$
$$\Lambda'_{(k)} := \{{t'_0}^{k_0} {t'_1}^{k_1} \dots {t'_m}^{k_m} \,|\, \textstyle\sum_{i=0}^{m} k_i = k,\ k_i \in \mathbb{Z} \setminus \{0\},\ k_i \leq k_{i+1}\ \forall i\} \tag{4.2}$$

In [**11**] it is shown that the sets $\Lambda_{(k)}$ and $\Lambda'_{(k)}$ are totally ordered and well ordered for all k ([**11**, Propositions 1 & 2]). Note that in [**16**] the exponents in the monomials of Λ are in decreasing order, while here the exponents are considered in increasing order, which is totally symmetric.

We finally define the set Λ^{aug}, which augments the basis Λ and its subset of level k, and we also introduce the notion of *homologous words*.

DEFINITION 4.12. We define the set:

$$\Lambda^{aug} := \{t_0^{k_0} t_1^{k_1} \dots t_n^{k_n},\ k_i \in \mathbb{Z}\setminus\{0\}\}. \tag{4.3}$$

and the *subset of level k*, $\Lambda^{aug}_{(k)}$, of Λ^{aug}:

$$\Lambda^{aug}_{(k)} := \{t_0^{k_0} t_1^{k_1} \dots t_m^{k_m} \,|\, \sum_{i=0}^{m} k_i = k,\ k_i \in \mathbb{Z}\setminus\{0\}\} \tag{4.4}$$

DEFINITION 4.13. We shall say that two words $w' \in \Lambda'$ and $w \in \Lambda$ are *homologous*, denoted $w' \sim w$, if w is obtained from w' by turning t'_i into t_i for all i.

EXAMPLE 4.14. The words $t^{-2} {t'_1}^2 t'_2$ and $t^{-2} t_1^2 t_2$ are homologous. Note also that $\Lambda' \ni t \sim t \in \Lambda$.

In order to relate the sets Λ' and Λ via a lower triangular matrix with invertible elements in the diagonal, we first express elements in Λ' to elements containing the t_i's. Indeed we have that (Theorem 7, [**11**]):

$$t^{k_0} {t'_1}^{k_1} \dots {t'_m}^{k_m} = q^{-\sum_{n=1}^{m} n k_n} \cdot t^{k_0} t_1^{k_1} \dots t_m^{k_m} + \sum_i f_i(q) \cdot t^{k_0} t_1^{k_1} \dots t_m^{k_m} \cdot w_i + \sum_j g_j(q) \tau_j \cdot u_j, \tag{4.5}$$

where $w_i, u_j \in \mathrm{H}_{m+1}(q), \forall i$, $\tau_j \in \Sigma_n$, such that $\tau_j < t^{k_0} t_1^{k_1} \dots t_m^{k_m}, \forall j$ and $f_i, g_j \in \mathbb{C}$, for all i, j.

EXAMPLE 4.15. We shall now give an example of a monomial in Λ' converted into sums of elements in Σ. We have that $t^2 t'_1 = q^{-1} t^2 t_1 + (q^{-1} - 1)\, t^3 \sigma_1$. For more details the interested reader should refer to the technical lemmas [**11**, Lemmas 3, 4, 5, 6, 9 & 11].

Equations (4.5) suggest that an element in Λ' can be written as a sum of elements in Λ, where one term is the homologous word, another term is the homologous word followed by a braiding "tail", and all other terms in the sum consist of lower order terms followed by braiding "tails". These elements belong to Σ_n since they may have gaps in their indices. We then manage the gaps in the indices (Theorem 8, [**11**]), namely:

$$\Sigma_n \ni \tau \,\widehat{=}\, \sum_i f_i(q)\, \tau_i \cdot w_i \ :\ \tau_i \in \Lambda^{aug},\ w_i \in \mathrm{H}_n(q),\ \forall i, \tag{4.6}$$

where $\widehat{=}$ denotes that conjugation is applied in the process.

Equation (4.6) is best demonstrated in the following example on a word with two gaps in the indices of the loop generators.

EXAMPLE 4.16. For the 2-gap word $t^{k_0}t_2t_4 \in \Sigma$ we have:

$$\begin{array}{lllll}
t^{k_0}\underline{t_2}t_4 & = & \underline{t^{k_0}\sigma_2\, t_1\, \sigma_2\, t_4} & = & \sigma_2 t^{k_0} t_1 t_4 \sigma_2 \quad \widehat{=} \\
 & \widehat{=} & t^{k_0} t_1 \underline{t_4} \sigma_2^2 & = & \underline{t^{k_0} t_1\, \sigma_4\sigma_3\, t_2\, \sigma_3\sigma_4}\, \sigma_2^2 \quad = \\
 & = & \sigma_4\sigma_3\, t^{k_0} t_1 t_2\, \sigma_3\sigma_4\, \sigma_2^2 & \widehat{=} & t^{k_0} t_1 t_2\, \sigma_3\sigma_4\, \sigma_2^2\, \sigma_4\sigma_3
\end{array}$$

We now deal with the elements in $\Lambda^{aug}_{(k)}$ that are followed by a braiding "tail" w in $H_n(q)$. More precisely we have (Theorem 9, [**11**]):

$$\tau \cdot w \,\widehat{\cong}\, \sum_j f_j(q,z) \cdot \tau_j,$$

such that $\tau_j \in \Lambda^{aug}_{(k)}$ and $\tau_j < \tau$, for all j.

EXAMPLE 4.17. In this example we demonstrate how to eliminate the 'braiding tail' in a word.

$$\begin{array}{lllllll}
t^k \underline{t_1^2 t_2 \sigma_1^{-1}} & & & & & & \\
= & \underline{t^k t_1 t_2 t_1 \sigma_1^{-1}} & = & t^k t_1 t_2 \sigma_1 \underline{t} & \widehat{=} & t^{k+1} \underline{t_1 t_2} \sigma_1 & = \ t^{k+1} t_2 \underline{t_1 \sigma_1} \\
= & (q-1)t^{k+1}\underline{t_2 t_1} & + & q t^{k+1} t_2 \sigma_1 \underline{t} & \widehat{=} & (q-1)t^{k+1} t_1 t_2 & +\ q t^{k+2} \underline{t_2}\sigma_1 \\
= & (q-1)t^{k+1} t_1 t_2 & + & q \underline{t^{k+2}\sigma_2}\, t_1\, \sigma_2\sigma_1 & = & (q-1)t^{k+1} t_1 t_2 & +\ q \underline{\sigma_2} t^{k+2}\, t_1\, \sigma_2\sigma_1 \\
\widehat{=} & (q-1)t^{k+1} t_1 t_2 & + & q t^{k+2}\, t_1\, \underline{\sigma_2\sigma_1\sigma_2} & = & (q-1)t^{k+1} t_1 t_2 & +\ q t^{k+2}\, t_1\, \underline{\sigma_2\sigma_1\sigma_2} \\
= & (q-1)t^{k+1} t_1 t_2 & + & q t^{k+2}\, t_1\, \sigma_1 \underline{\sigma_2} \sigma_1 & \cong & (q-1)t^{k+1} t_1 t_2 & +\ q z t^{k+2}\, t_1\, \underline{\sigma_1^2} \\
= & (q-1)t^{k+1} t_1 t_2 & + & q(q-1) z t^{k+2}\, \underline{t_1\, \sigma_1} & + & q^2 z t^{k+2}\, t_1 & \\
= & (q-1)t^{k+1} t_1 t_2 & + & q(q-1)^2 z t^{k+2}\, t_1 & + & q^2(q-1) z t^{k+2}\, \underline{\sigma_1 t} & +\ q^2 z t^{k+2}\, t_1 \\
\widehat{\cong} & (q-1)t^{k+1} t_1 t_2 & + & q(q-1)^2 z t^{k+2}\, t_1 & + & q^2(q-1) z^2 t^{k+3} & +\ q^2 z t^{k+2}\, t_1 \widehat{\cong}
\end{array}$$

One very important result in [**11**] is that one can change the order of the exponents by using conjugation and stabilization moves on elements in Λ and express them as sums of monomials in t_i's with arbitrary exponents and which are of lower order than the initial elements in Λ. Note that both conjugation and stabilization moves are captured by the trace rules, and that we translate here Theorem 10 in [**11**] using the trace:

$$tr(\tau_{0,m}^{k_{0,m}} \cdot w) \;=\; tr\left(\sum_j \tau_{0,j}^{\lambda_{0,j}} \cdot w_j\right),$$

where $\tau_{0,j}^{\lambda_{0,j}} \in \Lambda$ and $w, w_j \in \bigcup_{n\in\mathbb{N}} \mathrm{H}_n(q)$ for all j.

EXAMPLE 4.18. Consider the element $tt_1^2t_2 \in \Lambda^{aug}$. We have that:

$$\begin{aligned} t\underline{t_1^2}t_2 \;=\; t\,\sigma_1 t\underline{\sigma_1^2} t\sigma_1\, t_2 \;&=\; (q-1)\, t\,\underline{\sigma_1 t\sigma_1} t\sigma_1\, t_2 \;+\; q\, \underline{t\,\sigma_1} t^2\sigma_1\, t_2 \;\widehat{=} \\ &\widehat{=}\; (q-1)\, t\, t_1 t\sigma_1\, t_2 \;+\; q\, t^2 \underline{\sigma_1\, t_2} t\, \sigma_1 \;= \\ &=\; (q-1)\, t^2\, t_1\, t_2\, \sigma_1 \;+\; q\, t^2\, t_1\, t_2 \end{aligned}$$

Finally, in [**11**] it is shown that the infinite matrix converting elements of the basis Λ' of HOM(ST) to elements of the set Λ is a block diagonal matrix, where each block corresponds to a subset of Λ' of level k and it is an infinite lower triangular matrix with invertible elements in the diagonal. Using this infinite diagonal matrix, in [**11**, Theorem 11] it is shown that the set Λ is linearly independent and hence, it forms a basis for HOM(ST).

4.3. A new basis for KBSM(ST). KBSM(ST) is particularly important since, as shown in [**34**], in order to compute KBSM($L(p,1)$), it suffices to consider elements in KBSM(ST) and study the effect of band moves on these elements. Namely,

(4.7) $$\mathrm{KBSM}\,(L(p,1)) = \frac{\mathrm{KBSM(ST)}}{< a - bbm(a) >}, \quad \text{where } a \text{ basis element of KBSM(ST).}$$

Equivalently, in order to compute KBSM($L(p,1)$) we need to solve the infinite system of equations obtained by imposing on the universal invariant V, of the Kauffman bracket type, relations coming from the performance of braid band moves on elements in a basis of KBSM(ST). Toward that end, we first present a more suitable basis that involves elements with no crossings on the level of braids, and that it naturally describes bbm's (for an illustration see Figure 25). More precisely,

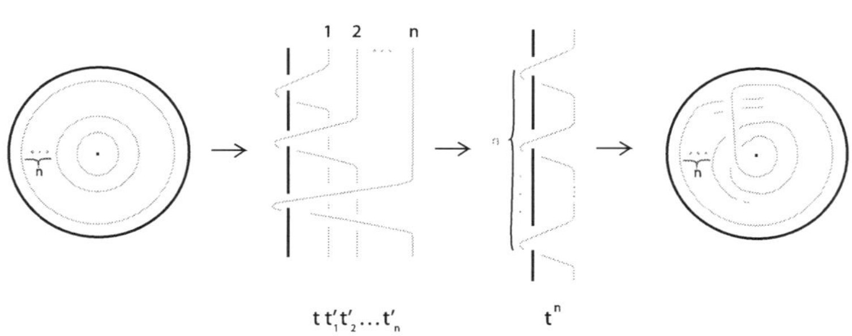

FIGURE 25. Elements in the bases B'_{ST} and B_{ST} of KBSM(ST).

we have the following result:

THEOREM 4.19. *The following set is a basis for KBSM(ST):*

$$B_{\mathrm{ST}} = \{t^n,\ n \in \mathbb{N}\}. \tag{4.8}$$

In order to prove Theorem 4.19 in [**1**] we follow the following procedure (see also [**2**]):

We first prove that the sets B'_{ST} and B_{ST} are totally ordered and well-ordered sets when equipped with the ordering relation of Definition 4.9.

We then pass from the set B'_{ST} consisting of monomials of the form $tt'_1t'_2\ldots t'_n$, $n \in \mathbb{N}$, to an augmented set ${B_{\mathrm{ST}}}^{aug}$, consisting of monomials of the form t^n, where $n \in \mathbb{Z}$. In particular in [**1**] it is shown that if $\tau \in B'_{\mathrm{ST}}$ such that $in(\tau) = n$, then

$$\tau \ \underset{\text{skein}}{\widehat{\cong}} \sum_{i\in\mathbb{Z}\,:\,|i|\leq n+1} a_i\, t^i,$$

where a_i coefficients for all i and where $\underset{\text{skein}}{\widehat{\cong}}$ signifies that the equivalence is established by employing a combination of conjugation, stabilization, and the skein relation.

EXAMPLE 4.20. In this example, we convert elements in the standard basis B'_{ST} to sums of elements in ${B_{\mathrm{ST}}}^{aug}$.

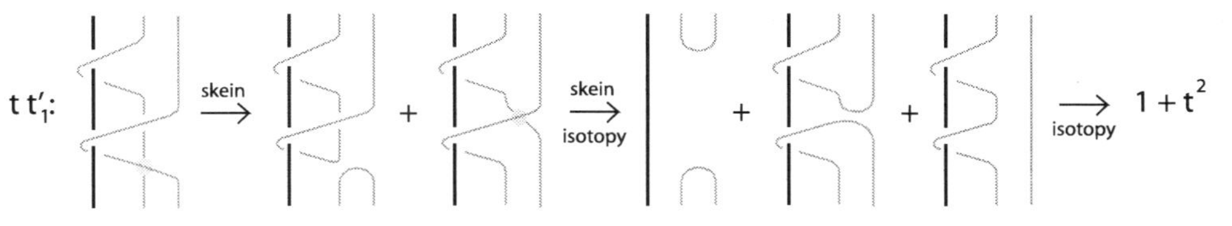

FIGURE 26. Expressing tt'_1 in terms of elements in ${B_{\mathrm{ST}}}^{aug}$.

i. The case tt'_1 is illustrated in Figure 26.

ii. $t\,t'_1\,t'_2 \ \underset{\text{skein}}{\widehat{\cong}}\ t^{-1} + t + t^3$

iii. $t\,t'_1\,t'_2\,t'_3\,t'_4\,t'_5 \ \underset{\text{skein}}{\widehat{\cong}}\ t^{-4} + t^{-2} + 1 + t^2 + t^4 + t^6$

iv. $t\,t'_1\,t'_2\,t'_3\,t'_4\,t'_5\,t'_6 \ \underset{\text{skein}}{\widehat{\cong}}\ t^{-5} + t^{-3} + t^{-1} + t + t^3 + t^5 + t^7$

We now express the elements $t^i \in {B_{\mathrm{ST}}}^{aug} \backslash B_{\mathrm{ST}}$ to sums of elements in B_{ST} (Proposition 2, [**1**]). In particular, we show that if $n \in \mathbb{N}$, then:

$$t^{-n} \ \underset{\text{skein}}{\widehat{\cong}} \sum_{i=0}^{n} a_i\, t^i.$$

In Figure 27 we illustrate some examples of converting elements in ${B_{\mathrm{ST}}}^{aug} \backslash B_{\mathrm{ST}}$ to sums of elements in B'_{ST}.

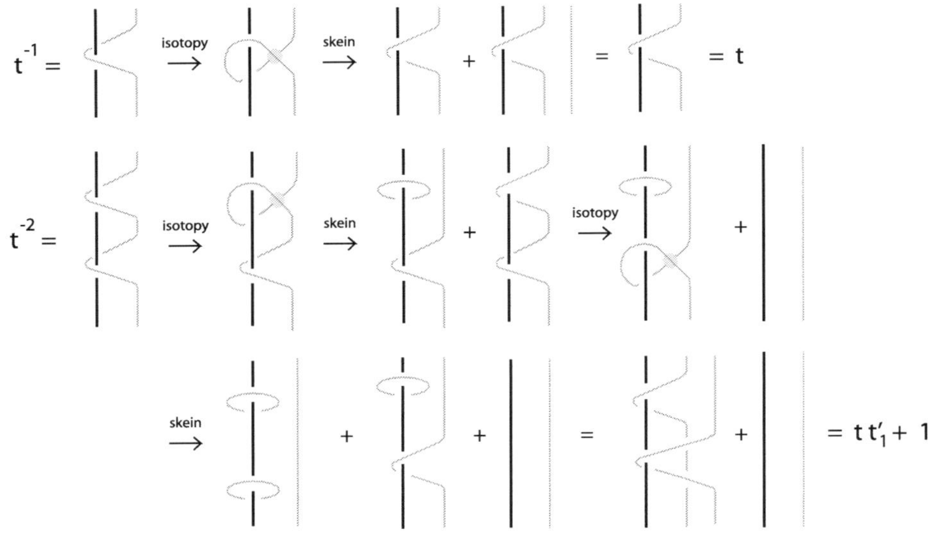

FIGURE 27. Converting t^{-1} and t^{-2} to elements in B'_{ST}.

The two (ordered) sets B'_{ST} and B_{ST} are finally related via a lower triangular infinite matrix with invertible elements on the diagonal, and we conclude that the set B_{ST} forms a basis for KBSM(ST).

4.4. Toward the HOMFLYPT skein module of the lens spaces $L(p, 1)$. In this subsection we go back to our original goal, which is the computation of the HOMFLYPT skein module of the lens spaces $L(p, 1)$. As explained before, in order to compute $\mathrm{HOM}(L(p, 1))$, we must normalize the universal invariant X of the HOMFLYPT type, by making it satisfy every possible braid band move. In [**16**] it is shown that the performance of a *bbm* on a mixed braid in $B_{1,n}$ reduces to performing *bbm*'s on elements in the canonical basis, Σ'_n, of the algebra $\mathrm{H}_{1,n}(q)$ and, in fact, on their first moving strand. Namely, if $\alpha \in B_{1,n}$ and $\beta_i \in \Sigma'_n$, for all i ([**16**, Proposition 1]). We have that:

$$X_{\widehat{\alpha}} = X_{\widehat{bbm(\alpha)}} \Leftrightarrow \sum_i \left(X_{\widehat{\beta_i}} = X_{\widehat{bbm_1(\beta_i)}} \right)$$

Namely, in order to compute $\mathrm{HOM}(L(p, 1))$, it suffices to consider the performance of braid band moves on the first strand of elements in the set Σ'. This simplifies the infinite system of equations. We then simplify the derived equations further, by showing that the equations obtained by performing bbm's on the first moving strand of elements in Σ'_n are equivalent to the equations obtained by performing bbm's on the first moving strand of elements in Σ_n ([**16**, Proposition 2]). That is,

$$X_{\widehat{s'}} = X_{\widehat{bbm_1(s')}} \Leftrightarrow \sum_i \left(X_{\widehat{s_i}} = X_{\widehat{bbm_1(s_i)}} \right),$$

where $s' \in \Sigma'_n$ and $s_i \in \Sigma_n$ for all i. The summation symbol indicates that the equation on the left-hand side is a combination of multiple equations on the right-hand side.

Recall now that elements in Σ consist of two parts: a monomial in t_i's with possible gaps in the indices and unordered exponents, followed by a 'braiding tail' in the basis of $\mathrm{H}_n(q)$. This allows us to further reduce the computation to elements in the basis Λ of HOM(ST). In order to obtain elements in the augmented $\mathrm{H}_n(q)$-module Λ^{aug} (followed by 'braiding tails'), we first handle the gaps in the indices of the looping generators of elements in Σ. It should be noted that the performance of a bbm is now seen as occurring on any moving strand ([**16**, Proposition 3]). Namely:

$$X_{\widehat{s}} = X_{\widehat{bbm_1(s)}} \Leftrightarrow \sum_i \left(X_{\widehat{h_i}} = X_{\widehat{bbm(h_i)}} \right),$$

where $s \in \Sigma_n$ and h_i in the $\mathrm{H}_n(q)$-module Λ^{aug}.

In [**16**], it shown that equations derived from elements in the Λ^{aug}-module $\mathrm{H}_n(q)$ are equivalent to the equations obtained by performing bbm's on an any moving strand of elements in the $\mathrm{H}_n(q)$-module Λ ([**16**, Proposition 4]). That is,

$$X_{\widehat{h}} = X_{\widehat{bbm(h)}} \Leftrightarrow \sum_i \left(X_{\widehat{k_i}} = X_{\widehat{bbm(k_i)}} \right),$$

where h in the $\mathrm{H}_n(q)$-module Λ^{aug} and k_i in the $\mathrm{H}_n(q)$-module Λ.

Finally, the 'braiding tails' are eliminated from the elements in the $\mathrm{H}_n(q)$-module Λ. This reduces the computations to the set Λ, where the bbm's are executed on any moving strand ([**16**, Theorem 8]). Thus, in order to compute $\mathrm{HOM}(L(p,1))$, it is sufficient to solve the infinite system of equations obtained by performing bbm's on every moving strand of elements in the set Λ. With a slight abuse of notation, we have that:

$$HOM(L(p,1)) = \frac{\Lambda}{< \lambda - bbm_i(\lambda) >}, \ \lambda \in \Lambda, \ \forall \ i.$$

Moreover, in [**13**] we consider the augmented set Λ^{aug} and show that the system of equations obtained from elements in Λ by performing bbm's on *any moving* strand, is equivalent to the system of equations obtained by performing *bbm*'s on the *first moving* strand of elements in Λ^{aug}. We show that by proving that the following diagram commutes:

$$\begin{array}{cccccc} \Lambda & \ni & \tau_i & \overset{bbm_m}{\longrightarrow} & bbm_m(\tau_i) \\ & & \updownarrow & & \updownarrow \\ \Lambda^{aug} & \ni & \sum_j \tau_j & \overset{bbm_1}{\longrightarrow} & \sum_j bbm_1(\tau_j) \end{array}$$

Note that the fact that the t_i's are not conjugate makes this procedure very non-trivial.

The above are summarized in the following sequence of equations:

$$\begin{array}{rclcl} \text{HOM}\,(L(p,1)) & = & \frac{\text{HOM(ST)}}{<bbm's>} & = & \frac{B_{1,n}}{<a-bbm_i(a)>,\ \forall\ i} & = & \frac{\Sigma'_n}{<s'-bbm_1(s')>} \\ & = & \frac{\Sigma_n}{<s-bbm_1(s)>} & = & \frac{\Lambda^{aug}|\text{H}_n}{<\lambda'-bbm_i(\lambda')>,\ \forall\ i} \\ & = & \frac{\Lambda|\text{H}_n}{<\lambda''-bbm_i(\lambda'')>,\ \forall\ i} & = & \frac{\Lambda}{<\lambda-bbm_i(\lambda)>,\ \forall\ i} \\ & = & \frac{\Lambda^{aug}}{<\mu-bbm_1(\mu)>}. \end{array}$$

Namely:

$$HOM(L(p,1)) \;=\; \frac{\Lambda^{aug}}{<\mu - bbm_1(\mu)>},\ \mu \in \Lambda^{aug}.$$

REMARK 4.21. It is worth mentioning that although $\Lambda^{aug} \supset \Lambda$, the advantage of considering elements in the augmented set Λ^{aug} is that we restrict the performance of the braid band moves only on the first moving strand and, thus, we obtain less equations and more control on the infinite system.

Moreover, in [**4**] we consider the skein module obtained by solving the infinite system of equations $X_{\widehat{\mu}} = X_{\widehat{bbm_i\mu}}$, where we only consider elements in Λ^{aug_+}, a set related to $\mathcal{B}^+$, and we perform braid band moves on their first moving strands, namely $\frac{\Lambda^{aug_+}}{<bbm_1>}$. Similarly, we consider the module $\frac{\Lambda^{aug_-}}{<bbm_1>}$ obtained by solving the infinite system of equations by considering elements in Λ^{aug_-}, and we perform braid band moves on their first moving strands. These modules are related via two maps f and I defined as follows:

DEFINITION 4.22.

(i) We define the automorphism $f : \Lambda^{aug} \to \Lambda^{aug}$ such that:

$$\begin{array}{rclcl} f(\tau_1 \cdot \tau_2) & = & f(\tau_1)\cdot f(\tau_2), & & \forall\, \tau_1, \tau_2 \in \Lambda^{aug} \\ t_i^k & \mapsto & t_i^{-k}, & & \forall\, i \in \mathbb{N}^*,\ \forall\, k \in \mathbb{Z}\setminus\mathbb{N} \\ \sigma_i & \mapsto & \sigma_i^{-1}, & & \forall\, i \in \mathbb{N}^* \end{array}$$

(ii) We define the map $I : R[z^{\pm 1}, s_k] \to R[z^{\pm 1}, s_k]$, $k \in \mathbb{Z}$ such that:

$$\begin{array}{rcll} I(\tau_1 + \tau_2) & = & I(\tau_1) + I(\tau_2), & \forall\, \tau_1, \tau_2 \\ I(\tau_1 \cdot \tau_2) & = & I(\tau_1) \cdot I(\tau_2), & \forall\, \tau_1, \tau_2 \end{array}$$

$$\begin{array}{rcll} s_{-k} & \mapsto & s_k, & \forall\, k \in \mathbb{N} \\ s_{p-k} & \mapsto & s_{p+k}, & \forall\, k\ :\ 0 \leq k \leq p \\ z & \mapsto & \lambda\cdot z & \\ q^{\pm 1} & \mapsto & q^{\mp 1} & \\ \frac{\lambda^k}{z} & \mapsto & \frac{1}{\lambda^{k+1} z}, & \forall\, k \end{array}$$

EXAMPLE 4.23.

i. $f(t^{-1}t_1^2\sigma_1^{-1}) = tt_1^{-2}\sigma_1$.

ii. For $p = 3$ we have that $I(z\,q^{-2}\,s_{-2}\,s_2) = \lambda z\,q^{-2}\,s_2\,s_4$, since $I(s_2) = I(s_{3-1}) := s_{3+1}$.

Using the two maps f and I it is proved that the solution of the system $\frac{\Lambda^{aug_-}}{<bbm_1>}$ can be derived from the solution of the system $\frac{\Lambda^{aug_+}}{<bbm_1>}$. Indeed we have the following result:

THEOREM 4.24. *The equations obtained by imposing on the invariant X relations coming from the performance of a $\pm$-bbm_1 on an element τ in Λ^{aug_+} are equivalent to the image of the equations obtained by performing $\mp$-bbm_1 on its corresponding element $f(\tau)$ in Λ^{aug_-} under I. That is:*

$$I\left(X_{\widehat{f(\tau)}} = X_{\widehat{bbm_{\mp 1}(f(\tau))}}\right) \Leftrightarrow X_{\widehat{\tau}} = X_{\widehat{bbm_{\pm 1}(\tau)}}$$

Equivalently we have that the following diagram commutes:

$$\begin{array}{ccccccc}
\Lambda^{aug}_{(k)} & \ni & \tau & \overset{bbm_{\pm}}{\rightarrow} & bbm_{\pm 1}(\tau) & \Rightarrow & X_{\widehat{\tau}} = X_{\widehat{bbm_1(\tau)}}, \qquad X_{\widehat{\tau}} = X_{\widehat{bbm_{-1}(\tau)}} \\
 & & \updownarrow f & & & & \uparrow I \qquad\qquad\qquad \uparrow I \\
\Lambda^{aug}_{(-k)} & \ni & f(\tau) & \overset{bbm_{\mp}}{\rightarrow} & bbm_{\mp 1}(\tau) & \Rightarrow & X_{\widehat{f(\tau)}} = X_{\widehat{bbm_{-1}(f(\tau))}}, \quad X_{\widehat{f(\tau)}} = X_{\widehat{bbm_{+1}(f(\tau))}}
\end{array}$$

We now demonstrate Theorem 4.24 with an example.

EXAMPLE 4.25. Consider the element $t \in \Lambda$ and perform a positive bbm to obtain $t^p\, t_1\, \sigma_1$. We have that

$$\begin{aligned}
tr(t^p\, \underline{t_1}\, \sigma_1) &= tr(t^p\, \sigma_1\, t\, \underline{\sigma_1^2}) = (q-1)\, tr(t^p\, \underline{t_1}) + q\, tr(t^p\, \sigma_1\, \underline{t}) = \\
&= q(q-1)\, tr(t^p\, t_1') + (q-1)^2\, tr(t^p\, \sigma_1\, t) + q\, tr(t^p\, \sigma_1\, t) = \\
&= q(q-1)\, s_1 s_p + (q^2-q+1)\, z\, s_{p+1}
\end{aligned}$$

Hence, we obtain the equation:

$$\begin{aligned}
X_{\widehat{t}} = X_{\widehat{t^p t_1 \sigma_1}} &\Leftrightarrow s_1 = -\frac{1-\lambda q}{\sqrt{\lambda}(1-q)} \sqrt{\lambda}^3 \cdot tr(t^p t_1 \sigma_1) \\
&\Leftrightarrow s_1 = \frac{\lambda}{z} \cdot \left[q(q-1)\, s_1 s_p + (q^2-q+1)\, z\, s_{p+1}\right]
\end{aligned}$$

We now consider the homologous word of t, $t^{-1} := f(t)$, and we perform a negative bbm, to obtain $bbm_{-1}(t^{-1}) = t^p\, t_1^{-1}\, \sigma_1^{-1}$. We now evaluate $tr(t^p\, t_1^{-1}\, \sigma_1^{-1})$:

$$\begin{aligned}
& tr(t^p\, t_1^{-1}\, \underline{\sigma_1^{-1}}) \\
&= q^{-1}\, tr(t^p\, \underline{t_1^{-1}\, \sigma_1}) + (q^{-1}-1)\, tr(t^p\, \underline{t_1^{-1}}) \\
&= q^{-1}\, tr(t^p\, \sigma_1^{-1}\, \underline{t^{-1}}) + (q^{-1}-1)\, tr(t^p\, \underline{\sigma_1^{-1}}\, t^{-1}\, \sigma_1^{-1}) \\
&= q^{-1}\, tr(t^{p-1}\, \underline{\sigma_1^{-1}}) + (q^{-1}-1)\, q^{-1}\, tr(t^p\, {t_1'}^{-1}) + (q^{-1}-1)^2\, tr(t^{p-1}\, \underline{\sigma_1^{-1}} \\
&= q^{-1}\, (q^{-1}-1)\, s_{-1}\, s_p + (\lambda\, z)\, (q^{-2}-q^{-1}+1)\, s_{p-1}
\end{aligned}$$

and we obtain the following equation:

$$\begin{aligned}
X_{\widehat{t^{-1}}} = X_{\widehat{t^p t_1^{-1} \sigma_1^{-1}}} &\Leftrightarrow s_{-1} = -\frac{1-\lambda q}{\sqrt{\lambda}(1-q)} \sqrt{\lambda}^{-3} \cdot tr(t^p t_1^{-1} \sigma_1^{-1}) \\
&\Leftrightarrow s_{-1} = \frac{1}{\lambda^2 z} \cdot [q^{-1}(q^{-1}-1)\, s_{-1} s_p \\
&\qquad\qquad + (q^{-2}-q^{-1}+1)\, \lambda\, z\, s_{p-1}]
\end{aligned}$$

We now consider $I\left(X_{\widehat{t^{-1}}}=X_{\widehat{t^p t_1^{-1}\sigma_1^{-1}}}\right)$ and we have that

$$s_1 = I\left(\frac{1}{\lambda^2 z}\right)\cdot\left[q(q-1)\,s_1 s_p + (q^2-q+1)\,I(\lambda\,z)\;s_{p+1}\right],$$

since $I(s_{-1}) = s_1$, $I(s_{-1}s_p) = s_1 s_p$, $I(s_{p-1}) = s_{p+1}$ and $I(q^{-1}) = q$. Finally, we have that

$$I\left(\frac{1}{\lambda^2 z}\right) = I\left(\frac{\lambda^{-2}}{z}\right) := \frac{\lambda}{z},\text{ and } I(\lambda\,z) = \lambda.$$

Hence, we obtain the equation $X_{\widehat{t}} = X_{\widehat{t^p t_1\sigma_1}}$.

Finally, in [**4**], potential bases for $\frac{\Lambda^{aug_+}}{<bbm_1>}$ and for $\frac{\Lambda^{aug_-}}{<bbm_1>}$ are presented. In particular, it is shown that the set

$$\left\{t'^{k_0}t'^{k_1}\dots t'^{\,k_n}_{n},\text{ where } n,\,k_i\in\mathbb{N}\;:\;0\le k_i\le p-1\right\} \tag{4.9}$$

is a generating set of the module $\frac{\Lambda^{aug_+}}{<bbm_1>}$, and since $\frac{\Lambda^{aug_+}}{<bbm_1>} = \frac{\Lambda^{+}}{<bbm_i>}$, we have that $\frac{\Lambda^{+}}{<bbm_i>} = \frac{\Lambda'^{+}}{<bbm_i>}$. We conjecture that the set in Eq. (4.9) forms a basis for $\mathcal{B}^+/<bbm_i>$. Similarly, the set

$$\left\{t'^{k_0}t'^{k_1}\dots t'^{\,k_n}_{n},\text{ where } n,\,k_i\in\mathbb{Z}\backslash\mathbb{N}\;:\;0\ge k_i\ge -p+1\right\}$$

is a generating set of the module $\frac{\Lambda^{aug_-}}{<bbm>}$. We conjecture that it also forms a basis for $\frac{\Lambda^{aug_-}}{<bbm>}$.

EXAMPLE 4.26. In this example we present some of the equations of the infinite system, a solution of which corresponds to the computation of HOM$(L(p,1))$:

i. For $k=0$ we have $1\in\Lambda^{aug_+}_{(0)}$ and $1\overset{bbm_\pm}{\to} t^p\sigma_1^{\pm1}\Rightarrow 1 := s_0 = s_p$.

ii. For $k=1$ we have $t\in\Lambda^{aug_+}_{(1)}$ and $t\overset{bbm_\pm}{\to} t^p t_1\sigma_1^{\pm1}\Rightarrow \begin{cases} s_{p+1} &= s_1\\ s_p s_1 &= a_0 s_1\end{cases}$

iii. For $k=2$ we have $t^2,\,tt_1\in\Lambda^{aug_+}_{(2)}$ and we obtain the following equations:

$$t^2\overset{bbm}{\to} t^p t_1^2\sigma_1^{\pm1}:\begin{cases} s_2 &= a_1 s_{p+2} + a_2 s_{p+1}s_1 + a_3 s_p s_2\\ s_2 &= b_1 s_{p+2} + b_2 s_{p+1} s_1\end{cases}$$

$$tt_1\overset{bbm}{\to} t^p t_1 t_2\sigma_1^{\pm1}:\begin{cases} s_2 + s_1^2 &= c_1 s_{p+2} + c_2 s_{p+1} s_1\\ s_2+s_1^2 &= d_1 s_{p+2} + d_2 s_{p+1}s_1 + d_3 s_p s_2 + d_4 s_p s_1^2\end{cases}$$

and thus:

$$\begin{cases} s_{p+2} &= A_1 s_2 + A_2 s_1^2\\ s_2 s_p &= B_1 s_2 + B_2 s_1^2\\ s_1 s_{p+1} &= C_1 s_2 + C_2 s_1^2\\ s_p s_1^2 &= D_1 s_2 + D_2 s_1^2\end{cases}$$

where $a_i, A_i, b_i, B_i, c_i, C_i, d_i, D_i\in\mathbb{C},\ \forall\, i$.

iv. For $k=-1$ we have $t^{-1} \in \Lambda^{aug_-}_{(-1)}$ and:

$$t^{-1} \overset{bbm_\pm}{\to} t^p t_1^{-1} \sigma_1^{\pm 1} \Leftrightarrow \begin{cases} s_{p-1} &= s_{-1} \\ s_p s_{-1} &= a'_0 s_1 \end{cases}$$

v. For the elements in $\Lambda^{aug_-}_{(-2)}$ we have:

$t^{-2} \overset{bbm}{\to} t^p t_1^{-2} \sigma_1^{\pm 1}$

$t^{-1} t_1^{-1} \overset{bbm}{\to} t^p t_1^{-1} t_{-2} \sigma_1^{\pm 1}$ and:

$$\begin{cases} s_{p-2} &= A'_1 s_{-2} + A'_2 s^2_{-1} \\ s_{-2} s_p &= B'_1 s_{-2} + B'_2 s^2_{-1} \\ s_{-1} s_{p-1} &= C'_1 s_{-2} + C'_2 s^2_{-1} \\ s_p s^2_{-1} &= D'_1 s_{-2} + D'_2 s^2_{-1} \end{cases}$$

where $A'_i, B'_i, C'_i, D'_i \in \mathbb{C}$, $\forall\ i$.

REMARK 4.27.

i. Results on the infinite system so far suggest that the infinite system of equations admits unique solution, which on the level of skein modules means that the skein module is torsion free, and that the set

$$\left\{ t'^{k_0} t'^{k_1} \dots t'^{\,k_n}_n, \text{ where } n, k_i \in \mathbb{Z} \ :\ -p/2 \leq k_i < p/2 \right\}$$

forms a basis for $\mathrm{HOM}(L(p,1))$.

ii. It is worth mentioning that this set was first presented in [**20**], where, using diagrammatic methods, it is shown that it forms a basis for $\mathrm{HOM}(L(p,1))$. In [**14**] we work toward proving this result using braids and techniques described within this paper.

iii. We have reasons to believe that the braid technique can be successfully applied in order to compute skein modules of other more complicated c.c.o. 3-manifolds (such as the lens spaces $L(p,q), q > 1$), where diagrammatic methods fail to do so ([**15**]).

4.5. The Kauffman bracket skein module of the lens spaces $L(p,1)$ for $p \neq 0$. In this subsection we present a new basis for $\mathrm{KBSM}(L(p,1))$, $p \neq 0$ first appeared in [**1**] (see also [**2**]). This basis is different from that presented in [**23**] and it so more natural on the level of braids, since elements in this basis have no crossings.

In [**1**, Theorem 3] the following result is proved:

THEOREM 4.28. *The set $\mathcal{B}_p := \{t^n,\ n \in I\}$, where $I = \{0, 1, \dots, \lfloor p/2 \rfloor\}$, is a basis for the Kauffman bracket skein module of the lens spaces $L(p,1)$.*

In order to prove Theorem 4.28 we first translate the relation $\mathrm{KBSM}(L(p,1)) = \frac{\mathrm{KBSM(ST)}}{<a - bm(a)>}$, for all a in a basis of KBSM(ST), in terms of braids. Namely, we impose on the universal invariant V of the Kauffman bracket type for knots and links in ST, relations coming from the performance of bbm's on elements in the B_{ST} basis of KBMS(ST). That is, in order to compute $\mathrm{KBSM}(L(p,1))$ we solve the infinite system of equations:

$$V_{\widehat{t^n}} = V_{\widehat{t^p t_1^n \sigma_1^{\pm 1}}}$$

Note that the unknowns of the infinite system are the s_i's that come from the fourth rule of the Markov trace.

For computing $tr\left(t^p t_1^n \sigma_1^{\pm 1}\right)$, we first express $t^p t_1^n \sigma_1^{\pm 1}$ to sums of elements of the form $t^m {t'_1}^k$. In particular, the following formulas hold in $TL_{1,n}$ for $n, p \in \mathbb{N}\backslash\{0\}$:

$$t^p\, t_1^n\, \sigma_1 \mathrel{\widehat{\cong}} \sum_{i=0}^{n} a_i\, t^{p+i}\, {t'_1}^{n-i} \qquad \text{and} \qquad t^p\, t_1^n\, \sigma_1^{-1} \mathrel{\widehat{\cong}} \sum_{i=1}^{n} b_i\, t^{p+i}\, {t'_1}^{n-i}$$

where a_i, b_i coefficients for all i and the symbol $\widehat{\cong}$ denotes that conjugation and stabilization moves are performed. Finally, we translate the resulting elements in terms of elements in the basis B_{ST} and we have that:

$$t^p\, {t'_1}^n \underset{\text{skein}}{\widehat{\cong}} \sum_{i=0}^{\lfloor n/2 \rfloor} a_i\, t^{p+n-2i}, \tag{4.10}$$

where a_i are coefficients for all i.

Hence, we have that $tr(t^p\, t_1^n\, \sigma_1) = \sum\limits_{i=0}^{\lfloor n/2 \rfloor} a_i\, s_{p+n-2i}$ for some coefficients a_i, since:

$$tr(t^p\, t_1^n\, \sigma_1) \;=\; tr\left(\sum_{i=0}^{n} a_i\, t^{p+i}\, {t'_1}^{n-i}\right) \;=\; tr\left(\sum_{i=0}^{n} \sum_{j=0}^{\lfloor \frac{n-i}{2} \rfloor} \left(a'_i\, t^{p+n-2j}\right)\right).$$

The following diagram summarizes the steps we followed in order to obtain the equations of the infinite system:

$$\begin{array}{ccccccc}
t^n & \overset{bbm}{\longrightarrow} & t^p t_1^n \sigma_1 & \widehat{\cong} & \sum\limits_{i=0}^{n} a_i\, t^{p+i}\, {t'_1}^{p-i} & \underset{\text{skein}}{\widehat{\cong}} & \sum\limits_{i=0}^{n} \sum\limits_{j=0}^{\lfloor \frac{n-i}{2} \rfloor} (s_{p+n-2j}) \\
\big\downarrow{\scriptstyle tr} & & & & & & \big\downarrow{\scriptstyle tr} \\
s_n & & & & & & \sum\limits_{i=0}^{n} s_{p-n+2i} \\
\big\downarrow{\scriptstyle V} & & & & & & \big\downarrow{\scriptstyle V} \\
s_n & & & = & & & \left(-\frac{1+u^2}{u}\right) u^{2e} \cdot \sum\limits_{i=0}^{n} s_{p-n+2i}
\end{array}$$

We now treat elements of the form s_{p+k} for $k \in \mathbb{N}$ and we show that these elements can be written as sums of elements in s_i's with $i \in \{0, 1, \ldots, p\}$. In particular, in [**1**, Proposition 3] the following formulas are proved:

$$s_{p+n} \;=\; \begin{cases} \sum\limits_{i=0}^{n/2} \left(c_i\, s_{2i} \,+\, d_i\, s_{p-2i}\right) & \text{, for } n \text{ even} \\ \sum\limits_{i=0}^{(n-1)/2} \left(c'_i\, s_{2i-1} \,+\, d'_i\, s_{p-2i-1}\right) & \text{, for } n \text{ odd} \end{cases},$$

where c_i, d_i, c'_i, d'_i are coefficients. Namely, for $p, k \in \mathbb{N}$ we have that:

$$s_{p+k} \;=\; \sum_{i<p} q_i\, s_i, \quad \text{where } i \in \mathbb{Z} \text{ and } q_i \text{ coefficients.}$$

Note now that the indices on some monomials of the resulting sum may be negative. As shown in [**1**], these monomials may be expressed in terms of s_i's where $i \in \mathbb{N}$. In particular we have that:

$$s_{-n} = \begin{cases} \sum\limits_{i=0}^{n/2} s_{2i} & , \text{for } n \text{ } even \\ \sum\limits_{i=0}^{(n-1)/2} s_{2i+1} & , \text{for } n \text{ } odd \end{cases}$$

Hence, we conclude that the unknowns of the infinite system of equations with indices greater than or equal to p can be written as sums of unknowns with non-negative indices and also that these indices are mod p. In other words, we conclude that the set $\mathcal{B}_p$ spans the Kauffman bracket skein module of the lens spaces $L(p,1)$. Finally § 4 [**1**], the set $\mathcal{B}_p$ is shown to be linearly independent and hence it forms a basis for KBSM$(L(p,1))$.

EXAMPLE 4.29. In this example we present equations of the infinite system and we demonstrate the above mentioned results. We have that:

$$t^0 \overset{bbm}{\rightarrow} t^p\, \sigma_1^{\pm 1} \overset{V}{\Rightarrow} 1 := s_0 = s_p \Rightarrow s_p = s_0$$

$$t^{-1} \overset{bbm}{\rightarrow} t^p\, t_1^{-1} \sigma_1^{\pm 1} \overset{V}{\Rightarrow} s_{-1} = s_{p-1} \underset{s_{-1}=s_1}{\Rightarrow} s_{p-1} = s_1$$

$$t^{-2} \overset{bbm}{\rightarrow} t^p\, t_1^{-2} \sigma_1^{\pm 1} \overset{V}{\Rightarrow} s_{-2} = s_{p-2} + s_p + s_{p+2} \underset{\left\{\begin{matrix} s_{-2} & = & s_0 + s_2 \\ s_{p+2} & = & \sum\limits_{i=0}^{p} s_i \end{matrix}\right\}}{\Rightarrow} s_{p-2} = \sum_{i=0}^{p-3} s_i$$

$$\vdots \qquad\qquad \vdots$$

$$t^{-\lfloor p/2 \rfloor + 1} \overset{bbm}{\rightarrow} t^p\, t_1^{-\lfloor p/2 \rfloor + 1} \sigma_1^{\pm 1} \overset{V}{\Rightarrow} s_{-\lfloor p/2 \rfloor + 1} = s_{p - \lfloor p/2 \rfloor - 1} + \ldots + s_{p + \lfloor p/2 \rfloor + 1}$$

$$\Rightarrow s_{p - \lfloor p/2 \rfloor + 1} = \sum_{i=0}^{p - \lfloor p/2 \rfloor} s_i$$

5. The Kauffman bracket skein module of $S^1 \times S^2$

$S^1 \times S^2$ may be obtained by 'gluing' two solid tori via some homeomorphism h on their boundaries, such that h takes a meridian curve m_1 on the first solid torus ST_1, to the corresponding meridian curve m_2 of the second solid torus ST_2. We may choose h to be the identity $i : \partial \mathrm{ST}_1 \rightarrow \partial \mathrm{ST}_2$. Since now $ST = S^1 \times D^2$, where D^2 denotes a disc, and since the gluing of two discs results in a 2-sphere S^2, the resulting 3-manifold can be considered as a family of 2-spheres parametrized by a circle (for an illustration see Figure 28).

In this subsection we compute the Kauffman bracket skein module of $S^1 \times S^2$, KBSM$\left(S^1 \times S^2\right)$, via braids. We follow the same procedure as in § 4.5 and we extend the universal Kauffman bracket type invariant V for knots and links in the Solid Torus to an invariant for knots and links in $S^1 \times S^2$. In particular, we show that KBSM$\left(S^1 \times S^2\right)$ is not torsion free and that its free part is generated by the unknot (or the empty knot).

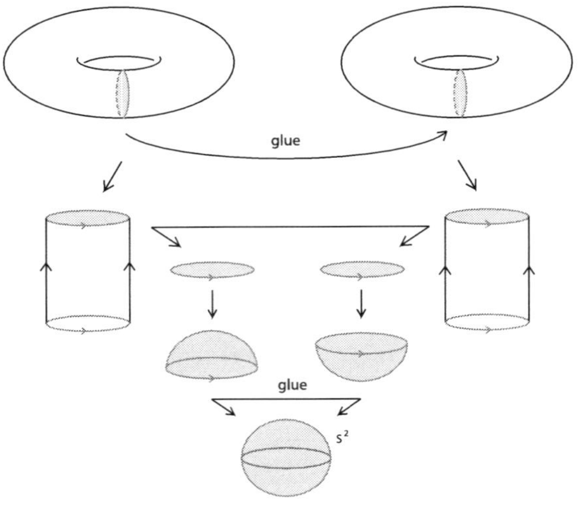

FIGURE 28. Gluing two solid tori to obtain $S^1 \times S^2$.

5.1. Topological set up. We consider $S^1 \times S^2$ as being obtained from S^3 by integral surgery along the unknot with coefficient zero, that is, $S^1 \times S^2$ is a special type of lens spaces $L(p, q)$ for $p = 0$ and $q = 1$ ($L(0, 1) \cong S^1 \times S^2$). The band moves, that reflect isotopy in $S^1 \times S^2$ are illustrated in Figure 29.

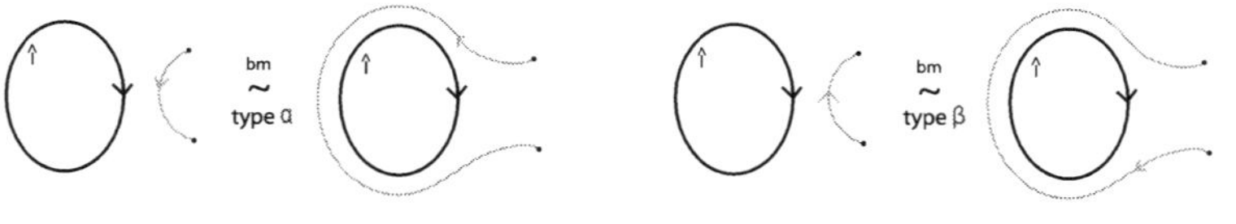

FIGURE 29. The two types of band moves for links in $S^1 \times S^2$.

Recall that a braid band move is defined as an equivalence move on mixed braids that corresponds to a band move on th closure of the mixed braid. In the case of $S^1 \times S^2$, these moves have the following algebraic expression:

$$\alpha \sim \alpha_+ \sigma_1^{\pm 1},$$

where $\alpha \in B_{1,n}$ and $\alpha_+ \in B_{1,n+1}$ is the word α with all indices shifted by $+1$. For an illustration see Figures 30 and 31.

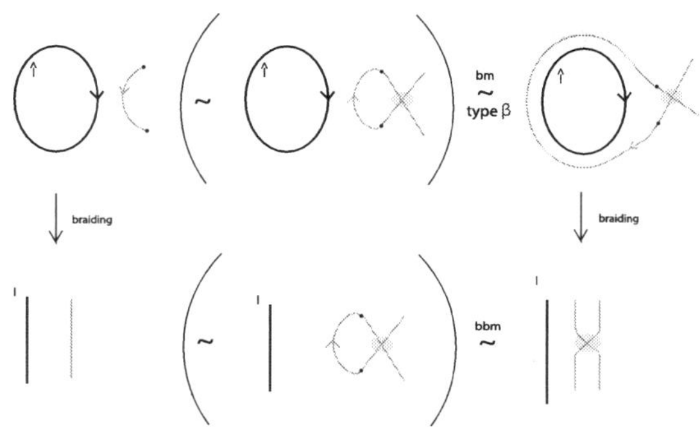

FIGURE 30. The two types of braid band moves for braids in $S^1 \times S^2$.

5.2. The Kauffman bracket skein module of $S^1 \times S^2$. In this subsection we solve the infinite system of equations obtained by performing braid band moves on elements in the basis B_{ST} of KBSM(ST) and by imposing to the generic invariant V for knots and links in ST relations of the form $V_{\widehat{t^n}} = V_{\widehat{bbm(t^n)}}$, for all $n \in \mathbb{N}$, where $bbm(t^n) = t_1^n \sigma_1^{\pm 1}$ (see Figure 31). Recall that the unknowns in the system are the s_i's, coming from the fourth rule of the trace function in Theorem 3.6, that is, $tr(t^n) = s_n$, for all $n \in \mathbb{Z}$.

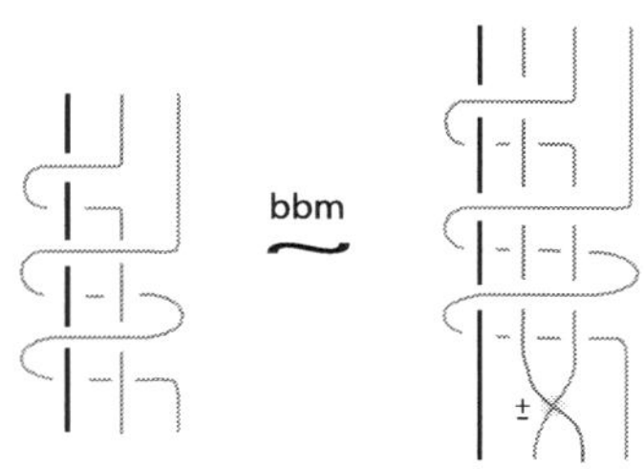

FIGURE 31. Braid band moves performed on $tt_1^2 \in B_{ST}$.

The method used for computing KBSM($S^1 \times S^2$) differs significantly from the one used in computing KBSM($L(p,1)$). More precisely, we start by considering the effect of negative bbm's on elements in B_{ST} and we obtain a spanning set for $B_-^{S^1\times S^2} := \frac{B_{\mathrm{ST}}}{< a - bbm_-(a) >}$. Then we study the effect of positive bbm's on elements in $B_-^{S^1\times S^2}$. With a slight abuse of notation, we denote the above as follows:

$$\mathrm{KBSM}\left(S^1 \times S^2\right)$$

$$= \frac{\mathrm{B}_{ST}}{< a - bbm_{\pm}(a) >} = \frac{\mathrm{B}_{ST}/ < a - bbm_-(a) >}{< a - bbm_+(a) >} = \frac{B_-^{S^1\times S^2}}{< a - bbm_+(a) >}.$$

Hence, we consider elements in B_{ST} and perform negative braid band moves, obtaining an infinite system of equations, the solution of which corresponds to a basis for $B^-_{S^1\times S^2}$.

EXAMPLE 5.1. In this example we show that t^0 and t are free in $B^-_{S^1\times S^2}$.

$$
\begin{aligned}
1 \overset{bbm_-}{\to} \sigma_1^{-1} \ \Rightarrow \ V_{\hat{1}} = V_{\widehat{\sigma_1^{-1}}} \ \Rightarrow \ 1 &= \left(-\tfrac{1+u^2}{u}\right) u^{-2}\, tr(\sigma_1^{-1}) \Rightarrow \\
1 &= -\tfrac{1+u^2}{u^3}\left(z-(u-u^{-1})\right) \overset{z=-\frac{1}{u(1+u^2)}}{\Rightarrow} \\
1 &= 1
\end{aligned}
$$

$$
\begin{aligned}
t \overset{bbm_-}{\to} t_1\,\sigma_1^{-1} \ \Rightarrow \ V_{\hat{t}} = V_{\widehat{t_1\,\sigma_1^{-1}}} \ \Rightarrow \ s_1 &= \left(-\tfrac{1+u^2}{u}\right) u^{2}\, tr(t_1\,\sigma_1^{-1}) \Rightarrow \\
s_1 &= --\tfrac{1+u^2}{u}\, u^2\, z\, s_1 \Rightarrow \\
s_1 &= s_1
\end{aligned}
$$

In [**3**], the following result is presented:

$$
V_{\widehat{t^n}} = V_{\widehat{t_1^n\,\sigma_1^{-1}}} \ \Rightarrow \ s_n = \begin{cases} \sum\limits_{i=0}^{(n-2)/2} s_{2i} & \text{, for } n \text{ even} \\ \sum\limits_{i=0}^{(n-3)/2} s_{2i+1} & \text{, for } n \text{ odd} \end{cases}
$$

Equivalently, the elements $t^n \in B_{\text{ST}}, n \geq 2$ can be written as sums of t^0 and t. This leads to the following theorem:

THEOREM 5.2. *$B^-_{S^1\times S^2}$ is generated by the unknot t^0 and t.*

From Theorem 5.2 and from the fact that the elements t^0 and t are linearly independent, we have that the set $\{t^0, t\}$ forms a basis for $B^-_{S^1\times S^2}$. We then consider the effect of positive braid band moves on the elements t^0 and t. We have that:

$$
\begin{aligned}
\text{For } t^0: \ 1 \overset{bbm_+}{\to} \sigma_1 \ \Rightarrow \ V_{\hat{1}} = V_{\hat{\sigma}_1} \ \Rightarrow \ 1 &= \left(-\tfrac{1+u^2}{u}\right) u^{-2}\, tr(\sigma_1) \Rightarrow \\
1 &= -(1+u^2)\, u\, z \overset{z=-\frac{1}{u(1+u^2)}}{\Rightarrow} \\
1 &= 1 \qquad \text{(free part)}
\end{aligned}
$$

Hence, the unknot is free in KBSM($S^1 \times S^2$. We now deal with the generator t and we have the following:

$$
\begin{aligned}
t \overset{bbm_+}{\to} t_1\,\sigma_1 \Rightarrow V_{\hat{t}} &= V_{\widehat{t_1\,\sigma_1}} \\
\Rightarrow s_1 &= \left(-\frac{1+u^2}{u}\right) u^6\, tr(t_1\,\sigma_1) \overset{tr(t_1\,\sigma_1)=(u^2-1+u^{-2})\,z\,s_1\,+\,(u-u^{-1})\,s_1}{\Rightarrow} \\
\Rightarrow (1-u^6)\,(1-u^2)\,s_1 &= 0.
\end{aligned}
$$

The discussion above leads to the following theorem:

THEOREM 5.3. *The free part of KBSM$(S^1 \times S^2)$ is generated by the unknot and the torsion part is generated by t.*

Equivalently, we have that

$$\text{KBSM}\left(S^1 \times S^2\right) \ = \ \mathbb{Z}[A^{\pm 1}] \oplus \text{Torsion}$$

and although we don't obtain a closed formula for the torsion part of $\text{KBSM}(S^1 \times S^2)$, Theorem 5.3 implies the following:

$$\pi(t^{2n+1}) \ = \ 0, \forall n \in \mathbb{N} \quad \text{and} \quad \text{KBSM}\left(S^1 \times S^2\right)/Tor \ = \ \mathbb{Z}[A^{\pm 1}],$$

where π denotes the natural projection

$$\pi : \text{KBSM}\left(S^1 \times S^2\right) \ \rightarrow \ \text{KBSM}\left(S^1 \times S^2\right)/Tor.$$

6. Conclusions

In this paper we present recent results on the computation of the HOMFLYPT and the Kauffman bracket skein modules of S^3, the Solid Torus, and the lens spaces. We first present new bases for HOM(ST) and KBSM(ST) in braid form and we then relate $\text{HOM}(L(p,1))$ to HOM(ST), and $\text{KBSM}(L(p,1))$ to KBSM(ST), by means of equations resulting from the performance of bbm's. For both skein modules, we arrive at infinite systems of equations, the solution of which coincides with the computation of the corresponding skein module of $L(p,1)$.

For the case of KBSM, we solve this system and we show that the set

$$\mathcal{B}_p \ := \{t^n, \ n \in I\}, \ \text{where } I = \{0, 1, \ldots, \lfloor p/2 \rfloor\}$$

forms a basis for the Kauffman bracket skein module of the lens spaces $L(p,1)$ (Theorem 4.28). For $\text{KBSM}\big(S^1 \times S^2\big)$, we show that it is generated by the unknot and the torsion part is generated by t (Theorem 5.3).

For the case of the HOMFLYPT skein module of the lens spaces $L(p,1)$, we show that in order to compute $\text{HOM}(L(p,1))$ it suffices to solve the infinite system of equations:

$$X_{\widehat{\tau}} \ = \ X_{\widehat{bbm_i(\tau)}},$$

where $bbm_i(\tau)$ is the result of the performance of bbm on the ith-moving strand of $\tau \in \Lambda$, for all $\tau \in \Lambda$ and for all i.
and we conclude that the set

$$\left\{{t'}^{k_0}{t'}^{k_1} \ldots {t'_n}^{k_n}, \ \text{where } n, \, k_i \in \mathbb{Z} \ : \ -p/2 \leq k_i < p/2\right\}$$

forms a potential basis for $\text{HOM}(L(p,1))$.

As a final note, it is worth mentioning that a different diagrammatic method based on braids has been successful in the computation of Kauffman bracket skein modules of 3-manifolds. More precisely, this method has been successfully applied for the case of the lens spaces $L(p,q)$ in [**1**], for $S^1 \times S^2$ in [**3**], for the handlebody of genus 2 in [**9**] and for the complement of $(2, 2p+1)$-torus knots in [**5**].

References

[1] Ioannis Diamantis, *The Kauffman bracket skein module of the lens spaces via unoriented braids*, Commun. Contemp. Math. **26** (2024), no. 2, Paper No. 2250076, 36, DOI 10.1142/S0219199722500766. MR4701452

[2] Ioannis Diamantis, *An alternative basis for the Kauffman bracket skein module of the solid torus via braids*, Knots, low-dimensional topology and applications, Springer Proc. Math. Stat., vol. 284, Springer, Cham, 2019, pp. 329–345, DOI 10.1007/978-3-030-16031-9_16. MR3986053

[3] Ioannis Diamantis, *The Kauffman bracket skein module of the handlebody of genus 2 via braids*, J. Knot Theory and Ramifications **28** (2019), no. 13, 1940020, DOI 10.1142/S0218216519400200.

[4] Ioannis Diamantis, *HOMFLYPT skein sub-modules of the lens spaces $L(p,1)$*, Topology Appl. **301** (2021), Paper No. 107500, 25, DOI 10.1016/j.topol.2020.107500. MR4312951

[5] Ioannis Diamantis, *The Kauffman bracket skein module of the complement of $(2,2p+1)$-torus knots via braids*, Topology Appl. **327** (2023), Paper No. 108433, 26, DOI 10.1016/j.topol.2023.108433.

[6] Ioannis Diamantis, *Pseudo links and singular links in the solid torus*, Commun. Math. **31** (2023), no. 1, 333–357. MR4527734

[7] Ioannis Diamantis, *Knotoids, pseudo knotoids, braidoids and pseudo braidoids on the torus*, Commun. Korean Math. Soc. **37** (2022), no. 4, 1221–1248, DOI 10.4134/CKMS.c210169. MR4505571

[8] Ioannis Diamantis, *Pseudo links in handlebodies*, Bull. Hellenic Math. Soc. **65** (2021), 17–34. MR4324469

[9] Ioannis Diamantis, *The Kauffman bracket skein module of $S^1 \times S^2$ via braids*, Axioms **13** (2024), no. 9, 617.

[10] Ioannis Diamantis and Sofia Lambropoulou, *Braid equivalences in 3-manifolds with rational surgery description*, Topology Appl. **194** (2015), 269–295, DOI 10.1016/j.topol.2015.08.009. MR3404617

[11] Ioannis Diamantis and Sofia Lambropoulou, *A new basis for the Homflypt skein module of the solid torus*, J. Pure Appl. Algebra **220** (2016), no. 2, 577–605, DOI 10.1016/j.jpaa.2015.06.014. MR3399379

[12] Ioannis Diamantis and Sofia Lambropoulou, *The braid approach to the HOMFLYPT skein module of the lens spaces $L(p,1)$*, Algebraic modeling of topological and computational structures and applications, Springer Proc. Math. Stat., vol. 219, Springer, Cham, 2017, pp. 143–176, DOI 10.1007/978-3-319-68103-0_7. MR3748894

[13] Ioannis Diamantis and Sofia Lambropoulou, *An important step for the computation of the HOMFLYPT skein module of the lens spaces $L(p,1)$ via braids*, J. Knot Theory Ramifications **28** (2019), no. 11, 1940007, 25, DOI 10.1142/S0218216519400078. MR4038332

[14] I. Diamantis and S. Lambropoulou, *The HOMFLYPT skein module of the lens spaces $L(p,1)$ via braids*, in preparation.

[15] I. Diamantis, B. Gabrovšek, S. Lambropoulou, and M. Mroczkowski, *Knot Theory of Lens Spaces*, De Gruyter. De Gruyter Expositions in Mathematics, 2024.

[16] Ioannis Diamantis, Sofia Lambropoulou, and Jozef H. Przytycki, *Topological steps toward the Homflypt skein module of the lens spaces $L(p,1)$ via braids*, J. Knot Theory Ramifications **25** (2016), no. 14, 1650084, 26, DOI 10.1142/S021821651650084X. MR3582885

[17] M. Flores and D. Goundaroulis, *Framization of a Temperley-Lieb algebra of type* B, J. Pure Appl. Algebra **224** (2020), no. 6, 106273, 29, DOI 10.1016/j.jpaa.2019.106273. MR4048529

[18] P. Freyd, D. Yetter, J. Hoste, W. B. R. Lickorish, K. Millett, and A. Ocneanu, *A new polynomial invariant of knots and links*, Bull. Amer. Math. Soc. (N.S.) **12** (1985), no. 2, 239–246, DOI 10.1090/S0273-0979-1985-15361-3. MR776477

[19] Boštjan Gabrovšek, *Tabulation of prime knots in lens spaces*, Mediterr. J. Math. **14** (2017), no. 2, Paper No. 88, 24, DOI 10.1007/s00009-016-0814-5. MR3621789

[20] Boštjan Gabrovšek and Maciej Mroczkowski, *The HOMFLYPT skein module of the lens spaces $L_{p,1}$*, Topology Appl. **175** (2014), 72–80, DOI 10.1016/j.topol.2014.07.003. MR3239212

[21] B. Gabrovšek and M. Mroczkowski, *Link diagrams and applications to skein modules*, in Algebraic Modeling of Topological and Computational Structures and Applications, Springer Proceedings in Mathematics & Statistics, 2017.

[22] Jim Hoste and Mark E. Kidwell, *Dichromatic link invariants*, Trans. Amer. Math. Soc. **321** (1990), no. 1, 197–229, DOI 10.2307/2001598. MR961623

[23] Jim Hoste and Józef H. Przytycki, *The $(2,\infty)$-skein module of lens spaces; a generalization of the Jones polynomial*, J. Knot Theory Ramifications **2** (1993), no. 3, 321–333, DOI 10.1142/S0218216593000180. MR1238877

[24] Jim Hoste and Józef H. Przytycki, *The Kauffman bracket skein module of $S^1 \times S^2$*, Math. Z. **220** (1995), no. 1, 65–73, DOI 10.1007/BF02572603. MR1347158

[25] Jim Hoste and Józef H. Przytycki, *A survey of skein modules of 3-manifolds*, Knots 90 (Osaka, 1990), de Gruyter, Berlin, 1992, pp. 363–379. MR1177433

[26] V. F. R. Jones, *Hecke algebra representations of braid groups and link polynomials*, Ann. of Math. (2) **126** (1987), no. 2, 335–388, DOI 10.2307/1971403. MR908150

[27] Vaughan F. R. Jones, *A polynomial invariant for knots via von Neumann algebras*, Bull. Amer. Math. Soc. (N.S.) **12** (1985), no. 1, 103–111, DOI 10.1090/S0273-0979-1985-15304-2. MR766964

[28] Louis H. Kauffman, *New invariants in the theory of knots*, Amer. Math. Monthly **95** (1988), no. 3, 195–242, DOI 10.2307/2323625. MR935433

[29] Kurt Reidemeister, *Knoten und Gruppen* (German), Abh. Math. Sem. Univ. Hamburg **5** (1927), no. 1, 7–23, DOI 10.1007/BF02952506. MR3069461

[30] Sofia S. F. Lambropoulou, *Solid torus links and Hecke algebras of $\mathcal{B}$-type*, Proceedings of the Conference on Quantum Topology (Manhattan, KS, 1993), World Sci. Publ., River Edge, NJ, 1994, pp. 225–245. MR1309934

[31] Sofia Lambropoulou, *Knot theory related to generalized and cyclotomic Hecke algebras of type B*, J. Knot Theory Ramifications **8** (1999), no. 5, 621–658, DOI 10.1142/S0218216599000419. MR1693556

[32] Sofia Lambropoulou and Colin P. Rourke, *Markov's theorem in 3-manifolds*, Topology Appl. **78** (1997), no. 1-2, 95–122, DOI 10.1016/S0166-8641(96)00151-4. Special issue on braid groups and related topics (Jerusalem, 1995). MR1465027

[33] S. Lambropoulou and C. P. Rourke, *Algebraic Markov equivalence for links in three-manifolds*, Compos. Math. **142** (2006), no. 4, 1039–1062, DOI 10.1112/S0010437X06002144. MR2249541

[34] Józef H. Przytycki, *Skein modules of 3-manifolds*, Bull. Polish Acad. Sci. Math. **39** (1991), no. 1-2, 91–100. MR1194712

[35] Józef H. Przytycki and Paweł Traczyk, *Invariants of links of Conway type*, Kobe J. Math. **4** (1988), no. 2, 115–139. MR945888

[36] V. G. Turaev, *The Conway and Kauffman modules of a solid torus* (Russian, with English summary), Zap. Nauchn. Sem. Leningrad. Otdel. Mat. Inst. Steklov. (LOMI) **167** (1988), no. Issled. Topol. 6, 79–89, 190, DOI 10.1007/BF01099241; English transl., J. Soviet Math. **52** (1990), no. 1, 2799–2805. MR964255

Department of Data Analytics and Digitalisation, Maastricht University, School of Business and Economics, P.O.Box 616, 6200 MD, Maastricht, The Netherlands

Email address: `i.diamantis@maastrichtuniversity.nl`

Contemporary Mathematics
Volume **827**, 2025
https://doi.org/10.1090/conm/827/16533

Topological symmetries of the Heawood family

Blake Mellor and Robin Wilson

Abstract. The *topological symmetry group* of an embedding Γ of an abstract graph γ in S^3 is the group of automorphisms of γ which can be realized by homeomorphisms of the pair (S^3, Γ). These groups are motivated by questions about the symmetries of molecules in space. In this paper, we find all the groups which can be realized as topological symmetry groups for each of the graphs in the Heawood family. This is an important collection of spatial graphs, containing the only intrinsically knotted graphs with 21 or fewer edges. As a consequence, we discover that the graphs in this family are all intrinsically chiral.

1. Introduction

The chemical properties of certain molecules are often strongly influenced by their symmetries in space. Molecules with the same chemical structure but whose embeddings in space are not equivalent are called **stereoisomers**, and can often have quite different properties. The subfield of *stereochemistry* focuses on the spatial structure of molecules, and particularly on the study of stereoisomers. Historically, chemists have been most interested in rigid symmetries of molecules, but as our ability to synthesize molecules advances there is increased interest in symmetries of more "flexible" molecules [**25**]. To describe the symmetries of these more complex molecules, Jonathan Simon [**27**] introduced the **topological symmetry group** of an embedded graph as the group of automorphisms of the graph induced by homeomorphisms of S^3.

Since Simon's original paper, considerable work has been done on topological symmetry groups. As a generalization of the study of symmetries and achirality of knots and links, this is a natural and important problem in low dimensional topology. In many cases, the motivating question is: given an abstract graph (e.g. chemical structure of a molecule), what are the possible topological symmetry groups over all embeddings of the graph in S^3? This problem has been addressed for several important families of graphs, including complete graphs [**2, 4, 14, 16, 19**], complete bipartite graphs [**13, 21, 23**] and Möbius ladders [**8, 11**], as well as for some individually interesting graphs [**3, 7, 12**]. There are also some broad results restricting possible topological symmetry groups for any 3-connected graph [**18**].

The Heawood family of graphs, shown in Figure 1, is an important family of graphs in low-dimensional topology, consisting of all graphs which can be derived

2020 *Mathematics Subject Classification.* Primary 57M15.

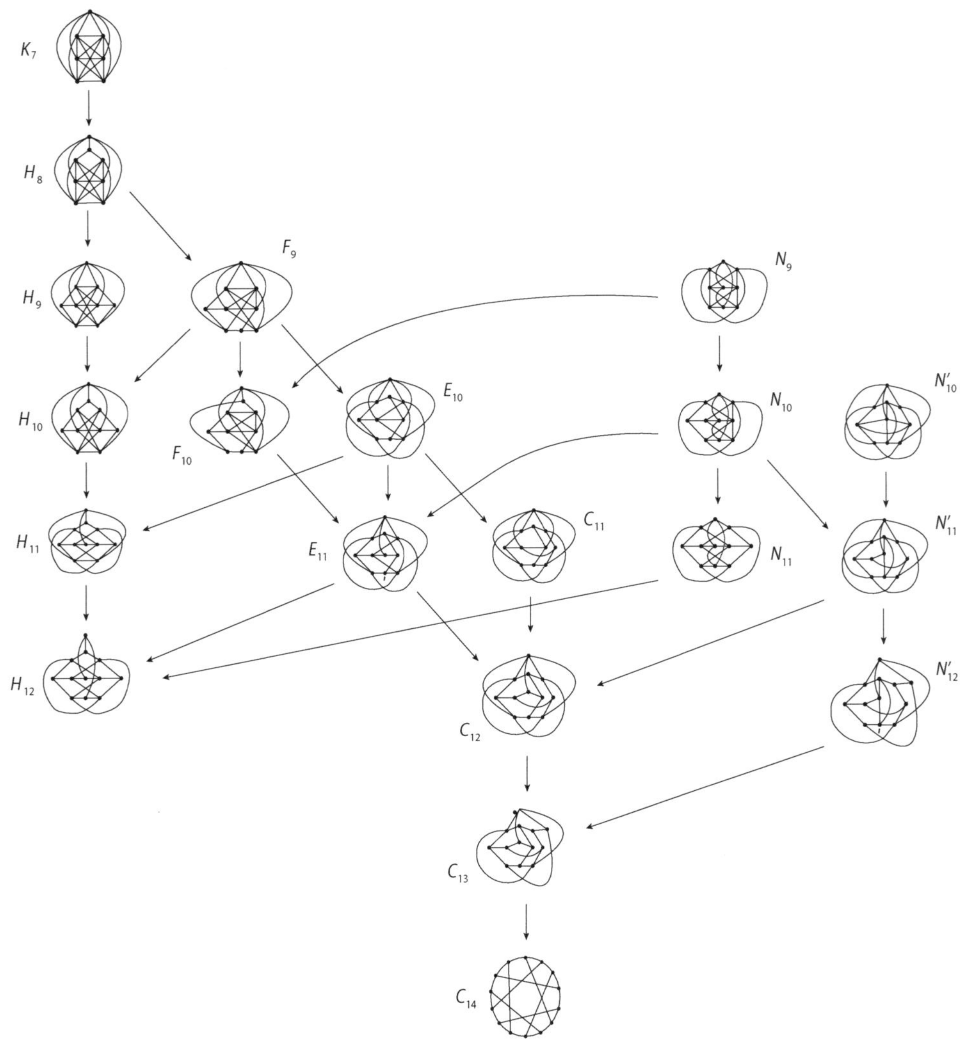

FIGURE 1. The Heawood family of graphs (adapted from [**22**]). Arrows indicate ∇Y moves between graphs.

from the complete graph K_7 by ∇Y or $Y\nabla$ moves. A ∇Y move is an operation in which a triangle in the graph is replaced by a degree-three vertex in the shape of a "Y", while a $Y\nabla$ move is the reverse, as shown in Figure 2. Aside from K_7, the best known graph in the family is the Heawood graph (denoted C_{14}). The 14 graphs of the family derived from K_7 only by ∇Y moves are all intrinsically knotted – in fact, they are the only intrinsically knotted graphs with 21 or fewer edges [**22**]. The other 6 graphs, denoted N_i and N_i', are not intrinsically knotted; they were the first examples demonstrating that $Y\nabla$ moves do not preserve intrinsic knottedness [**17**].

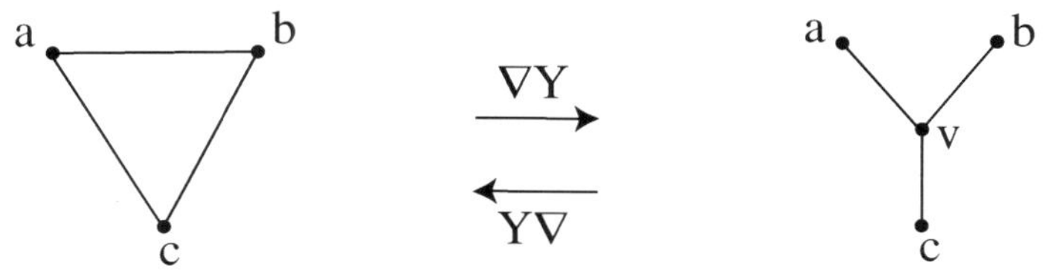

FIGURE 2. ∇Y and $Y\nabla$ moves.

TABLE 1. Realizable topological symmetry groups for graphs in the Heawood family. All realizable groups for these graphs are positively realizable.

Graph	Automorphism group	Positively realizable
K_7	S_7	$D_7, D_5, D_3, \mathbb{Z}_7, \mathbb{Z}_5, \mathbb{Z}_3, \mathbb{Z}_2$
H_8	$S_3 \times S_4$	$\mathbb{Z}_3, \mathbb{Z}_2$
H_9	$(D_3 \times D_3) \rtimes \mathbb{Z}_2$	$D_3, \mathbb{Z}_3, \mathbb{Z}_2$
H_{10}	D_4	$\mathbb{Z}_2$
H_{11}	D_2	$\mathbb{Z}_2$
H_{12}	$D_3 \times \mathbb{Z}_2$	$D_3, \mathbb{Z}_3, \mathbb{Z}_2$
F_9	$D_4 \times \mathbb{Z}_2$	$\mathbb{Z}_2$
F_{10}	$(\mathbb{Z}_2)^3 \rtimes D_3$	$D_3, \mathbb{Z}_3, \mathbb{Z}_2$
E_{10}	D_3	$\mathbb{Z}_3$
E_{11}	D_3	$\mathbb{Z}_3$
C_{11}	S_4	$\mathbb{Z}_3$
C_{12}	D_4	trivial
C_{13}	S_4	$\mathbb{Z}_3$
C_{14}	PGL(2, 7)	$D_7, D_3, \mathbb{Z}_7, \mathbb{Z}_6, \mathbb{Z}_3, \mathbb{Z}_2$
N_9	$(\mathbb{Z}_2)^3 \rtimes D_3$	$D_3 \times \mathbb{Z}_2, \mathbb{Z}_6, D_3, \mathbb{Z}_3, D_2, \mathbb{Z}_2$
N_{10}	D_3	$D_3, \mathbb{Z}_3, \mathbb{Z}_2$
N_{11}	$D_3 \times \mathbb{Z}_2$	$D_3 \times \mathbb{Z}_2, \mathbb{Z}_6, D_3, \mathbb{Z}_3, D_2, \mathbb{Z}_2$
N'_{10}	D_8	$D_2, \mathbb{Z}_2$
N'_{11}	D_2	$\mathbb{Z}_2$
N'_{12}	D_6	$D_6, D_3, D_2, \mathbb{Z}_6, \mathbb{Z}_3, \mathbb{Z}_2$

Our main purpose in this paper is to determine what groups can occur as topological symmetry groups for some embedding of each graph in the Heawood family of graphs. Aside from increasing the library of spatial graphs whose topological symmetry groups are known, the Heawood family provides an opportunity to explore how the ∇Y and $Y\nabla$ moves affect the symmetry groups. The classification of topological symmetry groups has been done for K_7 [**16**] and the Heawood graph C_{14} [**12**]. Baza [**1**] classified the topological symmetry group for the graphs we denote by H_8 and F_9, though we will include our own arguments here. Our names for the graphs are those used by Lee, Kim, Lee and Oh [**22**]. Table 1 summarizes our results, giving for each graph the automorphism group and the list of positively realizable topological symmetry groups (i.e., groups of automorphisms that are induced by orientation-preserving homeomorphisms of a single embedding of the graph in S^3). One of our most surprising observations is that, for the graphs in the Heawood family, every realizable group is also positively realizable. In fact, our methods allow us to show that all the graphs in the family are intrinsically chiral (i.e., for each graph in the family, there is no embedding of the graph that is ambient isotopic to its mirror image). In Section 2, we give the necessary definitions, and the results we will be using from previous work. Section 3 contains the details of the arguments for each graph, along with embeddings realizing the various groups. Finally, in Section 4 we connect our results to intrinsic chirality, and pose some questions for future work.

2. Background and definitions

Before we begin, we introduce some terminology, and some important tools from other papers. An *abstract graph* (or just a *graph*) γ is a pair (V, E), where V is a set of vertices and $E \subseteq V \times V$ is a set of edges. We will denote the edge between vertices v and w by $\overline{vw}$ (since our graphs are not directed, $\overline{vw} = \overline{wv}$). We should note that graph theorists usually represent an edge just as an unordered pair $\{v, w\}$; however, since we are looking at graph embeddings, we want to distinguish between a pair of vertices $\{v, w\}$ that will be embedded as two points, and an edge $\overline{vw}$ that will be embedded as an arc. An embedding $f : \gamma \to S^3$ means (1) an embedding of the vertices V in S^3 and (2) for each edge $\overline{vw}$, an embedding $f_{v,w} : [0,1] \to S^3$ such that $f_{v,w}(0) = f(v)$ and $f_{v,w}(1) = f(w)$, and the embeddings of distinct edges intersect only at the endpoints. The image $\Gamma = f(\gamma)$ is called an *embedded graph* or a *spatial graph*.

An *automorphism* of a graph γ is a bijection $\alpha : V \to V$ such that $\overline{vw} \in E$ if and only if $\overline{\alpha(v)\alpha(w)} \in E$. The automorphisms of a graph γ form a group (the *automorphism group*), denoted $\mathrm{Aut}(\gamma)$. To describe automorphism groups (and their subgroups), we will use the following standard terms:

- The *symmetric group* S_n is the group of permutations of a set of n objects.
- The *dihedral group* D_n is the group of size $2n$ with presentation $\langle m, r \mid m^2 = r^n = 1, rm = mr^{-1}\rangle$.
- The *cyclic group* $\mathbb{Z}_n$ is the group of size n with presentation $\langle r \mid r^n = 1\rangle$.
- The *direct product* $G \times H$ is the Cartesian product $G \times H$ with the group operation $(g_1, h_1)(g_2, h_2) = (g_1g_2, h_1h_2)$.
- The *semidirect product* $G \rtimes H$ is the Cartesian product $G \times H$ with the group operation $(g_1, h_1)(g_2, h_2) = (g_1\phi(h_1)(g_2), h_1h_2)$, for some homomorphism $\phi : H \to \mathrm{Aut}(G)$.

A homeomorphism of S^3 which takes an embedded graph Γ to itself (taking vertices to vertices and edges to edges) induces an automorphism on the underlying abstract graph. We are interested in which automorphisms can be induced in this way for some embedding of a given graph. In what follows, we will refer to a homeomorphism of S^3 taking an embedded graph Γ to itself (taking vertices to vertices and edges to edges) as a homeomorphism of the pair (S^3, Γ).

DEFINITION 2.1. Let γ be an abstract graph, and let Γ be an embedding of γ in S^3. We define the **topological symmetry group** $\mathrm{TSG}(\Gamma)$ as the subgroup of $\mathrm{Aut}(\gamma)$ induced by homeomorphisms of (S^3, Γ). We define the **orientation preserving topological symmetry group** $\mathrm{TSG}_+(\Gamma)$ as the subgroup of $\mathrm{Aut}(\gamma)$ induced by orientation preserving homeomorphisms of (S^3, Γ).

DEFINITION 2.2. Let G be a group and let γ denote an abstract graph. If there is some embedding Γ of γ in S^3 such that $\mathrm{TSG}(\Gamma) = G$ (resp. $\mathrm{TSG}_+(\Gamma) = G$), then we say that the group G is **realizable** (resp. **positively realizable**) for γ. We will also say that a particular automorphism $\sigma \in \mathrm{Aut}(\gamma)$ is **realizable** (resp. **positively realizable** or **negatively realizable**) if there is an embedding Γ of γ and a homeomorphism (resp. orientation-preserving homeomorphism or orientation-reversing homeomorphism) of (S^3, Γ) which induces σ.

Note that for any embedding Γ of an abstract graph, either $\mathrm{TSG}_+(\Gamma) = \mathrm{TSG}(\Gamma)$ or $\mathrm{TSG}_+(\Gamma)$ is an index two subgroup of $\mathrm{TSG}(\Gamma)$. Further, for graphs that contain at least one cycle (i.e. not trees), if a group G is positively realizable by some embedding Γ, then G is also realizable [**2**]. This is because we can add identical chiral knots to every edge of the embedding Γ to rule out any orientation-reversing homeomorphisms. This yields a new embedding Γ' with $\mathrm{TSG}(\Gamma') = \mathrm{TSG}_+(\Gamma') = \mathrm{TSG}_+(\Gamma) = G$. Similarly, we can add distinct knots to every edge of an embedding Γ to get a new embedding Γ' such that $\mathrm{TSG}(\Gamma') = \mathrm{TSG}_+(\Gamma')$ is the trivial group; so the trivial group is always both realizable and positively realizable. As a result, for any graph γ, we are interested in finding (1) the nontrivial groups that are positively realizable for γ and (2) the nontrivial groups that are realizable, but not positively realizable, for γ.

For example, these groups have been found for K_7 and the Heawood graph C_{14}. Both K_7 and C_{14} are **intrinsically chiral** [**10**], meaning that no automorphism of the graph is negatively realizable. So every group that is realizable for these graphs is positively realizable.

THEOREM 2.3 ([**16**]). *The nontrivial groups that can be (positively) realized for the complete graph K_7 are:*

$$D_7, D_5, D_3, \mathbb{Z}_7, \mathbb{Z}_5, \mathbb{Z}_3, \mathbb{Z}_2.$$

THEOREM 2.4 ([**12**]). *The nontrivial groups that can be (positively) realized for the Heawood graph C_{14} are:*

$$D_7, D_3, \mathbb{Z}_7, \mathbb{Z}_6, \mathbb{Z}_3, \mathbb{Z}_2.$$

We now turn to some of the important tools we will use in this paper. The following result is very useful in showing that certain groups can be realized as a topological symmetry group. In many cases, it tells us that if a group is realizable for a graph γ, then all of its subgroups are also realizable. (In [**15**], this is called the Subgroup Corollary.) Recall that a graph is *3-connected* if removing any two vertices leaves a connected graph.

SUBGROUP THEOREM. [**15**] *Let Γ be an embedding of a 3-connected graph in S^3. Suppose that Γ contains an edge e that is not pointwise fixed by any nontrivial element of* $\mathrm{TSG}_+(\Gamma)$. *Then for every $H \leq \mathrm{TSG}_+(\Gamma)$, there is an embedding Γ' of Γ with $H = \mathrm{TSG}_+(\Gamma')$.*

The following result is important for restricting which automorphisms or groups can be realized for a particular graph γ.

FINITE ORDER THEOREM. [**9**] *Let ϕ be a nontrivial automorphism of a 3-connected graph γ that is induced by a homeomorphism h of (S^3, Γ) for some embedding Γ of γ. Then there exists another embedding Γ' of γ such that the automorphism ϕ is induced by a* finite order *homeomorphism f of (S^3, Γ'), and f is orientation reversing if and only if h is orientation reversing.*

Due to the Finite Order Theorem, when considering individual automorphisms, we need only determine whether or not the automorphism is realizable by a finite order homeomorphism. In particular, this allows us to use the following result of P.A. Smith, which is one of our most powerful tools to determine that certain homeomorphisms are not positively or negatively realizable. Here fix(h) denotes the fixed point set of a homeomorphism $h : S^3 \to S^3$.

SMITH THEORY. [**28**] *Let h be a nontrivial finite order homeomorphism of S^3. If h is orientation preserving, then* fix(h) *is either the empty set or is homeomorphic to S^1. If h is orientation reversing, then* fix(h) *is homeomorphic to either S^0 (two distinct points) or S^2.*

In particular, we note that if h is orientation reversing with fix(h) $\cong S^2$, then h must have order 2 (since h^2 will be orientation preserving with fixed point set larger than S^1, and hence trivial). We can combine the Finite Order Theorem and Smith Theory to give criteria for when an automorphism is positively or negatively realizable.

DEFINITION 2.5. Let ϕ be an automorphism of a graph γ. Let $F = A \cup B$ where A is the subgraph of γ induced by the fixed vertices of ϕ, and B is the set of midpoints of edges whose endpoints are interchanged by ϕ (so B is a set of discrete points). We will call F the **fixed subgraph under** ϕ.

REALIZABILITY LEMMA. *Let ϕ be a nontrivial automorphism of a 3-connected nonplanar graph γ, and let F be the fixed subgraph under ϕ. If ϕ is positively realizable, then F embeds in S^1. If ϕ is negatively realizable, then ϕ has even order. Moreover, if ϕ is negatively realizable and F contains more than two points, then ϕ has order 2 and F is planar.*

PROOF. Suppose ϕ is realizable by a homeomorphism h of (S^3, Γ) for some embedding Γ of γ. Then, by the Finite Order Theorem, there exists another embedding Γ' of γ such that the automorphism ϕ is induced by a finite order homeomorphism f of (S^3, Γ'), and f is orientation reversing if and only if h is orientation reversing.

If f is orientation preserving, then fix(f) is either the empty set or homeomorphic to S^1 by Smith Theory. Since the fixed subgraph under ϕ is a subset of fix(f), it must embed in S^1.

If f is orientation reversing, then fix(f) is homeomorphic to either S^0 or S^2. Since the fixed subgraph F under ϕ is a subset of fix(f), if F contains more than two points, then fix(f) is homeomorphic to S^2. This implies F is planar. In addition, if fix(f) is homeomorphic to S^2, then f is a reflection with order 2. The order of ϕ must divide the order of f, so then ϕ also has order 2.

Finally consider the case when fix(f) is homeomorphic to S^0 (so F contains two or fewer points). Suppose f has order m, and ϕ has order n; there is some integer d such that $m = nd$. So f^n has order d. But f^n realizes $\phi^n = id$, so it fixes the embedded graph Γ'. Since γ is non-planar, this means the fixed set of f^n does not embed in S^2; hence, by Smith Theory, f^n is trivial. So $d = 1$, and $m = n$. Since f is orientation reversing, it has even order, so we conclude that ϕ also has even order. □

Note that all the graphs in the Heawood family are 3-connected and nonplanar, so the Realizability Lemma applies. From the Realizability Lemma, we can immediately conclude:

(1) If the fixed subgraph under $\phi \in \text{Aut}(\gamma)$ does not embed in S^1, then ϕ is not positively realizable.
(2) If the fixed subgraph under $\phi \in \text{Aut}(\gamma)$ is nonplanar, then ϕ is not positively or negatively realizable.
(3) If ϕ has odd order, then ϕ is not negatively realizable.
(4) If the fixed subgraph under $\phi \in \text{Aut}(\gamma)$ contains more than two points, and ϕ does not have order 2, then ϕ is not negatively realizable.

The following strengthening of the Finite Order Theorem allows us to assume that if a group of automorphisms is positively realizable, then they can all simultaneously be realized by finite order homeomorphisms of the same embedding of the graph in S^3.

ISOMETRY THEOREM. [**14**] *Let Γ be an embedded 3-connected graph, and let $H \leq \text{TSG}_+(\Gamma)$. Then Γ can be re-embedded as Γ' such that $H \leq \text{TSG}_+(\Gamma')$ and $\text{TSG}_+(\Gamma')$ is induced by an isomorphic finite subgroup of* SO(4) *(i.e., the group of orientation preserving isometries of S^3).*

The next lemma, the Subgraph Lemma, simply says if a group is realizable for a graph, it must be realizable for every stable subgraph. This lemma was given in [**7**], but the statement there was missing a necessary hypothesis (which was satisfied whenever the lemma was applied in that paper). We provide a corrected statement and proof here.

SUBGRAPH LEMMA. *Let γ be an abstract graph and γ' be a subgraph of γ so that every automorphism α of γ fixes γ' setwise. Further assume that the restriction map $\phi : \text{Aut}(\gamma) \to \text{Aut}(\gamma')$ defined by $\phi(\alpha) = \alpha|_{\gamma'}$ is injective (i.e., every automorphism of γ is determined by its action on γ'). Assume that Γ is an embedding of γ and let Γ' be the embedding of γ' induced by Γ. Then* $\text{TSG}(\Gamma) \leq \text{TSG}(\Gamma')$ *and* $\text{TSG}_+(\Gamma) \leq \text{TSG}_+(\Gamma')$.

PROOF. Every element $\alpha \in \text{TSG}(\Gamma)$ is realized by a homeomorphism of S^3 taking Γ to itself setwise. Hence, every such homeomorphism also takes Γ' to itself setwise and realizes $\phi(\alpha) \in \text{Aut}(\gamma')$. So ϕ induces a map from $\text{TSG}(\Gamma)$ to a subgroup of $\text{TSG}(\Gamma')$. Since ϕ is injective, $\text{TSG}(\Gamma)$ is isomorphic to a subgroup of $\text{TSG}(\Gamma')$. The same argument holds for $\text{TSG}_+(\Gamma)$. □

The Path Lemma is a new tool which provides an easy criterion for proving that an automorphism is not negatively realizable.

PATH LEMMA. *Suppose α is an automorphism of a graph γ whose fixed subgraph contains more than two points, and a and b are vertices of γ interchanged by α. Further suppose $P = av_1v_2 \dots v_kb$ is a path in γ such that no vertex v_i is fixed by α, and no edge $\overline{v_iv_{i+1}}$ is flipped by α. Then α is not negatively realizable.*

PROOF. Suppose α is negatively realizable by h; by the Finite Order Theorem we may assume h has finite order. Since α fixes more than two points, by Smith Theory h must have fixed point set homeomorphic to S^2, and h interchanges the inside and outside of the fixed S^2. Without loss of generality, suppose a is inside the sphere and b is outside the sphere. Then the embedding of the path P from a to b must cross the sphere at some point. If it crosses the sphere at a vertex v_i, then this vertex is fixed. If it crosses the sphere at a point inside the edge $\overline{v_i v_{i+1}}$, then this edge must be mapped to itself by h (since the graph is embedded, no two edges can share an interior point). If the endpoints of the edge are not fixed, then they must be interchanged. So either there are fixed vertices along the path, or a flipped edge. □

Our final observation allows us, in many case, to only consider automorphism up to conjugacy.

LEMMA 2.6. *If α and $\beta \in \mathrm{Aut}(\gamma)$ are conjugate, then β is positively (resp. negatively) realizable if and only if α is positively (resp. negatively) realizable.*

PROOF. Suppose α is realizable, and $\beta = \phi^{-1}\alpha\phi$ for some $\phi \in \mathrm{Aut}(\gamma)$. Then there is an embedding $\Gamma = f(\gamma)$ (where $f : \gamma \to S^3$) and a homeomorphism h of (S^3, Γ) that realizes α. Define a new embedding $\Gamma' = f'(\gamma)$ by (1) for each vertex v of γ, let $f'(v) = f(\phi(v))$ and (2) for each edge $\overline{vw}$ of γ, let $f'_{v,w} = f_{\phi(v)\phi(w)}$. Since ϕ is an automorphism, it is a bijection between vertices, and $\overline{vw}$ is an edge of γ if and only if $\overline{\phi(v)\phi(w)}$ is an edge of γ, so f' is well-defined. In fact, Γ' is the same subset of S^3 as Γ, except that the vertices have been relabeled, so the vertex labeled v in Γ' is labeled $\phi^{-1}(v)$ in Γ.

So, h is also a homeomorphism of (S^3, Γ'). For each vertex v in γ, we know that $h(f'(v)) = h(f(\phi(v))) = f(\alpha\phi(v))$ (since h realizes α on Γ). But since $\alpha\phi(v) = \phi\beta(v)$, this means $h(f'(v)) = f(\phi\beta(v)) = f'(\beta(v))$. So h realizes β on Γ'. Therefore β is also realizable, and is positively or negatively realizable as α is (since we are using the same homeomorphism of S^3). □

3. Computing the topological symmetry groups for the Heawood graphs

3.1. The graph H_8. The graph H_8 is obtained from the graph K_7 by performing a single ∇Y move, resulting in the graph shown in Figure 3.

The vertex a has degree 3, the vertices b, c, d have degree 5, and the vertices e, f, g, h have degree 6. So any automorphism of the graph must preserve these three sets of vertices (setwise). Since the vertices b, c, d all have the same neighbours, as do the vertices e, f, g, h, the vertices within each set can be permuted freely. Hence the automorphism group of H_8 is isomorphic to $S_3 \times S_4$. If we let $\alpha = (bcd)$, $\beta = (bc)$, $\gamma = (ef)$ and $\delta = (fgh)$, then α and β generate S_3 with relations $\alpha^3 = \beta^2 = (\alpha\beta)^2 = 1$, and γ and δ generate S_4 with relations $\gamma^2 = \delta^3 = (\gamma\delta)^4 = 1$.

Table 2 lists a representative of each conjugacy class of nontrivial automorphisms, its order, a diagram of the fixed subgraph, whether the fixed subgraph embeds in S^1 or S^2, and (if one exists) a path between two interchanged vertices that contains no fixed vertices or flipped edges (as in the Path Lemma). If the Path Lemma does not apply, we indicate "N/A", with a superscript 1 if no pairs of vertices are interchanged, 2 if there are two or fewer fixed points, or 3 if the fixed

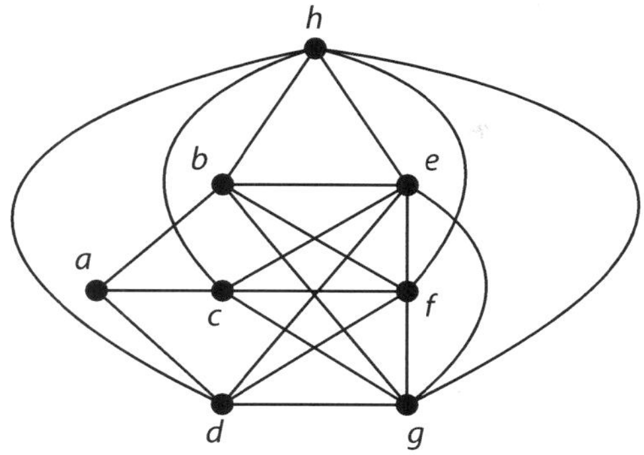

FIGURE 3. The graph H_8.

TABLE 2. Automorphisms of H_8.

Automorphism	Order	Fixed subgraph	$\hookrightarrow S^1$	$\hookrightarrow S^2$	Path Lemma
$\alpha = (bcd)$	3	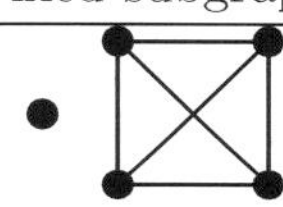	No	Yes	N/A[1]
$\beta = (bc)$	2	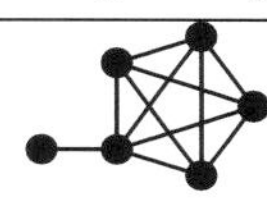	No	No	N/A[3]
$\gamma = (ef)$	2	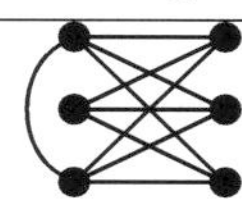	No	No	N/A[3]
$\delta = (fgh)$	3	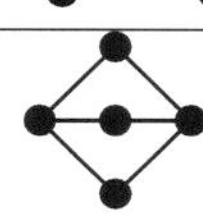	No	Yes	N/A[1]
$\gamma\delta = (efgh)$	4	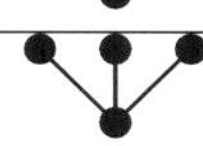	No	Yes	N/A[1]
$(\gamma\delta)^2 = (eg)(fh)$	2	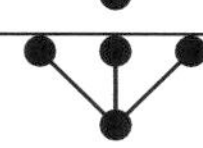	No	Yes	$e - f - g$
$\alpha\gamma = (bcd)(ef)$	6		Yes	Yes	$e - d - f$
$\alpha\delta = (bcd)(fgh)$	3		Yes	Yes	N/A[1]
$\alpha\gamma\delta = (bcd)(efgh)$	12		Yes	Yes	N/A[1]
$\alpha(\gamma\delta)^2 = (bcd)(eg)(fh)$	6		Yes	Yes	N/A[2]
$\beta\gamma = (bc)(ef)$	2	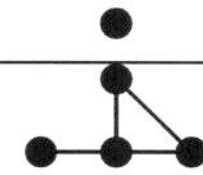	No	Yes	$b - e - c$
$\beta\delta = (bc)(fgh)$	6		Yes	Yes	$b - f - c$
$\beta\gamma\delta = (bc)(efgh)$	4		Yes	Yes	$b - e - c$
$\beta(\gamma\delta)^2 = (bc)(eg)(fh)$	2		Yes	Yes	$b - e - c$

subgraph does not embed in S^2. (In this last case, the Path Lemma might apply, but is redundant.)

THEOREM 3.1. *A nontrivial group is positively realizable for H_8 if and only if it is $\mathbb{Z}_3$ or $\mathbb{Z}_2$. Moreover, H_8 is intrinsically chiral, so these groups are only realizable by orientation-preserving homeomorphisms.*

PROOF. We can quickly see that the first 6 automorphisms in Table 2 are not realizable. They are not positively realizable, because their fixed subgraphs do not embed in S^1. The automorphisms α, δ and $\gamma\delta$ are not negatively realizable because they fix more than two points, and do not have order 2. Automorphisms β and γ are not negatively realizable because their fixed subgraphs do not embed in S^2, and $(\gamma\delta)^2$ is not negatively realizable by the Path Lemma.

Of the remaining automorphisms, $\alpha\gamma, \beta\gamma, \beta\delta, \beta\gamma\delta$ and $\beta(\gamma\delta)^2$ are not negatively realizable by the Path Lemma. Automorphism $\alpha\delta$ is not negatively realizable because it has odd order. Since $(\alpha\gamma\delta)^4 = \alpha$, which is not realizable, neither is $\alpha\gamma\delta$. Similarly, $[\alpha(\gamma\delta)^2]^3 = (\gamma\delta)^2$, so $\alpha(\gamma\delta)^2$ is not realizable either. So none of the automorphisms are negatively realizable, and hence H_8 is intrinsically chiral.

We already know that $\alpha, \beta, \gamma, \delta, \gamma\delta, (\gamma\delta)^2, \alpha\gamma\delta$ and $\alpha(\gamma\delta)^2$ are not positively realizable. Also, $\beta\gamma$ is not positively realizable by the Realizability Lemma. Since $(\alpha\gamma)^4 = \alpha$, $(\beta\delta)^3 = \beta$ and $(\beta\gamma\delta)^2 = (\gamma\delta)^2$, we also know $\alpha\gamma$, $\beta\delta$ and $\beta\gamma\delta$ are not positively realizable. This leaves only automorphisms conjugate to $\alpha\delta$ and $\beta(\gamma\delta)^2$.

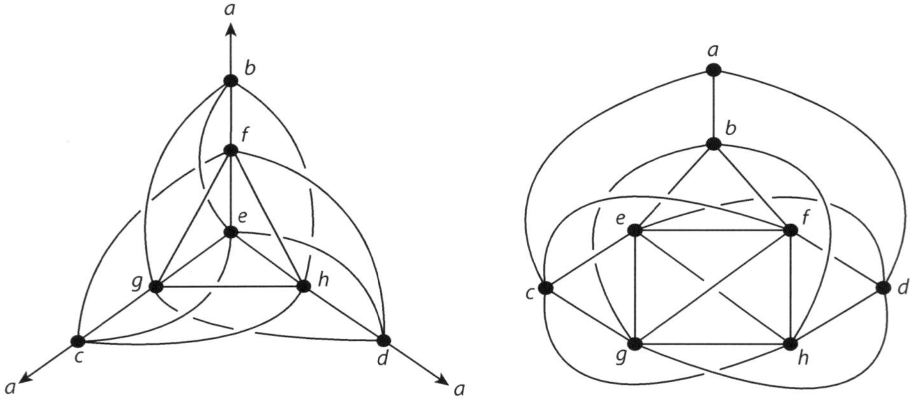

FIGURE 4. Embeddings of H_8 which positively realize $\mathbb{Z}_3$ (left) and $\mathbb{Z}_2$ (right).

Suppose there is an embedding Γ of H_8 such that automorphisms conjugate to $\alpha\delta$ and $\beta(\gamma\delta)^2$ are realized by orientation-preserving homeomorphisms h and g. By the Isometry Theorem, we may assume h and g are isometries. Since $\alpha\delta$ and $\beta(\gamma\delta)^2$ both have fixed vertices, we know h and g are both rotations around great circles of S^3. Since every automorphism of H_8 fixes vertex a, these great circles intersect at $a \in \Gamma$, so they must also intersect at the antipodal point. In addition, h fixes one of $\{e, f, g, h\}$ and g fixes one of $\{b, c, d\}$. Without loss of generality, suppose h fixes e and g fixes b. Since g has order 2, it takes the axis of h to itself (setwise), hence $g(e)$ must be another vertex on the axis of h. Since h fixes only two vertices, and $g(e) \neq a$, we must have $g(e) = e$. But this contradicts the fact that g does not fix any of $\{e, f, g, h\}$. So there is no embedding of H_8 that simultaneously realizes an automorphism conjugate to $\alpha\delta$ and one conjugate to $\beta(\gamma\delta)^2$. Moreover, the group generated by two automorphisms conjugate to $\alpha\delta$ (that are not powers of each other) includes automorphisms conjugate to δ or $(\gamma\delta)^2$, which are not realizable; so we cannot simultaneously realize two automorphisms conjugate to $\alpha\delta$. Similarly, two automorphisms conjugate to $\beta(\gamma\delta)^2$ have a product conjugate to α, $(\gamma\delta)^2$ or $\alpha(\gamma\delta)^2$, none of which are realizable. So the topological symmetry group can at most positively realize $\mathbb{Z}_3$ (generated by an automorphism conjugate to $\alpha\delta$) or $\mathbb{Z}_2$ (generated by an automorphism conjugate to $\beta(\gamma\delta)^2$).

The embeddings in Figure 4 positively realize $\mathbb{Z}_3$ and $\mathbb{Z}_2$. On the left, the axis of order 3 is perpendicular to the page, and e and a are placed at antipodal points on the axis (in S^3). On the right, the axis of rotation of order 2 is the vertical line through vertices a and b. So $\mathbb{Z}_3$ and $\mathbb{Z}_2$ are both positively realizable. □

3.2. The graph H_9. The graph H_9 is obtained from H_8 (as shown in Figure 3) by performing a ∇Y-move to the triangle efg, resulting in the graph in Figure 5.

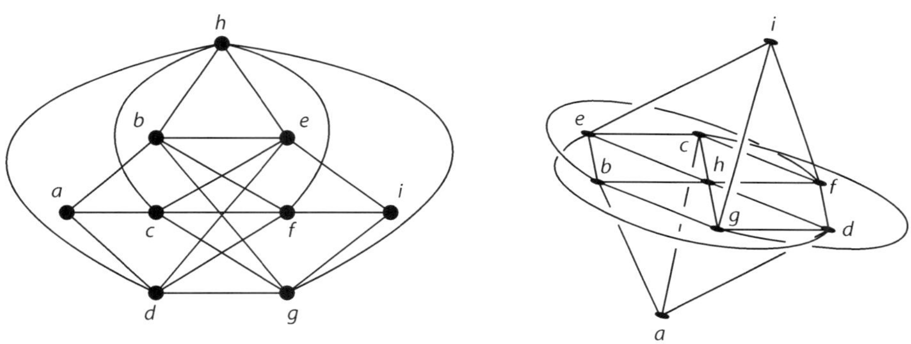

FIGURE 5. The graph H_9, and an embedding of H_9 which positively realizes D_3.

Observe that if we delete vertices a and i, the remaining graph is isomorphic to $K_{3,3,1}$. Moreover, since a and i are the only vertices with degree 3, every automorphism of H_9 preserves this $K_{3,3,1}$ subgraph setwise, and the induced automorphism of $K_{3,3,1}$ determines the automorphism of H_9. Hence $\mathrm{Aut}(H_9) = \mathrm{Aut}(K_{3,3,1}) = \mathrm{Aut}(K_{3,3}) = (D_3 \times D_3) \rtimes \mathbb{Z}_2$ [**7**]. The automorphism group is generated by $\alpha = (bcd)$, $\beta = (bc)$, $\gamma = (efg)$, $\delta = (ef)$ and $\phi = (ai)(be)(cf)(dg)$ (note that every automorphism fixes vertex h).

Table 3 lists a representative of each conjugacy class of nontrivial automorphisms, along with the same information given for H_8 in Table 2. Note that in the diagrams of the fixed subgraph, empty circles represent the midpoints of edges whose endpoints are interchanged by the automorphism. This is based on the conjugacy classes of automorphisms of $K_{3,3}$ found by Nikkuni and Taniyama [**24**].

By the Subgraph Lemma, given any embedding, the topological symmetry group for H_9 must be a subgroup of the topological symmetry group of the embedded $K_{3,3,1}$ subgraph. The possible topological symmetry groups for $K_{3,3,1}$ were classified in [**7**].

TABLE 3. Automorphisms of H_9.

Automorphism	Order	Fixed subgraph	$\hookrightarrow S^1$	$\hookrightarrow S^2$	Path Lemma
$\alpha = (bcd)$	3	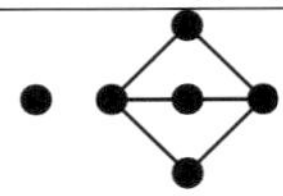	No	Yes	N/A[1]
$\beta = (bc)$	2	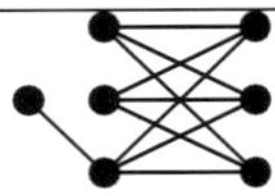	No	No	N/A[3]
$\alpha\gamma = (bcd)(efg)$	3	● ● ●	Yes	Yes	N/A[1]
$\beta\gamma = (bc)(efg)$	6		Yes	Yes	$b - e - c$
$\beta\delta = (bc)(ef)$	2	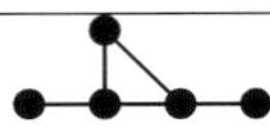	No	Yes	$b - e - c$
$\phi = (ai)(be)(cf)(dg)$	2	● ○ ○ ○	Yes	Yes	$a - b - f - i$
$\beta\phi = (ai)(becf)(dg)$	4	● ○	Yes	Yes	N/A[2]
$\alpha\phi = (ai)(becfdg)$	6	●	Yes	Yes	N/A[2]

THEOREM 3.2 ([**7**]).

(1) *A nontrivial group is positively realizable for $K_{3,3,1}$ if and only if it is one of: D_3, D_2, $\mathbb{Z}_3$, and $\mathbb{Z}_2$.*
(2) *The groups that are realizable but* **not** *positively realizable for $K_{3,3,1}$ are D_4 and $\mathbb{Z}_4$.*

We will use this result to classify the possible topological symmetry groups of H_9.

THEOREM 3.3. *A nontrivial group is positively realizable for H_9 if and only if it is one of: D_3, $\mathbb{Z}_3$, and $\mathbb{Z}_2$. Moreover, H_9 is intrinsically chiral, so there are no groups that are realizable but not positively realizable.*

PROOF. We can quickly see from Table 3 that automorphisms conjugate to β and $\beta\delta$ are neither positively nor negatively realizable. Since $(\beta\phi)^2 = \beta\delta$, $\beta\phi$ is also not realizable. Also, α and $\alpha\gamma$ are not negatively realizable because they have odd order, and $\beta\gamma$ and ϕ are not negatively realizable by the Path Lemma. Finally, $(\alpha\phi)^3 = (ai)(bf)(ed)(cg)$ is conjugate to ϕ; since ϕ is not negatively realizable, neither is $\alpha\phi$. Since none of the automorphisms are negatively realizable, H_9 is intrinsically chiral.

We already know β, $\beta\delta$ and $\beta\phi$ are not positively realizable; in addition, α is not positively realizable by the Realizability Lemma. Since $(\beta\gamma)^3 = \beta$, $\beta\gamma$ is also not positively realizable. Next, suppose $\alpha\phi$ is positively realizable; since it has a fixed point, by the Isometry Theorem it is realizable by a rotation about an axis containing h. Then $(\alpha\phi)^2 = (bcd)(efg)$ is also a rotation, with the same axis. But this means that a and i are also on the axis of rotation, and hence are fixed by $\alpha\phi$, which is a contradiction. Hence $\alpha\phi$ is not positively realizable.

So the only automorphisms which could be positively realizable are conjugate to $\alpha\gamma$ or ϕ. The embedding on the right in Figure 5 shows how to realize D_3 (based on an embedding for $K_{3,3,1}$ in [**7**]) by realizing both of these automorphisms together. Since the edge $\overline{ce}$ is not fixed by any symmetry of the embedding, the subgroups $\mathbb{Z}_3$ and $\mathbb{Z}_2$ are also realizable by the Subgroup Theorem.

By Theorem 3.2, the only groups that could be positively realized for H_9 are D_3, D_2, $\mathbb{Z}_3$, and $\mathbb{Z}_2$. It only remains to show that D_2 is not positively realizable. Suppose there is an embedding Γ of H_9 with $D_2 \leq \mathrm{TSG}_+(\Gamma)$. Then all three nontrivial elements in D_2 would be conjugate to ϕ (since these are the only positively realizable automorphisms of H_9 of order 2), and they would all commute with each other. Without loss of generality, suppose one of the automorphisms in D_2 is ϕ itself. There are five other automorphisms in the conjugacy class of ϕ; of these, three commute with ϕ, but none of those three commute with each other. So we cannot positively realize D_2. □

3.3. The graph H_{10}. The graph H_{10} is obtained from H_9 (as shown in Figure 5) by performing a ∇Y-move on the triangle beh, resulting in the graph shown in Figure 6.

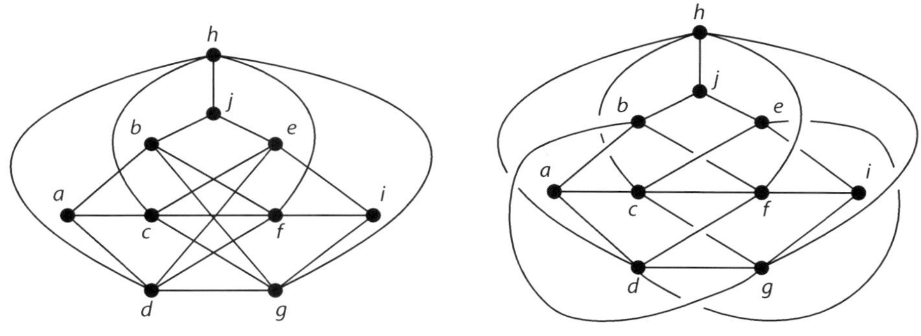

FIGURE 6. The graph H_{10}, and an embedding Γ of H_{10} with $\mathrm{TSG}(\Gamma) = \mathrm{TSG}_+(\Gamma) = \mathbb{Z}_2$.

Any automorphism of H_{10} must fix vertex j (the only vertex of degree 3 adjacent to two vertices of degree 4), vertex h (the only vertex of degree 5 *not* adjacent to a vertex of degree 4), the pair $\{a, i\}$ (the other vertices of degree 3), the pair $\{b, e\}$ (the vertices of degree 4), and the set $\{c, d, f, g\}$ (the other vertices of degree 5). Moreover, the two pairs $\{c, d\}$ and $\{f, g\}$ are either preserved or interchanged, so the square $c - f - d - g - c$ is preserved by every automorphism. Moreover, an automorphism of this square determines an automorphism of the entire graph, so $\mathrm{Aut}(H_{10}) = D_4$, generated by $\alpha = (ai)(be)(cfdg)$ and $\beta = (cd)$. The possible topological symmetry groups are D_4, $\mathbb{Z}_4$, D_2 and $\mathbb{Z}_2$ (the subgroups of D_4).

Table 4 lists a representative of each conjugacy class of nontrivial automorphisms, its order, a diagram of the fixed subgraph, whether the fixed subgraph embeds in S^1 or S^2, and (if one exists) a path between two interchanged vertices that contains no fixed vertices or flipped edges (as in the Path Lemma).

THEOREM 3.4. *The only nontrivial group that is positively realizable for H_{10} is $\mathbb{Z}_2$. Moreover, H_{10} is intrinsically chiral, so no other groups are realizable.*

PROOF. From the table above, none of the automorphisms of H_{10} are negatively realizable, either because the fixed subgraph does not embed in S^2 or due to the Path Lemma. Hence H_{10} is intrinsically chiral.

TABLE 4. Automorphisms of H_{10}.

Automorphism	Order	Fixed subgraph	$\hookrightarrow S^1$	$\hookrightarrow S^2$	Path Lemma
$\alpha = (ai)(be)(cfdg)$	4		Yes	Yes	$a - b - f - i$
$\alpha^2 = (cd)(fg)$	2	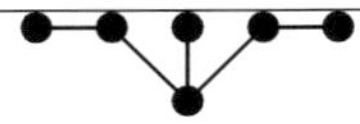	No	Yes	$c - f - d$
$\beta = (cd)$	2	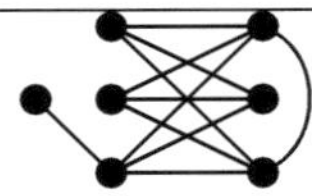	No	No	N/A[3]
$\alpha\beta = (ai)(be)(cg)(df)$	2		Yes	Yes	$a - b - f - i$

From the table, α^2 and β are not positively realizable (since their fixed subgraphs do not embed in S^1), which means α is also not positively realizable. So the only nontrivial automorphisms which could be positively realizable are those conjugate to $\alpha\beta$, namely $\alpha\beta$ and $\alpha^3\beta$. However, the product of these automorphisms is α^2, so no embedding of H_{10} can positively realize both of these automorphisms at the same time. Hence, $\mathbb{Z}_2$ is the only group which can possibly be positively realized.

The embedding Γ of H_{10} in Figure 6 is symmetric under the half-turn rotation around the axis through the edge $\overline{hj}$. Since no other nontrivial groups are realizable, this means $\mathrm{TSG}(\Gamma) = \mathrm{TSG}_+(\Gamma) = \mathbb{Z}_2$. □

3.4. The graph H_{11}. The graph H_{11} is obtained from H_{10} (as shown in Figure 6) by performing a ∇Y-move on the triangle hcf, resulting in the graph shown in Figure 7.

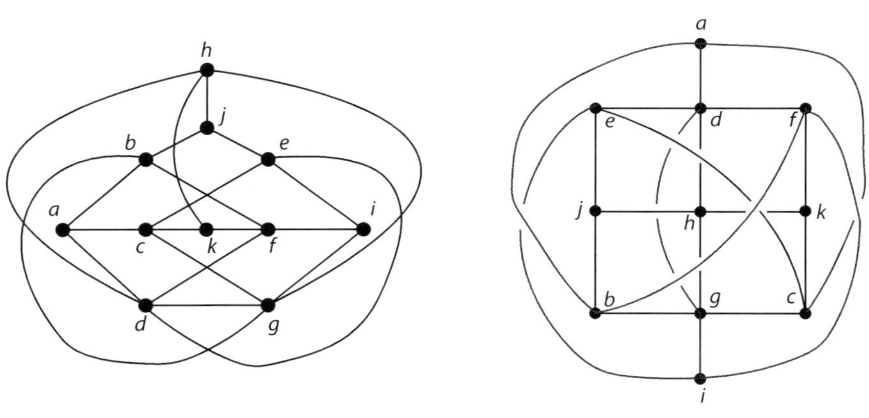

FIGURE 7. The graph H_{11}, and an embedding Γ of H_{11} with $\mathrm{TSG}(\Gamma) = \mathrm{TSG}_+(\Gamma) = \mathbb{Z}_2$.

Every automorphism of H_{11} fixes vertex h (the only vertex of degree 4 adjacent to two vertices of degree 3). Similarly, looking at the degrees of the vertices and their neighbors, every automorphism preserves each of the sets $\{a, i\}$, $\{j, k\}$, $\{d, g\}$ and $\{b, c, e, f\}$. In fact, the permutations of the sets $\{a, i\}$ and $\{j, k\}$ determine the permutation of the other vertices. Hence $\mathrm{Aut}(H_{11}) = \mathbb{Z}_2 \times \mathbb{Z}_2 = D_2$, and the possible

TABLE 5. Automorphisms of H_{11}.

Automorphism	Order	Fixed subgraph	$\hookrightarrow S^1$	$\hookrightarrow S^2$	Path Lemma
$\alpha = (ai)(be)(cf)(dg)$	2		Yes	Yes	$a - b - f - i$
$\beta = (jk)(bc)(ef)$	2	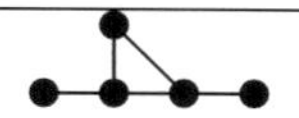	No	Yes	$j - b - f - k$
$\alpha\beta = (ai)(jk)(bf)(ce)(dg)$	2		Yes	Yes	$a - b - j - e - i$

nontrivial permutations of the vertices are: $\alpha = (ai)(be)(cf)(dg)$, $\beta = (jk)(bc)(ef)$, and $\alpha\beta = (ai)(jk)(bf)(ce)(dg)$. The properties of these automorphisms are listed in Table 5.

THEOREM 3.5. *The only nontrivial group that is positively realizable for H_{11} is $\mathbb{Z}_2$. Moreover, H_{11} is intrinsically chiral, so no other groups are realizable.*

PROOF. We can see from Table 5 that none of the automorphisms of H_{11} are negatively realizable, due to the Path Lemma, so H_{11} is intrinsically chiral. We also observe that β is not positively realizable, so the largest possible positively realizable group is $\mathbb{Z}_2$ (note that if α and $\alpha\beta$ were both positively realizable for one embedding, then β would also be positively realizable, so only one automorphism can be realized at once).

The embedding Γ of H_{11} in Figure 7 is symmetric under the half-turn rotation around the point h, positively realizing $\alpha\beta$. Here we imagine all the vertices are in the plane of the page, the edge $\overline{dg}$ is beneath the plane, with its midpoint directly below vertex h, and the edges $\overline{ec}$ and $\overline{bf}$ are above the plane, with their midpoints directly above vertex h (and with $\overline{bf}$ above $\overline{ec}$). Since no other nontrivial groups are realizable, this means $\mathrm{TSG}(\Gamma) = \mathrm{TSG}_+(\Gamma) = \mathbb{Z}_2$. □

3.5. The graph H_{12}. The graph H_{12} is obtained from H_{11} (as shown in Figure 7) by performing a ∇Y-move on the triangle hdg, resulting in the graph shown in Figure 8.

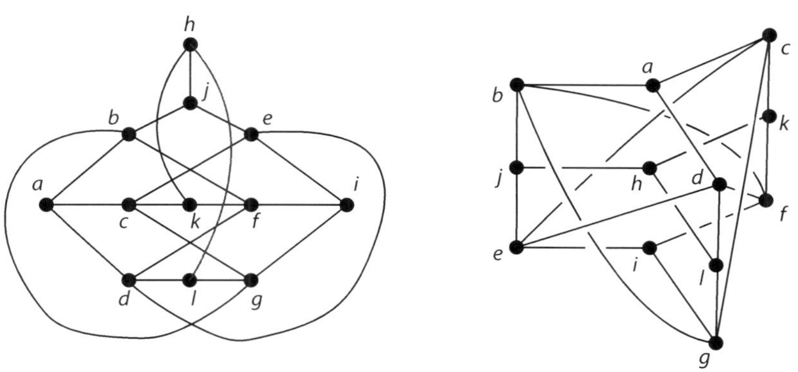

FIGURE 8. The graph H_{12}, and an embedding Γ of H_{12} with $\mathrm{TSG}(\Gamma) = \mathrm{TSG}_+(\Gamma) = D_3$.

Note that any automorphism of H_{12} fixes h (the only vertex of degree 3 all of whose neighbors are also degree 3), so it permutes $\{j, k, l\}$, and also permutes

TABLE 6. Automorphisms of H_{12}.

Automorphism	Order	Fixed subgraph	$\hookrightarrow S^1$	$\hookrightarrow S^2$	Path Lemma
$\alpha = (jkl)(bcd)(efg)$	3		Yes	Yes	N/A[1]
$\beta = (ai)(jk)(bf)(ec)(dg)$	2		Yes	Yes	$a - c - g - i$
$\gamma = (ai)(be)(cf)(dg)$	2	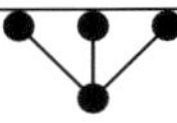	No	Yes	$a - b - f - i$
$\alpha\gamma = (ai)(bfdecg)(jkl)$	6		Yes	Yes	N/A[2]
$\beta\gamma = (bc)(ef)(jk)$	2	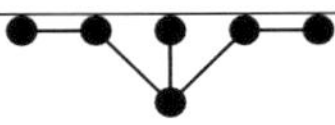	No	Yes	$j - b - f - k$

$\{a, i\}$. The result determines the permutation of the remaining 6 vertices. So $\text{Aut}(H_{12}) = D_3 \times \mathbb{Z}_2$, generated by $\alpha = (jkl)(bcd)(efg)$, $\beta = (ai)(jk)(bf)(ec)(dg)$ and $\gamma = (ai)(be)(cf)(dg)$ (α and β generate a subgroup isomorphic to D_3, and γ commutes with both). The properties of the automorphisms are given in Table 6 (we list one automorphism from each conjugacy class).

THEOREM 3.6. *The groups D_3, $\mathbb{Z}_3$ and $\mathbb{Z}_2$ are positively realizable for H_{12}. Moreover, H_{12} is intrinsically chiral, and there are no other nontrivial groups that are realizable for H_{12}.*

PROOF. From Table 6, γ, $\beta\gamma$ and their conjugates are neither positively nor negatively realizable. Since $(\alpha\gamma)^3 = \gamma$, $\alpha\gamma$ and its conjugates are not realizable either. The remaining automorphisms are not negatively realizable (either by the Path Lemma or because they have odd order), so H_{12} is intrinsically chiral. Since only automorphisms conjugate to α or β can be realizable (namely, $\alpha, \alpha^2, \beta, \alpha\beta$ and $\alpha^2\beta$), the only possible topological symmetry groups are subgroups of D_3: D_3, $\mathbb{Z}_3$ and $\mathbb{Z}_2$.

Figure 8 shows an embedding of H_{12} which positively realizes D_3. Here α is realized by the rotation of order 3 around the axis through a, h and i, and β is realized by the rotation of order 2 around the axis through the edge $\overline{hl}$. Since the edge $\overline{ab}$ is not fixed by any nontrivial element of the topological symmetry group, we can also positively realize $\mathbb{Z}_3$ and $\mathbb{Z}_2$ by the Subgroup Theorem. □

3.6. The graph F_9. The graph F_9 is obtained from H_8 (as shown in Figure 3) by performing a ∇Y-move on the triangle dfg, resulting in the graph shown in Figure 9.

Every automorphism of F_9 fixes d (the only vertex of degree 4), $\{a, i\}$ (the vertices of degree 3), $\{h, e\}$ (the vertices of degree 6) and $\{b, c, f, g\}$ (the vertices of degree 5). Note that the subgraph of F_9 induced by $\{b, c, f, g\}$ is the 4-cycle (or square) $bfcg$. So every automorphism of F_9 must preserve this square. A permutation of $\{b, c, f, g\}$ determines a permutation of $\{a, i\}$; since h and e are both adjacent to every vertex in the square (along with d and each other), they can be interchanged independent of the permutation of $\{b, c, f, g\}$. So $\text{Aut}(F_9) \cong D_4 \times \mathbb{Z}_2$, generated by $\alpha = (bfcg)(ai)$, $\beta = (bc)$ and $\gamma = (eh)$. The properties of the automorphisms are given in Table 7 (we list one automorphism from each conjugacy class).

THEOREM 3.7. *The only nontrivial group that is positively realizable for F_9 is $\mathbb{Z}_2$. Moreover, F_9 is intrinsically chiral, so no other groups are realizable.*

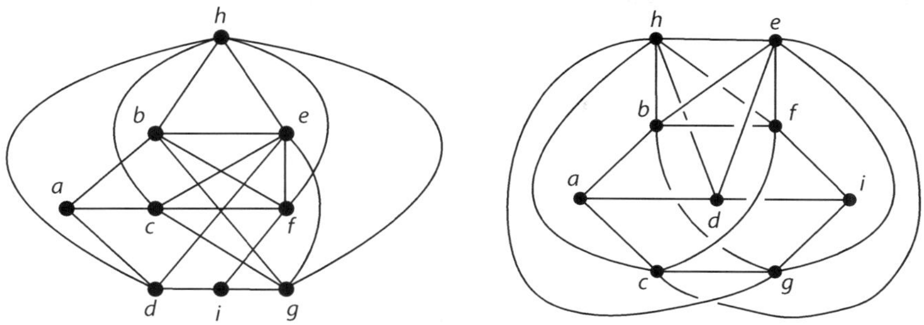

FIGURE 9. The graph F_9, and an embedding Γ of F_9 with $\mathrm{TSG}(\Gamma) = \mathrm{TSG}_+(\Gamma) = \mathbb{Z}_2$.

TABLE 7. Automorphisms of F_9.

Automorphism	Order	Fixed subgraph	$\hookrightarrow S^1$	$\hookrightarrow S^2$	Path Lemma
$\alpha = (bfcg)(ai)$	4	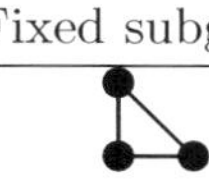	Yes	Yes	$a-b-f-i$
$\alpha^2 = (bc)(fg)$	2	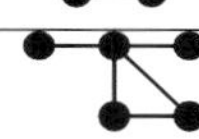	No	Yes	$b-f-c$
$\beta = (bc)$	2	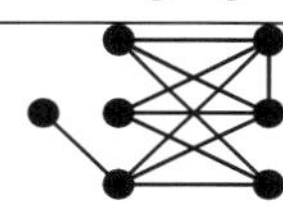	No	No	N/A[3]
$\alpha\beta = (bg)(cf)(ai)$	2	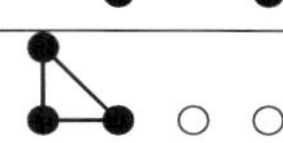	No	Yes	$a-c-g-i$
$\gamma = (eh)$	2	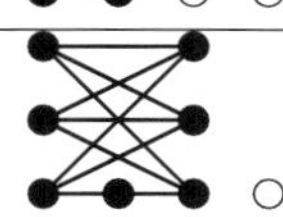	No	No	N/A[3]
$\alpha\gamma = (bfcg)(ai)(eh)$	4		Yes	Yes	N/A[2]
$\alpha^2\gamma = (bc)(fg)(eh)$	2		Yes	Yes	$b-f-c$
$\beta\gamma = (bc)(eh)$	2	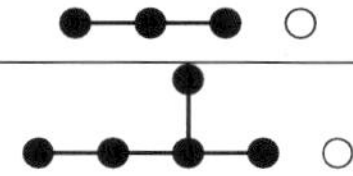	No	Yes	$b-e-c$
$\alpha\beta\gamma = (bg)(cf)(ai)(eh)$	2	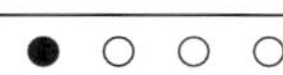	Yes	Yes	$b-f-i-g$

PROOF. We first observe from Table 7 that β and γ are not realizable by the Realizability Lemma. Moreover, α^2 is not positively realizable by the Realizability Lemma, and is not negatively realizable by the Path Lemma; similarly, $\alpha\beta$ and $\beta\gamma$ are not realizable. Since $(\alpha\gamma)^2 = \alpha^2$, neither $\alpha\gamma$ nor α are realizable either. Finally, we see that $\alpha^2\gamma$ and $\alpha\beta\gamma$ are not negatively realizable by the Path Lemma. So none of the automorphisms are negatively realizable, and hence F_9 is intrinsically chiral.

We already know α, α^2, β, $\alpha\beta$, γ, $\alpha\gamma$ and $\beta\gamma$ (and their conjugates) are not positively realizable. We are left with three automorphisms that may be positively realizable: $\alpha\beta\gamma$, its conjugate $\alpha^3\beta\gamma$, and $\alpha^2\gamma$ (which is only conjugate to itself), all of order 2. The product of any two of these yields an unrealizable automorphism, so the only group which may be realizable is $\mathbb{Z}_2$. Figure 9 shows an embedding of

F_9 which positively realizes $\mathbb{Z}_2$ (specifically, it realizes the automorphism $\alpha\beta\gamma$ by a 180° rotation about the vertical axis). □

3.7. The graph F_{10}. The graph F_{10} is obtained from F_9 (as shown in Figure 9) by performing a ∇Y-move on the triangle deh, resulting in the graph shown in Figure 10.

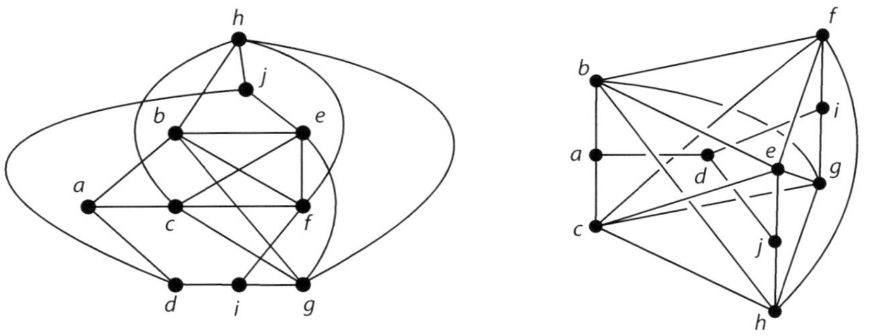

FIGURE 10. The graph F_{10}, and an embedding Γ of F_{10} with $\mathrm{TSG}(\Gamma) = \mathrm{TSG}_+(\Gamma) = D_3$.

Every automorphism of F_{10} fixes d (the only vertex of degree 3 that is adjacent to the other three vertices of degree 3), $\{a, i, j\}$ (the other vertices of degree 3), and $\{b, c, e, h, f, g\}$ (the vertices of degree 5). Moreover, a is adjacent to $\{b, c\}$, i is adjacent to $\{f, g\}$ and j is adjacent to $\{e, h\}$; so the permutation of $\{a, i, j\}$ determines a permutation of these three pairs of vertices. Finally, the vertices within each pair can be interchanged (since they have the same neighbours). So the group of automorphisms is generated by $\alpha = (aij)(bfe)(cgh)$, $\beta = (ai)(bf)(cg)$, $\gamma_a = (bc)$, $\gamma_i = (fg)$ and $\gamma_j = (eh)$. The automorphisms α and β generate a subgroup isomorphic to D_3, while γ_a, γ_i, and γ_j generate $(\mathbb{Z}_2)^3$. The automorphism group is a semidirect product of these two subgroups, with the homeomorphism $\phi : D_3 \to \mathrm{Aut}((\mathbb{Z}_2)^3)$ defined by $\phi(g)(\gamma_x) = g\gamma_x g^{-1} = \gamma_{g(x)}$. So $\mathrm{Aut}(F_{10}) \cong (\mathbb{Z}_2)^3 \rtimes D_3$. The 47 nontrivial automorphisms fall into 9 conjugacy classes (where $x, y \in \{a, i, j\}$):

- $[\alpha] = \{\alpha, \alpha^2, \alpha\gamma_a\gamma_i, \alpha\gamma_a\gamma_j, \alpha\gamma_i\gamma_j, \alpha^2\gamma_a\gamma_i, \alpha^2\gamma_a\gamma_j, \alpha^2\gamma_i\gamma_j\}$
- $[\alpha\gamma_a] = \{\alpha\gamma_a, \alpha\gamma_i, \alpha\gamma_j, \alpha^2\gamma_a, \alpha^2\gamma_i, \alpha^2\gamma_j, \alpha\gamma_a\gamma_i\gamma_j, \alpha^2\gamma_a\gamma_i\gamma_j\}$
- $[\beta] = \{\beta, \alpha\beta, \alpha^2\beta, \beta\gamma_a\gamma_i, \alpha\beta\gamma_a\gamma_j, \alpha^2\beta\gamma_i\gamma_j\}$
- $[\beta\gamma_a] = \{\beta\gamma_a, \beta\gamma_i, \alpha\beta\gamma_a, \alpha\beta\gamma_j, \alpha^2\beta\gamma_i, \alpha^2\beta\gamma_j\}$
- $[\beta\gamma_j] = \{\beta\gamma_j, \alpha\beta\gamma_i, \alpha^2\beta\gamma_a, \beta\gamma_a\gamma_i\gamma_j, \alpha\beta\gamma_a\gamma_i\gamma_j, \alpha^2\beta\gamma_a\gamma_i\gamma_j\}$
- $[\beta\gamma_a\gamma_j] = \{\beta\gamma_a\gamma_j, \beta\gamma_i\gamma_j, \alpha\beta\gamma_a\gamma_i, \alpha\beta\gamma_i\gamma_j, \alpha^2\beta\gamma_a\gamma_i, \alpha^2\beta\gamma_a\gamma_j\}$
- $[\gamma_a] = \{\gamma_a, \gamma_i, \gamma_j\}$
- $[\gamma_a\gamma_i] = \{\gamma_a\gamma_i, \gamma_a\gamma_j, \gamma_i\gamma_j\}$
- $[\gamma_a\gamma_i\gamma_j] = \{\gamma_a\gamma_i\gamma_j\}$

The properties of the automorphisms are given in Table 8 (we list one automorphism from each conjugacy class).

THEOREM 3.8. *The only nontrivial group that is positively realizable for F_{10} is $\mathbb{Z}_3$. Moreover, F_{10} is intrinsically chiral, so no other nontrivial group is realizable.*

PROOF. From Table 8, we easily see that none of β, $\beta\gamma_a$, γ_a, $\gamma_a\gamma_i$ and $\gamma_a\gamma_i\gamma_j$ (and their conjugates) are positively or negatively realizable (by the Realizability

TABLE 8. Automorphisms of F_{10}.

Automorphism	Order	Fixed subgraph	$\hookrightarrow S^1$	$\hookrightarrow S^2$	Path Lemma
$\alpha = (aij)(bfe)(cgh)$	3	•	Yes	Yes	N/A[1]
$\alpha\gamma_a = (aij)(bghcfe)$	6	•	Yes	Yes	N/A[1]
$\beta = (ai)(bf)(cg)$	2	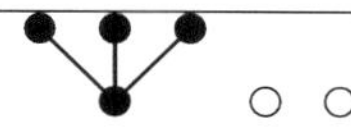	No	Yes	$a-b-g-i$
$\beta\gamma_a = (ai)(bgcf)$	4	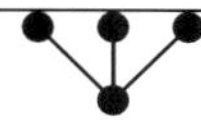	No	Yes	$a-b-g-i$
$\beta\gamma_j = (ai)(bf)(cg)(eh)$	2		Yes	Yes	$a-b-g-i$
$\beta\gamma_a\gamma_j = (ai)(bgcf)(eh)$	4		Yes	Yes	$a-b-g-i$
$\gamma_a = (bc)$	2	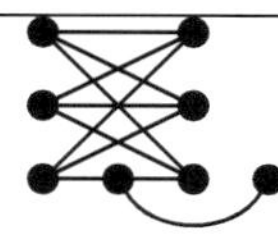	No	No	N/A[3]
$\gamma_a\gamma_i = (bc)(fg)$	2	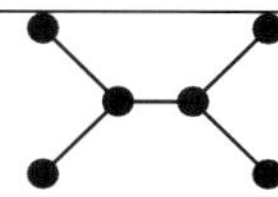	No	Yes	$b-f-c$
$\gamma_a\gamma_i\gamma_j = (bc)(fg)(eh)$	2	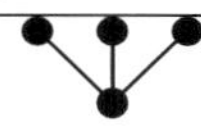	No	Yes	$b-f-c$

Lemma and the Path Lemma). Also, $\alpha\gamma_a$ is not realizable, since $(\alpha\gamma_a)^3 = \gamma_a\gamma_i\gamma_j$, which is not realizable. Similarly, $(\beta\gamma_a\gamma_j)^2 = \gamma_a\gamma_i$, so $\beta\gamma_a\gamma_j$ is not realizable. Finally, $\beta\gamma_j$ is not negatively realizable by the Path Lemma, and α is not negatively realizable since it has odd order. So no automorphisms are negatively realizable, and F_{10} is intrinsically chiral.

The automorphisms conjugate to α and $\beta\gamma_j$ are positively realizable. The embedding on the right in Figure 10 realizes both α (by a rotation of order 3 around the vertical axis through d) and $\beta\gamma_a\gamma_i\gamma_j = (ai)(bg)(cf)(eh)$ (by a rotation of order 2 around the axis through $\overline{dj}$); together, these generate a group isomorphic to D_3. However, including any other automorphism conjugate to either α or $\beta\gamma_a\gamma_i\gamma_j$ (that is not already in the group) generates an automorphism in one of the other conjugacy classes, which are not realizable. So this embedding positively realizes D_3. Since the edge $\overline{ab}$ (among others) is not pointwise fixed by any automorphism, the subgroups $\mathbb{Z}_3$ and $\mathbb{Z}_2$ are also positively realizable by the Subgroup Theorem.

Any combination of automorphisms from the conjugacy classes of α and $\beta\gamma_j$ which do not generate D_3, $\mathbb{Z}_3$ or $\mathbb{Z}_2$ generate groups containing non-realizable automorphisms conjugate to either $\beta\gamma_a$ or $\gamma_a\gamma_i$, so no other groups are realizable. □

3.8. The graph E_{10}. The graph E_{10} is obtained from F_9 (as shown in Figure 9) by performing a ∇Y-move on the triangle bef, resulting in the graph shown in Figure 11.

Any automorphism of E_{10} will fix vertex h (the only vertex of degree 6), and will permute the sets $\{a, i, j\}$ of degree 3 vertices, $\{b, d, f\}$ of degree 4 vertices, and $\{c, e, g\}$ of degree 5 vertices. Moreover, since each of c, e, g is adjacent to exactly one vertex of degree 3 and one of degree 4, the permutation of $\{c, e, g\}$ determines

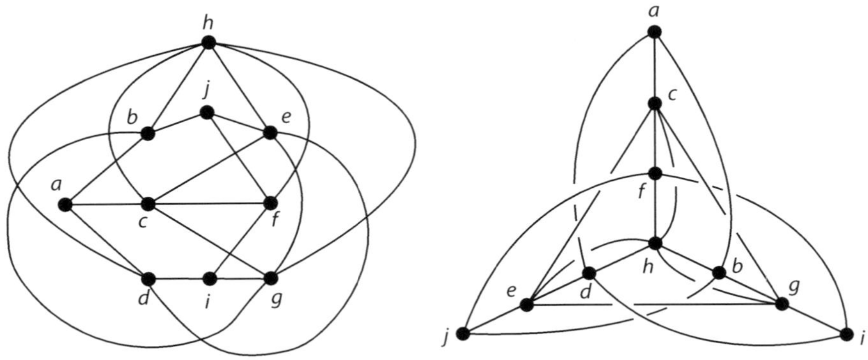

FIGURE 11. The graph E_{10}, and an embedding Γ of E_{10} with $\mathrm{TSG}(\Gamma) = \mathrm{TSG}_+(\Gamma) = \mathbb{Z}_3$.

TABLE 9. Automorphisms of E_{10}.

Automorphism	Order	Fixed subgraph	$\hookrightarrow S^1$	$\hookrightarrow S^2$	Path Lemma
$\alpha = (ceg)(aji)(fdb)$	3	●	Yes	Yes	N/A[1]
$\beta = (ce)(aj)(df)$	2	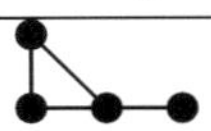 ○	No	Yes	$a - c - f - j$

the permutation of the remaining vertices. Hence $\mathrm{Aut}(E_{10}) \cong D_3$, generated by $\alpha = (ceg)(aji)(fdb)$ and $\beta = (ce)(aj)(df)$. The properties of the automorphisms are given in Table 9 (every nontrivial automorphism is conjugate to α or β).

THEOREM 3.9. *The only nontrivial group that is positively realizable for E_{10} is $\mathbb{Z}_3$. Moreover, E_{10} is intrinsically chiral, so no other nontrivial group is realizable.*

PROOF. From Table 9, β is not positively realizable (by the Realizability Lemma) or negatively realizable (by the Path Lemma). Also, α is not negatively realizable, since it has odd order, showing that E_{10} is intrinsically chiral.

Figure 11 shows an embedding of E_{10} which realizes the automorphism α by the rotation around the vertex h. Since the only realizable automorphisms are the powers of α, this shows $\mathbb{Z}_3$ is positively realizable, but no other nontrivial group is realizable. □

3.9. The graph E_{11}. The graph E_{11} is obtained from E_{10} (as shown in Figure 11) by performing a ∇Y-move on the triangle cfh, resulting in the graph shown in Figure 12.

The vertices of degree 3 are $\{a, f, i, k, j\}$, the vertices of degree 4 are $\{b, c, d\}$ and the vertices of degree 5 are $\{e, g, h\}$. Observe that a is the only vertex of degree 3 adjacent to three vertices of degree 4, and f is the only vertex of degree 3 adjacent to three other vertices of degree 3. Hence a and f are both fixed by any automorphism of E_{11}. Since i, j, k are each adjacent to (in addition to f) one vertex of degree 4 and one of degree 5, the permutation of $\{i, j, k\}$ determines the permutations of $\{b, c, d\}$ and $\{e, g, h\}$. So $\mathrm{Aut}(E_{11}) \cong D_3$, generated by $\alpha = (bcd)(geh)(ijk)$ and $\beta = (bd)(ge)(ij)$. The properties of the automorphisms are given in Table 10 (every nontrivial automorphism is conjugate to α or β).

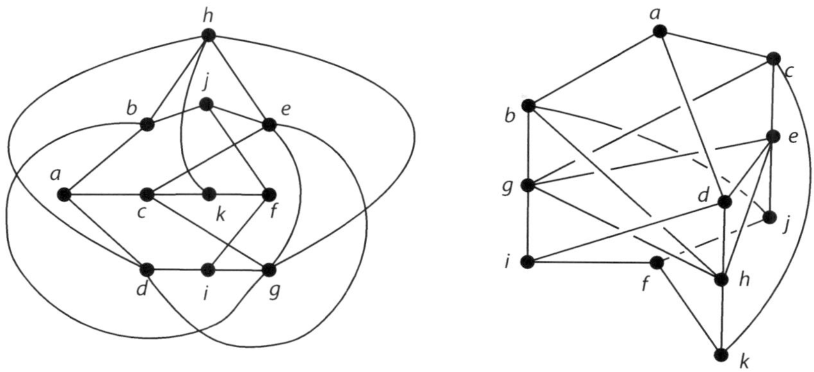

FIGURE 12. The graph E_{11}, and an embedding Γ of E_{11} with $\text{TSG}(\Gamma) = \text{TSG}_+(\Gamma) = \mathbb{Z}_3$.

TABLE 10. Automorphisms of E_{11}.

Automorphism	Order	Fixed subgraph	$\hookrightarrow S^1$	$\hookrightarrow S^2$	Path Lemma
$\alpha = (bcd)(geh)(ijk)$	3	● ●	Yes	Yes	N/A[1]
$\beta = (bd)(ge)(ij)$	2	●–●–●–● (with ● attached above) ○	No	Yes	$b - g - i - d$

THEOREM 3.10. *The only nontrivial group that is positively realizable for E_{11} is $\mathbb{Z}_3$. E_{11} is intrinsically chiral, so no other nontrivial group is realizable.*

PROOF. From Table 9, β is not postively realizable (by the Realizability Lemma) or negatively realizable (by the Path Lemma). Moreover, α is not negatively realizable, since it has odd order, showing that E_{11} is intrinsically chiral.

Figure 12 shows an embedding of E_{11} which realizes the automorphism α by the rotation around the vertical axis through a and f. Since the only realizable automorphisms are the powers of α, this shows $\mathbb{Z}_3$ is positively realizable, but no other nontrivial group is realizable. □

3.10. The graph C_{11}. The graph C_{11} is obtained from E_{10} (as shown in Figure 11) by performing a ∇Y-move on the triangle ceg, resulting in the graph shown in Figure 13.

Since h is the only vertex of degree 6, it is fixed by every automorphism of C_{11}. Similarly, the sets $\{b, c, d, e, f, g\}$ of degree 4 vertices and $\{a, i, j, k\}$ of degree 3 vertices are fixed setwise. Since each pair of degree 3 vertices have one degree 4 vertex adjacent to both, and it is a different common neighbor for each pair, the permutation of $\{a, i, j, k\}$ determines the permutation of $\{b, c, d, e, f, g\}$. Since all the transpositions of $\{a, i, j, k\}$ give automorphisms (e.g. the transposition (ai) induces the automorphism $(ai)(bf)(cg)$), we conclude that $\text{Aut}(C_{11}) \cong S_4$, with generators $\alpha = (ai)(bf)(cg)$ and $\beta = (ijk)(dbc)(feg)$, and with relations $\alpha^2 = \beta^3 = (\alpha\beta)^4 = 1$. The properties of the nontrivial automorphisms are given in Table 11 (listing one automorphism from each conjugacy class).

THEOREM 3.11. *The only nontrivial group that is positively realizable for C_{11} is $\mathbb{Z}_3$. Moreover, C_{11} is intrinsically chiral, and no other nontrivial group is realizable.*

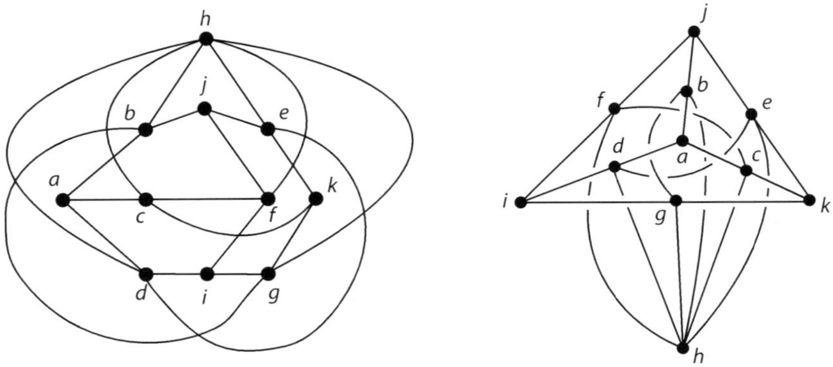

FIGURE 13. The graph C_{11}, and an embedding Γ of C_{11} with $\mathrm{TSG}(\Gamma) = \mathrm{TSG}_+(\Gamma) = \mathbb{Z}_3$.

TABLE 11. Automorphisms of C_{11}.

Automorphism	Order	fixed subgraph	$\hookrightarrow S^1$	$\hookrightarrow S^2$	Path Lemma
$\alpha = (ai)(bf)(cg)$	2	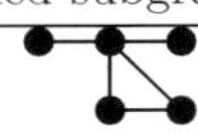	No	Yes	$a - c - f - i$
$\beta = (ijk)(dbc)(feg)$	3	● ●	Yes	Yes	N/A[1]
$\alpha\beta = (aijk)(bg)(dfec)$	4	● ○	Yes	Yes	N/A[2]
$(\alpha\beta)^2 = (aj)(ik)(de)(fc)$	2	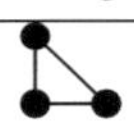 ○ ○	No	Yes	$a - c - k - e - j$

PROOF. From the Realizability Lemma and the Path Lemma, α and $(\alpha\beta)^2$ are neither positively nor negatively realizable. Since $(\alpha\beta)^2$ is not realizable, neither is $\alpha\beta$. And β is not negatively realizable, since it has odd order, so C_{11} is intrinsically chiral.

The only subgroups of S_4 which do not contain transpositions, products of two disjoint transpositions or 4-cycles are the subgroups isomorphic to $\mathbb{Z}_3$ induced by a 3-cycle, so the only possibly realizable group is $\mathbb{Z}_3$. Figure 13 shows an embedding of C_{11} that positively realizes $\mathbb{Z}_3$, where the automorphism β is realized by the rotation around the vertical axis through a and h. □

3.11. The graph C_{12}. The graph C_{12} is obtained from C_{11} (as shown in Figure 13) by performing a ∇Y-move on the triangle cfh, resulting in the graph shown in Figure 14.

Every automorphism of C_{12} fixes h, the only vertex of degree 5. Since l is the only vertex of degree 3 adjacent to h, it is also fixed by every automorphism. So the set $\{c, f\}$ (the other vertices adjacent to l) is fixed setwise. The remaining vertices of the graph are contained in a unique 8-cycle $b - a - d - i - g - k - e - j - b$; the subgraph they induce is homeomorphic (in the graph theoretic sense) to a square, where b, d, g, e are the vertices (connected by diagonals $\overline{bg}$ and $\overline{de}$), and a, i, k, j are the midpoints of the sides. This subgraph has an automorphism group isomorphic to D_4. Since each automorphism of the square subgraph determines the permutation of $\{c, f\}$, the automorphism group of C_{12} is also isomorphic to

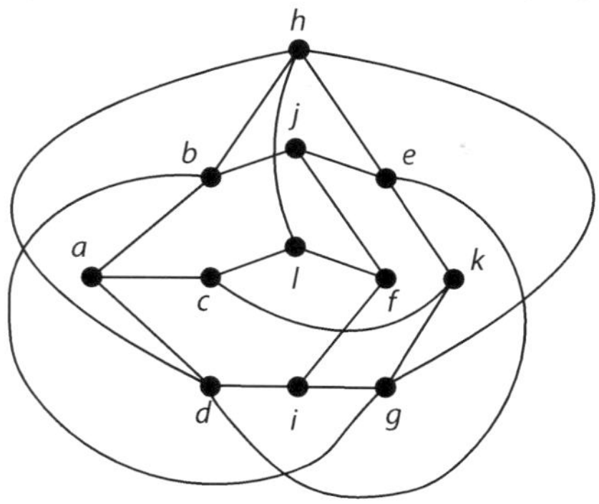

FIGURE 14. The graph C_{12}.

TABLE 12. Automorphisms of C_{12}.

Automorphism	Order	Fixed subgraph	$\hookrightarrow S^1$	$\hookrightarrow S^2$	Path Lemma
$\alpha = (aikj)(bdge)(cf)$	4		Yes	Yes	$c - k - e - j - f$
$\alpha^2 = (ak)(ij)(bg)(de)$	2	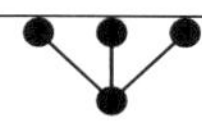	No	Yes	$a - d - i - g - k$
$\beta = (aj)(de)(ik)(cf)$	2	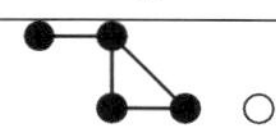	No	Yes	$a - c - k - e - j$
$\alpha\beta = (bd)(eg)(ij)$	2	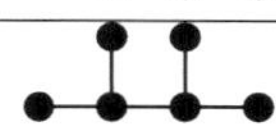	No	Yes	$b - j - e - d$

D_4, generated by $\alpha = (aikj)(bdge)(cf)$ and $\beta = (aj)(de)(ik)(cf)$. The properties of the nontrivial automorphisms are given in Table 12 (listing one automorphism from each conjugacy class).

THEOREM 3.12. *No nontrivial group is positively realizable or realizable for C_{12}. Hence C_{12} is intrinsically chiral.*

PROOF. From Table 12, we easily see that α^2, β and $\alpha\beta$ are neither positively realizable (by the Realizability Lemma) or negatively realizable (by the Path Lemma). Since α^2 is not realizable, neither is α. So no nontrivial automorphism of C_{12} is realizable. Hence C_{12} is intrinsically chiral, and no nontrivial group is realizable. □

3.12. The graph C_{13}. The graph C_{13} is obtained from C_{12} (as shown in Figure 14) by performing a ∇Y-move on the triangle deh, resulting in the graph shown in Figure 15.

The graph C_{13} has three vertices of degree 4, $\{b, g, h\}$, and the other ten have degree 3. Of these ten, a, i, j, k, l, m are adjacent to a vertex of degree 4, while c, d, e, f are only adjacent to vertices of degree 3. So these two sets of vertices are preserved setwise by every automorphism. In fact, the permutation of $\{c, d, e, f\}$ determines the permutation of $\{a, i, j, k, l, m\}$ (each of which is adjacent to a different pair of vertices from $\{c, d, e, f\}$), and these determine the permutation of $\{b, g, h\}$. Since every permutation of $\{c, d, e, f\}$ induces an automorphism on C_{13}, we conclude that $\mathrm{Aut}(C_{13}) \cong S_4$, with generators $\alpha = (cd)(il)(km)(gh)$ and

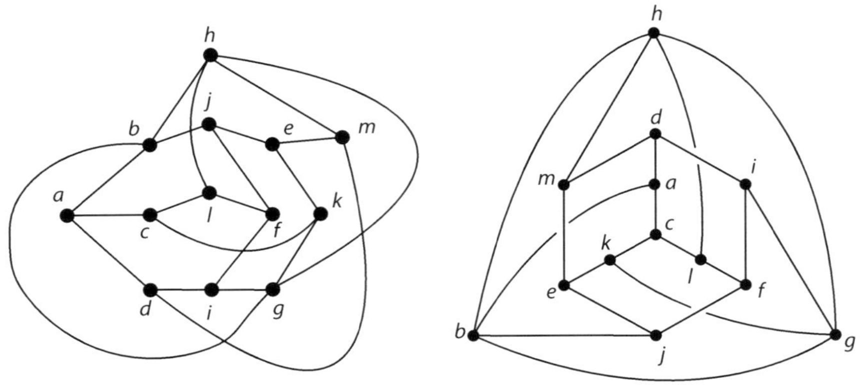

FIGURE 15. The graph C_{13}, and an embedding Γ of C_{13} with $\mathrm{TSG}(\Gamma) = \mathrm{TSG}_+(\Gamma) = \mathbb{Z}_3$.

TABLE 13. Automorphisms of C_{13}.

Automorphism	Order	Fixed subgraph	$\hookrightarrow S^1$	$\hookrightarrow S^2$	Path Lemma
$\alpha = (cd)(il)(km)(gh)$	2	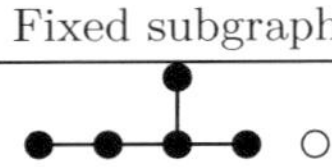	No	Yes	$c - l - h - m - d$
$\beta = (def)(imj)(akl)(bgh)$	3		Yes	Yes	N/A[1]
$\alpha\beta = (cdef)(amjl)(bh)$	4		Yes	Yes	$b - j - e - m - h$
$(\alpha\beta)^2 = (ce)(df)(aj)(ml)$	2	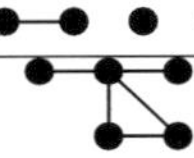	No	Yes	$a - c - l - f - j$

$\beta = (def)(imj)(akl)(bgh)$, and with relations $\alpha^2 = \beta^3 = (\alpha\beta)^4 = 1$. The properties of the nontrivial automorphisms are given in Table 13 (listing one automorphism from each conjugacy class).

THEOREM 3.13. *The only nontrivial group that is positively realizable for C_{13} is $\mathbb{Z}_3$. No other nontrivial group is realizable; moreover, C_{13} is intrinsically chiral.*

PROOF. From Table 13, we see that α and $(\alpha\beta)^2$ are not realizable (by the Realizability Lemma and the Path Lemma). Hence $\alpha\beta$ is also not realizable. Since β has odd order, it is not negatively realizable, so C_{13} is intrinsically chiral.

The only subgroups of S_4 which do not contain transpositions, products of two disjoint transpositions or 4-cycles are the subgroups isomorphic to $\mathbb{Z}_3$ induced by a 3-cycle, so the only possibly realizable group is $\mathbb{Z}_3$. Figure 15 shows an embedding of C_{13} that positively realizes $\mathbb{Z}_3$, generated by the automorphism β. □

3.13. The graph N_{11}. The graph N_{11} is obtained from H_{12} (as shown in Figure 8) by performing a $Y\nabla$-move removing vertex h and adding triangle jkl, resulting in the graph shown in Figure 16.

The automorphisms of N_{11} permute the set $\{a, i\}$ of degree 3 vertices, permute the set $\{j, k, l\}$ of vertices not adjacent to a or i, and permute or interchange the sets $\{b, c, d\}$ and $\{e, f, g\}$. The automorphisms are completely determined by the permutation of $\{j, k, l\}$ and the permutation of $\{a, i\}$. These permutations commute, so $\mathrm{Aut}(N_{11}) \cong D_3 \times \mathbb{Z}_2$. The properties of the automorphisms are given in Table 14.

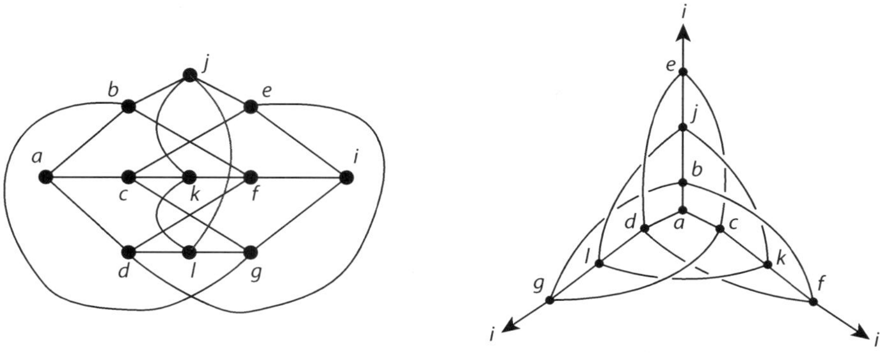

FIGURE 16. The graph N_{11}.

TABLE 14. Automorphisms of N_{11}.

Automorphism	Order	Fixed subgraph	$\hookrightarrow S^1$	$\hookrightarrow S^2$	Path Lemma
$\alpha = (bcd)(efg)(jkl)$	3	● ●	Yes	Yes	N/A[1]
$\beta = (bc)(ef)(jk)$	2	●–●–●–●–● ○	Yes	Yes	$j - b - f - k$
$\gamma = (ai)(be)(cf)(dg)$	2	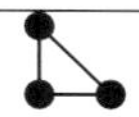	Yes	Yes	$a - b - f - i$
$\alpha\gamma = (ai)(bfdecg)(jkl)$	6	$\emptyset$	Yes	Yes	N/A[2]
$\beta\gamma = (ai)(bf)(ec)(dg)(jk)$	2	● ○ ○ ○	Yes	Yes	$a - c - g - i$

THEOREM 3.14. *The only groups that are positively realizable for* N_{11} *are* $D_3 \times \mathbb{Z}_2$ *and its subgroups* $\mathbb{Z}_6, D_3, \mathbb{Z}_3, D_2$ *and* $\mathbb{Z}_2$. *Moreover,* N_{11} *is intrinsically chiral, and no other nontrivial group is realizable.*

PROOF. From Table 14 we see that α is not negatively realizable since it has odd order, and β, γ and $\beta\gamma$ are not negatively realizable by the Path Lemma. Since $(\alpha\gamma)^3 = \gamma$, the automorphism $\alpha\gamma$ is also not negatively realizable. So none of the automorphisms are negatively realizable, and N_{11} is intrinsically chiral.

The embedding of N_{11} on the right in Figure 16 positively realizes $D_3 \times \mathbb{Z}_2$. In this embedding, i is embedded at the point in S^3 antipodal to a, and they are interchanged by the order 2 rotation around the triangle jkl, which realizes γ. Then α is realized by rotation around vertex a, and β by rotation about the axis through vertices d, g, l. Since the edge $\overline{de}$ is not fixed by any automorphism, the subgroups of $D_3 \times \mathbb{Z}_2$ are also positively realizable by the Subgroup Theorem. Since $D_3 \times \mathbb{Z}_2$ is the full automorphism group, there are no other realizable groups. □

3.14. The graph N_{10}. The graph N_{10} is obtained from N_{11} (as shown in Figure 16) by performing a $Y\nabla$-move removing vertex i and adding triangle efg, resulting in the graph shown in Figure 17.

Every automorphism of N_{10} fixes vertex a (the only vertex of degree 3), the set $\{b, c, d\}$ (the vertices adjacent to a), the set $\{j, k, l\}$ (the other vertices of degree 4) and the set $\{e, f, g\}$ (the vertices of degree 5). The permutation of $\{j, k, l\}$ determines the permutation of the other vertices, so $\mathrm{Aut}(N_{10}) \cong D_3$. The properties of the automorphisms are given in Table 15.

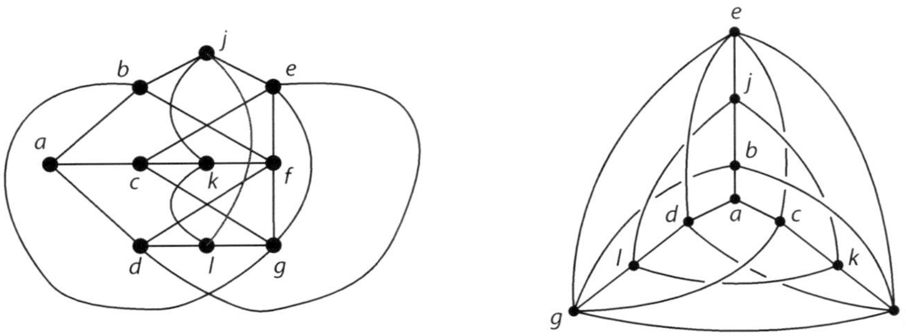

FIGURE 17. The graph N_{10}, and an embedding Γ of N_{10} with $\mathrm{TSG}(\Gamma) = \mathrm{TSG}_+(\Gamma) = D_3$.

TABLE 15. Automorphisms of N_{10}.

Automorphism	Order	Fixed subgraph	$\hookrightarrow S^1$	$\hookrightarrow S^2$	Path Lemma
$\alpha = (bcd)(efg)(jkl)$	3	●	Yes	Yes	N/A[1]
$\beta = (bc)(ef)(jk)$	2	●—●—●—● ○ ○	Yes	Yes	$b - f - k - c$

THEOREM 3.15. *The only groups that are positively realizable for N_{10} are D_3 and its subgroups $\mathbb{Z}_3$ and $\mathbb{Z}_2$. N_{10} is intrinsically chiral, and no other nontrivial group is realizable.*

PROOF. From Table 15, we see that α is not negatively realizable because it has odd order, and β is not negatively realizable by the Path Lemma. Hence N_{10} is intrinsically chiral. The embedding of N_{10} on the right in Figure 17 positively realizes D_3. Since the edge $\overline{de}$ is not fixed by any automorphism, the subgroups of D_3 are also positively realizable by the Subgroup Theorem. Since D_3 is the full automorphism group, there are no other realizable groups. □

3.15. The graph N_9. The graph N_9 is obtained from N_{10} (as shown in Figure 17) by performing a $Y\nabla$-move removing vertex a and adding triangle bcd, resulting in the graph shown in Figure 17.

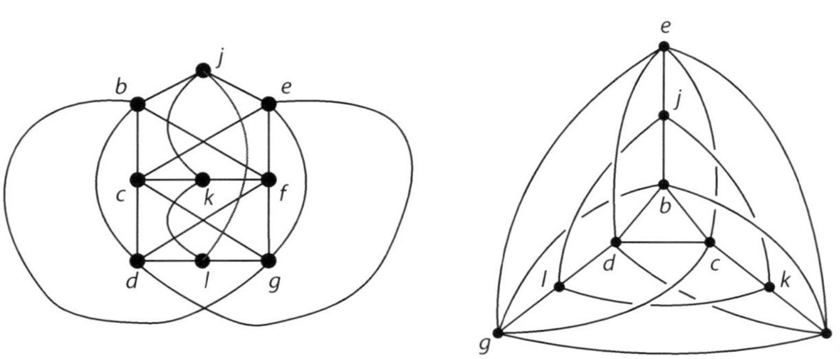

FIGURE 18. The graph N_9, and an embedding Γ of N_9 with $\mathrm{TSG}(\Gamma) = \mathrm{TSG}_+(\Gamma) = D_3 \times \mathbb{Z}_2$.

TABLE 16. Automorphisms of N_9.

Automorphism	Order	Fixed subgraph	$\hookrightarrow S^1$	$\hookrightarrow S^2$	Path Lemma
$\alpha = (jkl)(bcd)(efg)$	3	$\emptyset$	Yes	Yes	N/A[1]
$\alpha\gamma_j = (jkl)(bfgecd)$	6	$\emptyset$	Yes	Yes	N/A[1]
$\beta = (cd)(lk)(gf)$	2		Yes	Yes	$c - g - l - d$
$\beta\gamma_k = (cgfd)(lk)$	4		Yes	Yes	$l - d - c - k$
$\beta\gamma_j = (be)(cd)(lk)(gf)$	2		Yes	Yes	$c - g - l - d$
$\beta\gamma_j\gamma_k = (be)(cgfd)(lk)$	4		Yes	Yes	N/A[2]
$\gamma_j = (be)$	2	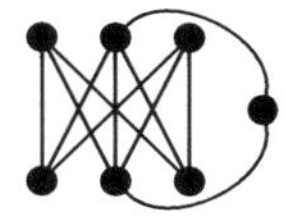	No	No	N/A[3]
$\gamma_j\gamma_k = (be)(cf)$	2	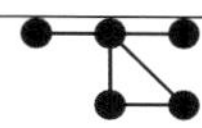	No	Yes	$b - c - e$
$\gamma_j\gamma_k\gamma_l = (be)(cf)(dg)$	2	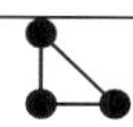	Yes	Yes	$b - c - e$

Every automorphism of N_9 preserves the set $\{j, k, l\}$ of degree 4 vertices and the set $\{b, c, d, e, f, g\}$ of degree 5 vertices. Morever, the permutation of $\{j, k, l\}$ permutes the three pairs $\{b, e\}$, $\{c, f\}$, $\{d, g\}$; the vertices in each pair can also be interchanged. So the automorphism group is generated by the permutations $\alpha = (jkl)(bcd)(efg)$, $\beta = (cd)(lk)(gf)$, $\gamma_j = (be)$, $\gamma_k = (cf)$ and $\gamma_l = (dg)$. Automorphisms α and β generate a subgroup isomorphic to D_3, while γ_j, γ_k, and γ_l generate $(\mathbb{Z}_2)^3$. The automorphism group is a semidirect product of these two subgroups, with the homeomorphism $\phi : D_3 \to \mathrm{Aut}((\mathbb{Z}_2)^3)$ defined by $\phi(g)(\gamma_x) = g\gamma_x g^{-1} = \gamma_{g(x)}$. So $\mathrm{Aut}(N_9) \cong (\mathbb{Z}_2)^3 \rtimes D_3$. The properties of the automorphisms are given in Table 16 (we list one automorphism from each conjugacy class, as in Table 8).

THEOREM 3.16. *The only groups that are positively realizable for N_9 are $D_3 \times \mathbb{Z}_2$ and its subgroups $\mathbb{Z}_6, D_3, \mathbb{Z}_3, D_2$ and $\mathbb{Z}_2$. Moreover, N_9 is intrinsically chiral, and no other nontrivial group is realizable.*

PROOF. From Table 16, we see that γ_j and $\gamma_j\gamma_k$ are not realizable (by the Realizability Lemma and the Path Lemma), β, $\beta\gamma_k$, $\beta\gamma_j$ and $\gamma_j\gamma_k\gamma_l$ are not negatively realizable by the Path Lemma, and α is not negatively realizable because it has odd order. Since $(\alpha\gamma_j)^3 = \gamma_j\gamma_k\gamma_l$, it is also not negatively realizable. Finally, $(\beta\gamma_k)^2 = (\beta\gamma_j\gamma_k)^2 = \gamma_k\gamma_l$, which is in the conjugacy class of $\gamma_j\gamma_k$, so it is not realizable; therefore $\beta\gamma_k$ and $\beta\gamma_j\gamma_k$ are also not realizable. Hence no automorphisms are negatively realizable, so N_9 is intrinsically chiral.

The only positively realizable automorphisms are those conjugate to α, $\alpha\gamma_j$, β, $\beta\gamma_j$, and $\gamma_j\gamma_k\gamma_l$. The embedding shown on the right in Figure 18 positively realizes $D_3 \times \mathbb{Z}_2$, generated by α (the rotation of order 3 around the center), β (the rotation of order 2 around the axis through vertices e, j, b), and $\gamma_j\gamma_k\gamma_l$ (the rotation of order 2 around the triangle jkl). Including any other automorphisms generates a group that contains non-realizable automorphisms, so the topological symmetry group is $D_3 \times \mathbb{Z}_2$. Since the edge $\overline{de}$ is not fixed by any element of the group, every subgroup of $D_3 \times \mathbb{Z}_2$ is positively realizable by the Subgroup Theorem. Using *Sage* [**26**], we verified that any subgroup of $\mathrm{Aut}(N_9)$ not isomorphic to a subgroup of $D_3 \times \mathbb{Z}_2$ must contain non-realizable automorphisms, so these are the only realizable topological symmetry groups. □

3.16. The graph N'_{12}. The graph N'_{12} is obtained from C_{13} (as shown in Figure 15) by performing a $Y\nabla$-move removing vertex f and adding triangle ijl, resulting in the graph shown in Figure 19.

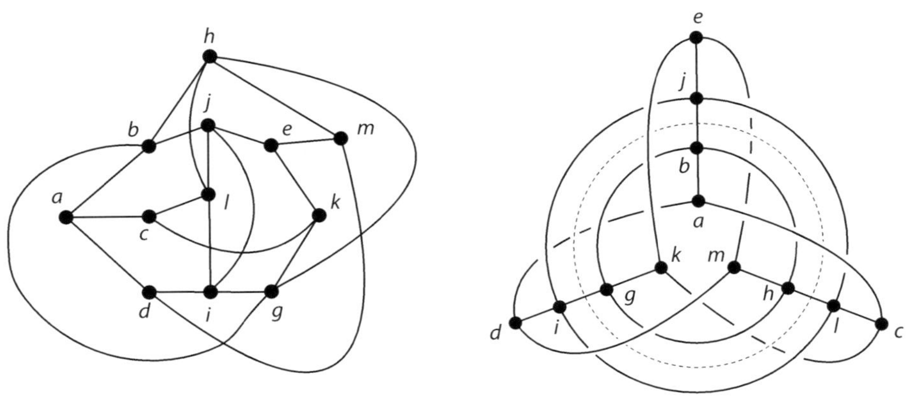

FIGURE 19. The graph N'_{12}, and an embedding Γ of N'_{12} with $\mathrm{TSG}(\Gamma) = \mathrm{TSG}_+(\Gamma) = D_6$.

TABLE 17. Automorphisms of N'_{12}.

Automorphism	Order	Fixed subgraph	$\hookrightarrow S^1$	$\hookrightarrow S^2$	Path Lemma
$\alpha = (ackemd)(blgjhi)$	6	$\emptyset$	Yes	Yes	N/A[1]
$\alpha^2 = (akm)(ced)(bgh)(lji)$	3	$\emptyset$	Yes	Yes	N/A[1]
$\alpha^3 = (ae)(cm)(kd)(bj)(lh)(gi)$	2	○ ○ ○	Yes	Yes	$a - c - k - e$
$\beta = (cd)(li)(hg)(mk)$	2	●–●–●–●	Yes	Yes	$c - k - g - i - d$
$\alpha\beta = (ac)(dk)(em)(bl)(gi)(hj)$	2	○ ○ ○	Yes	Yes	$d - a - b - g - k$

The graph N'_{12} has six vertices of degree 3. These form a 6-cycle in the graph: $a - c - k - e - m - d - a$. This 6-cycle must be preserved by any automorphism of the graph, so the automorphism group of the graph is a subgroup of D_6. Since every automorphism of the 6-cycle can be realized by an automorphism of the graph, we conclude $\mathrm{Aut}(N'_{12}) \cong D_6$, generated by $\alpha = (ackemd)(blgjhi)$ and $\beta = (cd)(li)(hg)(mk)$. The properties of the automorphisms are given in Table 17.

THEOREM 3.17. *The only groups that are positively realizable for N'_{12} are D_6 and its subgroups $\mathbb{Z}_6$, D_3, $\mathbb{Z}_3$, D_2 and $\mathbb{Z}_2$. N'_{12} is intrinsically chiral, and no other nontrivial group is realizable.*

PROOF. From Table 17, we can see that α^3, β and $\alpha\beta$ are not negatively realizable by the Path Lemma, and α^2 is not negatively realizable because it has odd order. Finally, since α^3 is not negatively realizable, neither is α, so N'_{12} is intrinsically chiral.

The embedding on the right in Figure 19 positively realizes $\mathrm{Aut}(N'_{12}) \cong D_6$: α is realized by the glide rotation consisting of an order 3 rotation around the center of the figure, and an order 2 rotation about the dashed circle in the diagram (going through the midpoints of $\overline{bj}$, $\overline{hl}$ and $\overline{gi}$); β is realized by the order 2 rotation about the vertical axis through a, b, j, e.

Since the edge $\overline{bh}$ is not fixed pointwise by any automorphism of the graph, we can also positively realize every subgroup of D_6, by the Subgroup Theorem. □

3.17. The graph N'_{11}. The graph N'_{11} is obtained from N'_{12} (as shown in Figure 19) by performing a $Y\nabla$-move removing vertex m and adding triangle deh, resulting in the graph shown in Figure 20.

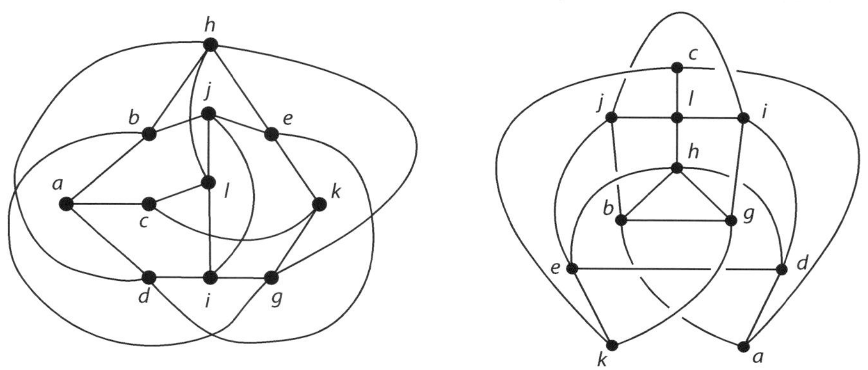

FIGURE 20. The graph N'_{11}, and an embedding Γ of N'_{11} with $\mathrm{TSG}(\Gamma) = \mathrm{TSG}_+(\Gamma) = \mathbb{Z}_2$.

TABLE 18. Automorphisms of N'_{11}.

Automorphism	Order	Fixed subgraph	$\hookrightarrow S^1$	$\hookrightarrow S^2$	Path Lemma
$\alpha = (ak)(be)(dg)$	2	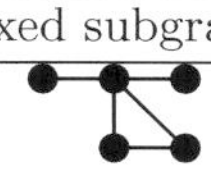	No	Yes	$a - d - e - k$
$\beta = (ji)(bd)(eg)$	2	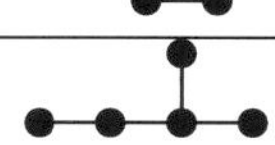	No	Yes	$b - g - i - d$
$\alpha\beta = (ak)(ji)(bg)(de)$	2		Yes	Yes	$a - d - i - g - k$

Every automorphism of N'_{11} fixes h (the only vertex of degree 5), c (the only vertex of degree 3 with two neighbors of degree 3) and l (the only vertex adjacent to both h and c). Moreover, every automorphism fixes (setwise) the sets $\{a, k\}$ (the other vertices of degree 3), $\{j, i\}$ (the other vertices adjacent to l) and $\{b, d, e, g\}$ (the remaining vertices). The permutations of $\{a, k\}$ and $\{j, i\}$ determine the permutation of the remaining vertices, so $\mathrm{Aut}(N'_{11}) \cong \mathbb{Z}_2 \times \mathbb{Z}_2 = D_2$, generated by $\alpha = (ak)(be)(dg)$ and $\beta = (ji)(bd)(eg)$. The properties of the automorphisms are given in Table 18.

THEOREM 3.18. *The only group that is positively realizable for N'_{11} is $\mathbb{Z}_2$. N'_{11} is intrinsically chiral, and no other nontrivial group is realizable.*

PROOF. From Table 18, none of the automorphisms are negatively realizable, by the Path Lemma, so N'_{11} is intrinsically chiral. Also, by the Realizability Lemma, neither α nor β is positively realizable. However, the product $\alpha\beta = (ak)(ji)(bg)(ed)$ is positively realizable, as shown on the right in Figure 20. So the group $\mathbb{Z}_2$ is positively realizable, and no other nontrivial groups are realizable. □

3.18. The graph N'_{10}. The graph N'_{10} is obtained from N'_{11} (as shown in Figure 20) by performing a $Y\nabla$-move removing vertex c and adding triangle akl, resulting in the graph shown in Figure 21.

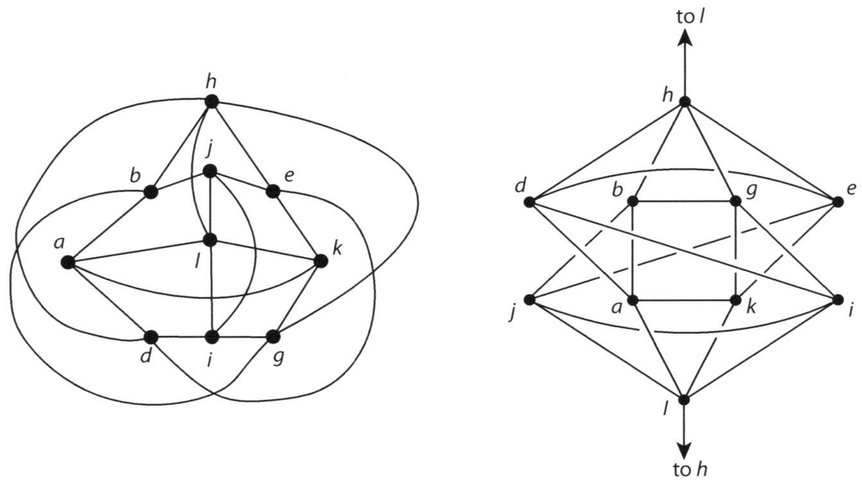

FIGURE 21. The graph N'_{10}, and an embedding Γ of N'_{10} with $\mathrm{TSG}(\Gamma) = \mathrm{TSG}_+(\Gamma) = D_2$.

TABLE 19. Automorphisms of N'_{10}.

Automorphism	Order	Fixed subgraph	$\hookrightarrow S^1$	$\hookrightarrow S^2$	Path Lemma
$\alpha = (hl)(abjekgid)$	8	○	Yes	Yes	N/A[2]
$\alpha^2 = (ajki)(begd)$	4	●—●	Yes	Yes	N/A[1]
$\alpha^3 = (hl)(aeibkdjg)$	8	○	Yes	Yes	N/A[2]
$\alpha^4 = (ak)(bg)(ji)(ed)$	2	○ ○ ○	Yes	Yes	$a-d-i-g-k$
$\beta = (hl)(ab)(dj)(ei)(gk)$	2	○ ○ ○	Yes	Yes	$d-i-l-j$
$\alpha\beta = (aj)(de)(ik)$	2	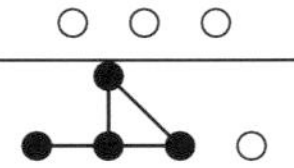	No	Yes	$a-d-i-j$

The graph N'_{10} has two vertices of degree 5, $\{h, l\}$, and eight vertices of degree 4. The subgraph induced by the vertices of degree 4 is isomorphic to a Möbius ladder M_4, which has automorphism group D_8 [**11**]. These automorphisms extend to the rest of N'_{10}, so $\mathrm{Aut}(N'_{10}) \cong D_8$, generated by $\alpha = (hl)(abjekgid)$ and $\beta = (hl)(ab)(dj)(ei)(gk)$. The properties of the automorphisms are given in Table 19.

THEOREM 3.19. *The only groups that are positively realizable for N'_{10} are D_2 and $\mathbb{Z}_2$. Moreover, N'_{10} is intrinsically chiral, and no other nontrivial groups are realizable.*

PROOF. From Table 19, we see that α^4, β and $\alpha\beta$ are not negatively realizable by the Path Lemma, and $\alpha\beta$ is also not positively realizable. Since α^2 fixes more than two points (h, l and the edge between them), and does not have order 2, it is not negatively realizable by the Realizability Lemma.

Towards contradiction, suppose the automorphism $\alpha^2 = (ajki)(begd)$ is positively realizable for an embedding Γ of N'_{10}. Since α^2 fixes h and l, it must be realized by a rotation; hence α^4 is also a rotation. Since α^4 interchanges a and k, it must fix a point on the edge $\overline{ak}$, so $\overline{ak}$ crosses the axis of rotation. But then α^2 would also fix a point on $\overline{ak}$, which contradicts the fact that α^2 sends $\overline{ak}$ to the disjoint edge $\overline{ji}$. Hence α^2 is not positively realizable (and so not realizable). Since α^2 is a power of each of $\alpha, \alpha^3, \alpha^5, \alpha^6$ and α^7, none of these are realizable. So the only power of α which may be realizable is α^4. Since none of the automorphisms are negatively realizable, N'_{10} is intrinsically chiral.

The only automorphisms which could be realizable are α^4 and β (and their conjugates). Observe that α^4 is only conjugate to itself, and multiplying β by any conjugate other than $\alpha^4\beta$ gives a power of α which is not realizable. So the largest possible positively realizable group is isomorphic to D_2, with α^4, β and $\alpha^4\beta$. The embedding shown on the right in Figure 21 positively realizes D_2 by realizing α^4 by an order 2 rotation around the axis through h and l, and realizing β by an order two rotation about the horizontal line through the middle of the diagram. Since the edge $\overline{hd}$ is not fixed by any of these automorphisms, we can also positively realize $\mathbb{Z}_2$ by the Subgroup Theorem. Hence the only groups that are realizable for N'_{10} are D_2 and $\mathbb{Z}_2$. □

4. Conclusions and Questions

Aside from adding to our library of graphs for which we know the topological symmetry groups, the purpose of this project was to explore whether there are any relationships among the topological symmetry groups of graphs related by ∇Y and $Y\nabla$ moves. The most interesting observation is that none of the graphs in the Heawood family have any topological symmetry groups that are realizable but not positively realizable; in fact, every graph in the Heawood family is intrinsically chiral! This is in contrast to the Petersen family (another family related by ∇Y moves, shown in Figure 22), where most of the graphs do have topological symmetry groups that are realizable but not positively realizable [**7**].

Putting together the results of all our theorems in Sections 2 and 3, we have proved the following theorem.

THEOREM 4.1. *The graphs of the Heawood family are all intrinsically chiral.*

Intrinsically chiral graphs are of particular interest in both chemistry and topology, and substantial work has been done on proving particular graphs are intrinsically chiral [**5, 6, 10, 20**]. These proofs are often difficult and performed only for small graphs or very special families. In light of our results for the Heawood family, it is natural to ask whether ∇Y and $Y\nabla$ moves always preserve intrinsic chirality.

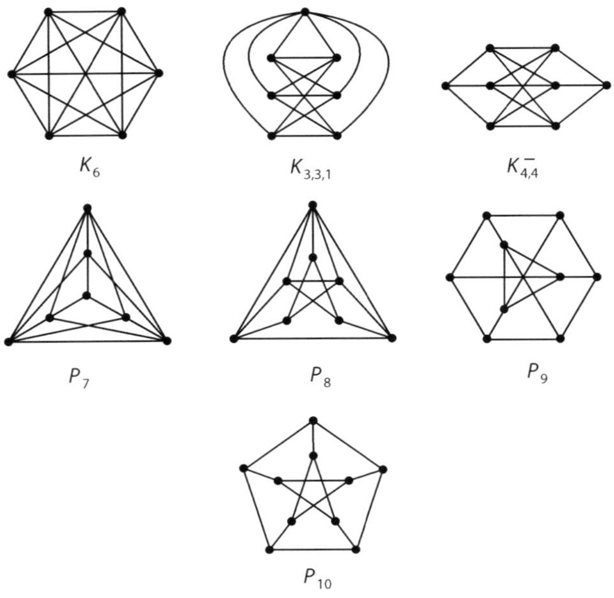

FIGURE 22. The Petersen family of graphs.

In general, however, they do not, as we can see in the following examples:

- Let γ_1 be the wedge of two copies of K_7 joined at a single vertex v, and Γ_1 be an embedding where one copy of K_7 is embedded as the mirror image of the other. Then γ_1 is not intrinsically chiral. However, if we apply a ∇Y move to one of the copies of K_7, then the resulting graph γ_2 will be intrinsically chiral (since K_7 is intrinsically chiral [**10**]). A similar construction can be done with the wedge of two copies of C_{14} using a $Y\nabla$ move (C_{14} is also shown to be intrinsically chiral in [**10**]). So neither ∇Y nor $Y\nabla$ moves preserve intrinsic chirality in general.
- We can also note that, in the Petersen family of graphs, the graph P_8 is intrinsically chiral from the results in [**7**], but neither P_7 (obtained from P_8 by a $Y\nabla$ move) nor P_9 (obtained from P_8 by a ∇Y move) are intrinsically chiral.

So is the result for the Heawood family simply a coincidence, or is something deeper involved?

QUESTION. Under what conditions do ∇Y and $Y\nabla$ moves preserve intrinsic chirality?

We hope to investigate this question further in future work.

Acknowledgements. The authors are grateful to Erica Flapan for valuable conversations and insight during the writing of this paper.

References

[1] M. Baza, *The positively realizable groups of H_8 and G_3*, Master's thesis, California State University, Fresno, 2021.

[2] Dwayne Chambers and Erica Flapan, *Topological symmetry groups of small complete graphs*, Symmetry **6** (2014), no. 2, 189–209, DOI 10.3390/sym6020189. MR3230775
[3] D. Chambers, E. Flapan, D. Heath, E. Lawrence, C. Thatcher, and R. Vanderpool, *Topological symmetry groups of the Petersen graph*, J. Knot Theory Ramifications **30** (2021), no. 12, Paper No. 2141004, 11, DOI 10.1142/S0218216521410042. MR4394063
[4] Dwayne Chambers, Erica Flapan, and John D. O'Brien, *Topological symmetry groups of K_{4r+3}*, Discrete Contin. Dyn. Syst. Ser. S **4** (2011), no. 6, 1401–1411, DOI 10.3934/dcdss.2011.4.1401. MR2754164
[5] Howon Choi, Hyoungjun Kim, and Sungjong No, *The minor minimal intrinsically chiral graphs*, Discrete Appl. Math. **291** (2021), 237–245, DOI 10.1016/j.dam.2020.12.009. MR4192050
[6] Howon Choi, Hyoungjun Kim, and Sungjong No, *Chirality for simple graphs of size up to 12*, J. Math. Chem. **60** (2022), no. 10, 2013–2028, DOI 10.1007/s10910-022-01398-9. MR4496428
[7] D. Elzie, S. Fridhi, B. Mellor, D. Silva, and R. Wilson, *Topological symmetry groups of the Petersen graphs*, Symmetry **15** (2023), no. 6, 1267.
[8] Erica Flapan, *Symmetries of Möbius ladders*, Math. Ann. **283** (1989), no. 2, 271–283, DOI 10.1007/BF01446435. MR980598
[9] Erica Flapan, *Rigidity of graph symmetries in the 3-sphere*, J. Knot Theory Ramifications **4** (1995), no. 3, 373–388, DOI 10.1142/S0218216595000181. MR1347360
[10] Erica Flapan, Will Fletcher, and Ryo Nikkuni, *Reduced Wu and generalized Simon invariants for spatial graphs*, Math. Proc. Cambridge Philos. Soc. **156** (2014), no. 3, 521–544, DOI 10.1017/S0305004114000073. MR3181638
[11] Erica Flapan and Emille Davie Lawrence, *Topological symmetry groups of Möbius ladders*, J. Knot Theory Ramifications **23** (2014), no. 14, 1450077, 13, DOI 10.1142/S0218216514500771. MR3312620
[12] E. Flapan, E. Lawrence, and R. Wilson, *Topological symmetry groups of the Heawood graph*, Symmetry **12** (2020), no. 4.
[13] Erica Flapan, Nicole Lehle, Blake Mellor, Matt Pittluck, and Xan Vongsathorn, *Symmetries of embedded complete bipartite graphs*, Fund. Math. **226** (2014), no. 1, 1–16, DOI 10.4064/fm226-1-1. MR3208292
[14] Erica Flapan, Blake Mellor, and Ramin Naimi, *Complete graphs whose topological symmetry groups are polyhedral*, Algebr. Geom. Topol. **11** (2011), no. 3, 1405–1433, DOI 10.2140/agt.2011.11.1405. MR2821428
[15] Erica Flapan, Blake Mellor, and Ramin Naimi, *Spatial graphs with local knots*, Rev. Mat. Complut. **25** (2012), no. 2, 493–510, DOI 10.1007/s13163-011-0072-9. MR2931423
[16] Erica Flapan, Blake Mellor, Ramin Naimi, and Michael Yoshizawa, *Classification of topological symmetry groups of K_n*, Topology Proc. **43** (2014), 209–233. MR3104921
[17] Erica Flapan and Ramin Naimi, *The Y-triangle move does not preserve intrinsic knottedness*, Osaka J. Math. **45** (2008), no. 1, 107–111. MR2416651
[18] Erica Flapan, Ramin Naimi, James Pommersheim, and Harry Tamvakis, *Topological symmetry groups of graphs embedded in the 3-sphere*, Comment. Math. Helv. **80** (2005), no. 2, 317–354, DOI 10.4171/CMH/16. MR2142245
[19] Erica Flapan, Ramin Naimi, and Harry Tamvakis, *Topological symmetry groups of complete graphs in the 3-sphere*, J. London Math. Soc. (2) **73** (2006), no. 1, 237–251, DOI 10.1112/S0024610705022490. MR2197381
[20] Erica Flapan and Nikolai Weaver, *Intrinsic chirality of complete graphs*, Proc. Amer. Math. Soc. **115** (1992), no. 1, 233–236, DOI 10.2307/2159591. MR1112490
[21] Kathleen Hake, Blake Mellor, and Matt Pittluck, *Topological symmetry groups of complete bipartite graphs*, Tokyo J. Math. **39** (2016), no. 1, 133–156, DOI 10.3836/tjm/1459367261. MR3543135
[22] Minjung Lee, Hyoungjun Kim, Hwa Jeong Lee, and Seungsang Oh, *Exactly fourteen intrinsically knotted graphs have 21 edges*, Algebr. Geom. Topol. **15** (2015), no. 6, 3305–3322, DOI 10.2140/agt.2015.15.3305. MR3450762
[23] Blake Mellor, *Complete bipartite graphs whose topological symmetry groups are polyhedral*, Tokyo J. Math. **37** (2014), no. 1, 135–158, DOI 10.3836/tjm/1406552435. MR3264518
[24] R. Nikkuni and K. Taniyama, *Symmetries of spatial graphs and Simon invariants*, Fund. Math. **205** (2009), 219–236.

[25] Vladimir R. Rosenfeld, *Looking into the future of molecules with novel topological symmetries*, J. Math. Chem. **57** (2019), no. 7, 1850–1867, DOI 10.1007/s10910-019-01042-z. MR3983458

[26] *SageMath*, https://www.sagemath.org. Accessed: 2024-03-24.

[27] Jonathan Simon, *Topological chirality of certain molecules*, Topology **25** (1986), no. 2, 229–235, DOI 10.1016/0040-9383(86)90041-8. MR837623

[28] P. A. Smith, *Transformations of finite period. II*, Ann. of Math. (2) **40** (1939), 690–711, DOI 10.2307/1968950. MR177

Loyola Marymount University, 1 LMU Drive, Los Angeles, CA 90045
Email address: blake.mellor@lmu.edu

Loyola Marymount University, 1 LMU Drive, Los Angeles, CA 90045
Email address: robin.wilson@lmu.edu

Contemporary Mathematics
Volume **827**, 2025
https://doi.org/10.1090/conm/827/16534

On the cohomology of two stranded braid varieties

Tonie Scroggin

ABSTRACT. We compute the cohomologies of two strand braid varieties using the two-form present in cluster structures. We confirm these results with proof using Alexander and Poincaré duality. Further, we consider products of braid varieties and their interactions with the cohomologies.

CONTENTS

1. Introduction
2. Braid varieties
3. Ring structure on cohomology using (algebraic) deRham cohomology
4. Performing cuts

1. Introduction

In this paper, we will study the relationship between braid varieties and their associated cluster structure in order to compute their cohomologies. Braid varieties are a class of affine algebraic varieties associated to positive braids [**1**–**3**]. Braid varieties are closely related to augmentation varieties of Legendrian links [**4**] and also include interesting geometric spaces such as positroid varieties, open Richardson varieties and double Bruhat cells.

To define the braid variety, we use the braid group on n strands,

$$\mathrm{Br}_n = \langle \sigma_1, \ldots, \sigma_{n-1} : \sigma_i\sigma_{i+1}\sigma_i = \sigma_{i+1}\sigma_i\sigma_{i+1},\ \sigma_i\sigma_j = \sigma_j\sigma_i \text{ if } |i-j| > 1\rangle$$

and restrict to positive crossings σ_i between the i and $i+1$ strand. At each crossing σ_i of the positive braid we assign a complex variable z, see Figure 2, and a matrix

$$B_i(z) := \begin{pmatrix} 1 & \cdots & & & \cdots & 0 \\ \vdots & \ddots & & & & \vdots \\ 0 & \cdots & z & -1 & \cdots & 0 \\ 0 & \cdots & 1 & 0 & \cdots & 0 \\ \vdots & & & & \ddots & \vdots \\ 0 & \cdots & & & \cdots & 1 \end{pmatrix}$$

2020 *Mathematics Subject Classification.* Primary 14M15, 14F40.

This work was partially supported by NSF grant DMS-2302305.

where the 2×2 embedded matrix is at the i and $i+1$ row and column. Let $\beta = \sigma_{i_1} \dots \sigma_{i_k} \in \mathrm{Br}_n^+$ be a positive braid word, then the braid variety $X(\beta)$ is defined by

$$X(\beta) := \left\{ (z_1, \dots, z_k) : \begin{pmatrix} 0 & \dots & 1 \\ \vdots & \iddots & \vdots \\ 1 & \dots & 0 \end{pmatrix} B_{i_1}(z_1) \cdots B_{i_k}(z_k) \text{ is upper-triangular.} \right\}$$

There is an alternative geometric definition of braid varieties using certain configurations of flags, however, it will not be relevant to this paper, see [**3**, **11**] for further information. As indicated by the title, this paper will solely focus on two-strand braid varieties and we denote such a braid with k crossings as σ^k.

THEOREM 1.1 (Hughes[**15**], Chantraine-Ng-Sivek[**5**]). *The braid variety $X(\sigma^k)$ is defined in $\mathbb{C}^k$ by the equation $F_k(z_1, \dots, z_k) = 0$ where F_k is given by the recursion*

$$B_\beta(z_1, \dots, z_k) = \begin{pmatrix} F_k(z_1, \dots, z_k) & -F_{k-1}(z_1, \dots, z_{k-1}) \\ F_{k-1}(z_2, \dots, z_k) & -F_{k-2}(z_2, \dots, z_{k-1}) \end{pmatrix}$$

where

$$F_k(z_i, \dots, z_{i+k}) = z_k F_{k-1}(z_i, \dots, z_{i+k-1}) - F_{k-2}(z_i, \dots, z_{i+k-2}) \tag{1}$$

with initial values $F_1(z_i) = z_i$, $F_0 \equiv 1$ and $F_{-1} \equiv 0$. Moreover, if $F_k(z_1, \dots, z_k) = 0$, then $F_{k-1}(z_1, \dots, z_{k-1}) \neq 0$ and

$$z_k = \frac{F_{k-2}(z_1, \dots, z_{k-2})}{F_{k-1}(z_1, \dots, z_{k-1})}.$$

REMARK 1.2. As a corollary, we have $X(\sigma^k) \cong \{(z_1, \dots, z_{k-1}) \in \mathbb{C}^{k-1} : F_{k-1}(z_1, \dots, z_{k-1}) \neq 0\}$ as algebraic varieties. In particular, $X(\beta)$ is smooth of complex dimension $k-1$.

We construct an explicit isomorphism between the two-strand braid variety and positroid varieties in the Grassmannian $\mathrm{Gr}(2, k+1)$. By Scott [**18**], these admit a cluster structure of type A.

We define the open positroid variety as the set of elements in the Grassmannian such that there is a representative $k \times n$-matrix such that all cyclically consecutive $k \times k$ minors don't vanish, i.e.,

$$\Delta_{i,\dots,i+k-1} = \det(v_i, \dots, v_{i+k-1}) \neq 0.$$

This condition does not depend on the representative, so the positroid is well-defined.

THEOREM 1.3. *Let $\Pi^\circ_{2,k+1}$ be the open positroid variety defined by the condition that all consecutive 2×2 minors do not vanish, i.e., $\Delta_{i,i+1} = \det(v_i, v_{i+1}) \neq 0$, and $\Pi^{\circ,1}_{2,k+1}$ be the subset of the open positroid variety such that for all $\Delta_{i,i+1} = 1$ for all $1 \leq i \leq k$. Then*

a) $\Pi^{\circ,1}_{2,k+1}$ is isomorphic to $X(\sigma^k)$.
b) $\Pi^\circ_{2,k+1}$ is isomorphic to $X(\sigma^k) \times (\mathbb{C}^*)^k$.

One of the main motivations for studying the homologies of braid varieties is their relation to the Khovanov-Rozansky homology of the corresponding link.

THEOREM 1.4 (Trinh [**20**]). *For all r-strand braids $\beta \in Br_W^+$ we have*

$$\mathrm{HHH}^{r,r+j,k}(\beta\Delta)^\vee \simeq \mathrm{gr}^w_{j+2(r-N)} \mathrm{H}^{!,G}_{-(j+k+2(r-N))}(X(\beta)).$$

Equivalently, by Gorsky-Hogancamp-Mellit-Nakagane [**14**],

$$\mathrm{H}^*(X(\beta)) \simeq \mathrm{HHH}^{0,*,*}(\beta\Delta^{-1})^\vee,$$

where Δ is the half-twist (aka longest word). Here gr^w denotes the associated graded with respect to the weight filtration in cohomology.

On two strands this equivalence simplifies to

$$H^*(X(\sigma^k)) \simeq \mathrm{HHH}^{0,*,*}(\sigma^{k-1})$$

where the braid σ^{k-1} closes up to the torus link $T(2, k-1)$.

The cohomology of $X(\sigma^k)$ was computed by Lam and Speyer in [**17**] using cluster algebra machinery. Here we give a simpler and more direct proof.

First, we describe the cohomology of $X(\sigma^k)$ as a vector space.

THEOREM 1.5. *Let $\beta = \sigma^n$, then the cohomology of the two-strand braid variety is given by:*

$$H^i(X(\beta); \mathbb{C}) = \begin{cases} \mathbb{C} & \text{for } 0 \le i \le n-1 \\ 0 & \text{otherwise.} \end{cases}$$

Next, we identify the ring structure in cohomology using algebraic forms with (algebraic) de Rham cohomology. For this, we introduce in Section 3.1 a regular one-form α and a regular two-form ω on $X(\beta)$. We write explicit formulas for α and ω in terms of both z_i and the independent Plücker coordinates in Theorems 4.9 and 3.4.

THEOREM 1.6. *The 1-form α and 2-form ω generate $H^*(X(\sigma^k))$ as a $\mathbb{C}$-algebra, modulo the following relations:*

1) *If k is even, the only relation is $\omega^{\frac{k}{2}} = 0$. The basis in cohomology is given by:*

$$1, \alpha, \omega, \alpha\omega, \ldots, \omega^{\frac{k}{2}-1}, \alpha\omega^{\frac{k}{2}-1}.$$

2) *If k is odd, the relations are $\alpha\omega^{\frac{k-1}{2}} = \omega^{\frac{k+1}{2}} = 0$. The basis in cohomology is given by:*

$$1, \alpha, \omega, \alpha\omega, \ldots, \alpha\omega^{\frac{k-3}{2}}, \omega^{\frac{k-1}{2}}.$$

Next, we study the relation between different two-strand braid varieties. We show that the product of two braid varieties $X(\sigma^a) \times X(\sigma^b)$ can be embedded as an open subset into a larger braid variety $X(\sigma^{a+b-1})$. All such embeddings are parametrized by the diagonals in the $(a+b)$-gon (we refer to them as to diagonal cuts), and we write them explicitly in coordinates.

THEOREM 1.7. *Performing one diagonal cut on P along D_{ij} defines an injective map*

$$\Phi_{ij} : X(\sigma^a) \times X(\sigma^b) \longrightarrow X(\sigma^{a+b-1}).$$

By Theorem 1.3 we identify $X(\sigma^{a+b-1})$ with $\Pi^{o,1}_{2,a+b}$ and the image of the map is the open subset $\{\Delta_{ij} \neq 0\}$ in $\Pi^{o,1}_{2,a+b}$.

We can study the corresponding maps in cohomology of braid varieties.

THEOREM 1.8. *We have*

$$\Phi_{ij}^* \alpha = \alpha_2 + (-1)^{k-j}\alpha_1, \quad \Phi_{ij}^* \omega = \omega_1 + \omega_2 + (-1)^{k-j}\alpha_1 \wedge \alpha_2. \tag{2}$$

The pullback map in cohomology

$$\Phi_{ij}^* : H^*(X(\sigma^k)) \to H^*(X(\sigma^{j-i})) \otimes H^*(X(\sigma^{k-j+i+1}))$$

is injective. and can be described by (2).

THEOREM 1.9. *The map Φ_{ij} defines a quasi-equivalence of cluster varietes $\{\Delta_{ij} \neq 0\} \subset X(\sigma^{a+b-1})$ and $X(\sigma^a) \times X(\sigma^b)$. The latter has a cluster structure obtained by freezing Δ_{ij} in the cluster structure from $X(\sigma^{a+b-1})$.*

REMARK 1.10. By Gorsky and Hogancamp [13], on the level of knot homology, the maps $X(\sigma^a) \times X(\sigma^b) \to X(\sigma^{a+b-1})$ correspond via Theorem 1.4 to the maps

$$\mathrm{HHH}(\sigma^{a-1}) \otimes \mathrm{HHH}(\sigma^{b-1}) \to \mathrm{HHH}(\sigma^{a+b-2})$$

induced by the cobordism between the closures of the corresponding braids $T(2, a-1) \sqcup T(2, b-1)$ and $T(2, a+b-2)$.

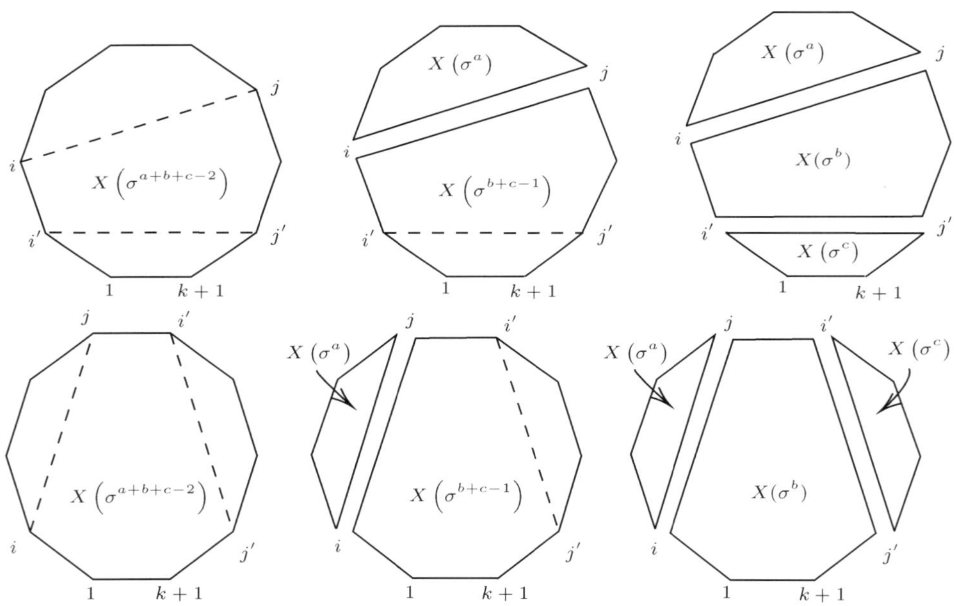

FIGURE 1. Examples of two diagonal cuts. The top is shows a Type A cut and the bottom shows a Type B cut.

Finally, we study the interactions between the maps Φ_{ij} associated to different cuts, which can be thought of as associativity of "gluing" of different braid varieties

$$X(\sigma^a) \times X(\sigma^b) \times X(\sigma^c) \to X(\sigma^{a+b+c-2}).$$

Actually, it happens that there are two different cases which we call "Type A" and "Type B" cuts (see Figure 1 and 7).

THEOREM 1.11. *Performing two diagonal cuts on P along Δ_{ij} and $\Delta_{i'j'}$ we have two commutative diagrams*

(i) *For Type A cuts*

$$\begin{array}{ccc} X(\sigma^a)\times X(\sigma^b)\times X(\sigma^c) & \xrightarrow{\Phi_{ij}\times \mathrm{Id}} & X(\sigma^{a+b-1})\times X(\sigma^c) \\ \Big\downarrow{\scriptstyle \mathrm{Id}\times\Phi_{i'j'}} & & \Big\downarrow{\scriptstyle \Phi_{i'j'}} \\ X(\sigma^a)\times X(\sigma^{b+c-1}) & \xrightarrow[\Phi_{ij}]{} & X(\sigma^{a+b+c-2}) \end{array}$$

(ii) *For Type B cuts*

$$\begin{array}{ccccc} X(\sigma^a)\times X(\sigma^b)\times X(\sigma^c) & \xrightarrow{\mathrm{Id}\times\Phi_{i'j'}} & X(\sigma^a)\times X(\sigma^{b+c-1}) & \xrightarrow{\Phi_{ij}} & X(\sigma^{a+b+c-2}) \\ \Big\downarrow{\scriptstyle \mathrm{Id}\times\mathrm{Id}\times T_{\Delta_{ij}}} & & & & \Big\uparrow{\scriptstyle \Phi_{i'j'}} \\ X(\sigma^a)\times X(\sigma^b)\times X(\sigma^c) & & \xrightarrow{\Phi_{ij}\times\mathrm{Id}} & & X(\sigma^{a+b-1})\times X(\sigma^c) \end{array}$$

Here $T_{\Delta_{ij}}$ preserves $\Pi^{\circ,1}_{2,c+1}$ and defines a $\mathbb{C}^*$ action on $\Pi^{\circ}_{2,c+1}$ and $\Pi^{\circ,1}_{2,c+1}$ as defined in Lemma 4.7 with $\lambda = \Delta_{ij}$. Informally, we can say that the gluing P from smaller polygons is associative only up to the additional transformation $T_{\Delta_{ij}}$.

2. Braid varieties

2.1. Definition of braid varieties. We consider the standard definition of the braid group on n strands, Br_n, given by the presentation

$$\mathrm{Br}_n = \langle \sigma_1,\dots,\sigma_{n-1} : \sigma_i\sigma_{i+1}\sigma_i = \sigma_{i+1}\sigma_i\sigma_{i+1},\ \sigma_i\sigma_j = \sigma_j\sigma_i \text{ if } |i-j|>1\rangle$$

where σ_i is the positive crossing defined by

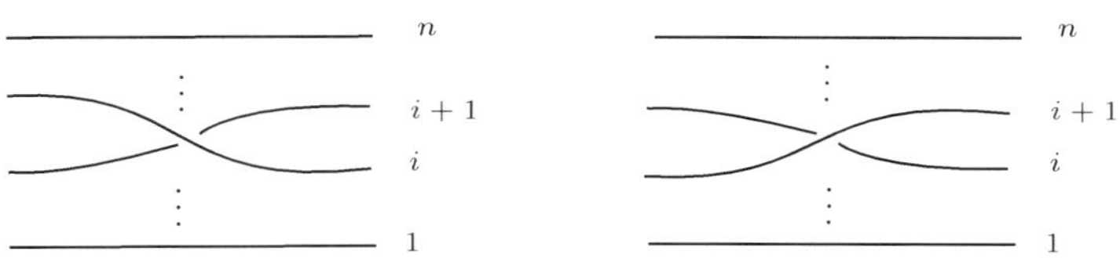

σ_i $\qquad\qquad$ σ_i^{-1}

We consider the positive braid monoid $\mathrm{Br}_n^+ \subseteq \mathrm{Br}_n$ which is generated by the nonnegative powers of the generators σ_i, for $i \in [1, n-1]$. We follow the notations in [**3**].

DEFINITION 2.1. Let $n \in \mathbb{N}$, $i \in [1, n-1] \in \mathbb{N}$ and z a (complex) variable. Then the braid matrix $B_i(z) \in GL(n, \mathbb{C}[z])$ is defined

$$(B_i(z))_{jk} := \begin{cases} 1 & j=k \text{ and } j\neq i, i+1 \\ -1 & (j,k)=(i,i+1) \\ 1 & (i+1,i) \\ z & j=k=i \\ 0 & \text{otherwise} \end{cases}, \quad \text{i.e. } B_i(z) := \begin{pmatrix} 1 & \cdots & & & \cdots & 0 \\ \vdots & \ddots & & & & \vdots \\ 0 & \cdots & z & -1 & \cdots & 0 \\ 0 & \cdots & 1 & 0 & \cdots & 0 \\ \vdots & & & & \ddots & \vdots \\ 0 & \cdots & & & \cdots & 1 \end{pmatrix}$$

Given a positive braid word $\beta = \sigma_{i_1} \cdots \sigma_{i_r} \in \mathrm{Br}_n^+$ and $z_1, \ldots, z_r$ complex variables, define the braid matrix

$$B_\beta(z_1, \ldots, z_r) = B_{i_1}(z_1) \cdots B_{i_r}(z_r) \in GL(n, \mathbb{C}[z_1, \ldots, z_r]).$$

Braid matrices satisfy the braid relations up to a change of variables given as

$$\begin{aligned} B_i(z_1)B_{i+1}(z_2)B_i(z_3) &= B_{i+1}(z_3)B_i(z_1z_3 - z_2)B_{i+1}(z_1), \quad &&\text{for all } i \in [1, n-2] \\ B_i(z_1)B_j(z_z) &= B_j(z_2)B_i(z_1), &&\text{for } |i - j| > 1. \end{aligned}$$

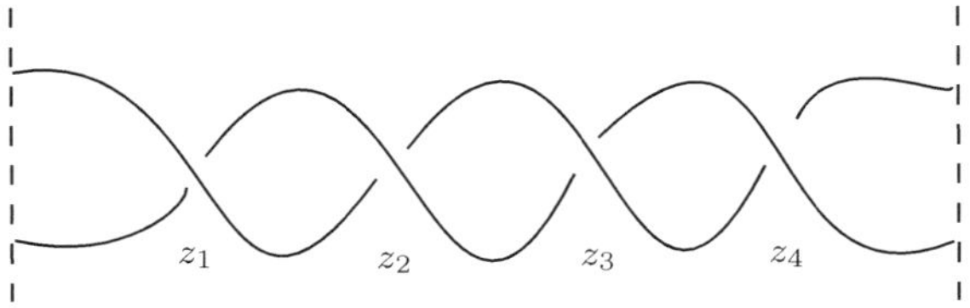

FIGURE 2. The braid σ_1^4 with each crossing j labeled with a complex variable z_j.

This paper concerns only braids on two strands, for the remainder of the paper we will refer to a two strand braid with k crossings as σ^k. The braid matrices on two strands are given by

$$B(z) = \begin{pmatrix} z & -1 \\ 1 & 0 \end{pmatrix}$$

DEFINITION 2.2. The braid variety $X(\sigma^k)$ on two strands is defined by the equation

$$X(\sigma^k) := \left\{ (z_1, \ldots, z_k) : \begin{pmatrix} 0 & -1 \\ 1 & 0 \end{pmatrix} B(z_1) \cdots B(z_k) \text{ is upper-triangular.} \right\}$$

If β and β' are related by braid moves then $X(\beta) \simeq X(\beta')$, this isomorphism arises from the invariance of braid matrices. It is easy to see that the use of -1 does not affect that definition of $X(\beta)$.

To develop a general understanding of these varieties we first consider cases of a small number of crossings.

EXAMPLE 2.3. Let $\beta = \sigma^1 \in \mathrm{Br}_2^+$, the braid matrix is given by

$$B_\beta(z_1) = \begin{pmatrix} z_1 & -1 \\ 1 & 0 \end{pmatrix}$$

Therefore, the braid variety is defined as

$$\begin{aligned} X(\sigma^1) &= \left\{ z_1 : \begin{pmatrix} 0 & -1 \\ 1 & 0 \end{pmatrix} \begin{pmatrix} z_1 & -1 \\ 1 & 0 \end{pmatrix} \text{ is upper-triangular.} \right\} \\ &= \left\{ z_1 : \begin{pmatrix} -1 & 0 \\ z_1 & -1 \end{pmatrix} \text{ is upper triangular} \right\} = \{z_1 = 0\}. \end{aligned}$$

More precisely, $X(\sigma^1)$ is a point.

EXAMPLE 2.4. Let $\beta = \sigma^2 \in \mathrm{Br}_2^+$ with braid matrix

$$B_\beta(z_1, z_2) = \begin{pmatrix} z_1 z_2 - 1 & -z_1 \\ z_2 & -1 \end{pmatrix}$$

then the braid variety associated to β is

$$\begin{aligned} X(\sigma^2) &= \left\{ (z_1, z_2) : \begin{pmatrix} 0 & -1 \\ 1 & 0 \end{pmatrix} \begin{pmatrix} z_1 z_2 - 1 & -z_1 \\ z_2 & -1 \end{pmatrix} \text{ is upper-triangular.} \right\} \\ &= \{(z_1, z_2) \in \mathbb{C}^2 : z_1 z_2 - 1 = 0\} \cong \{z_1 \in \mathbb{C} : z_1 \neq 0\} \end{aligned}$$

It is important to note that the choice of coordinate z_1 on $X(\sigma^2) = \{z_1 \neq 0\}$ is not unique in this case, we may have also chosen $X(\sigma^2) = \{z_2 \neq 0\}$. However, the choice of $X(\sigma^2) = \{z_1 \neq 0\}$ is helpful when developing an inductive way to describe the braid variety in order to compute its cohomology.

EXAMPLE 2.5. Let $\beta = \sigma^3 \in \mathrm{Br}_2^+$ with braid matrix

$$B_\beta(z_1, z_2, z_3) = \begin{pmatrix} z_1 z_2 z_3 - z_3 - z_1 & 1 - z_1 z_2 \\ z_2 z_3 - 1 & -z_2 \end{pmatrix}$$

then the braid variety associated to β is

$$\begin{aligned} X(\sigma^3) &= \left\{ (z_1, z_2, z_3) : \begin{pmatrix} 0 & -1 \\ 1 & 0 \end{pmatrix} \begin{pmatrix} z_1 z_2 z_3 - z_3 - z_1 & 1 - z_1 z_2 \\ z_2 z_3 - 1 & -z_2 \end{pmatrix} \text{ is upper-triangular.} \right\} \\ &= \{(z_1, z_2, z_3) \in \mathbb{C}^3 : z_1 z_2 z_3 - z_3 - z_1 = 0\} \cong \{(z_1, z_2) \in \mathbb{C}^2 : z_1 z_2 - 1 \neq 0\} \end{aligned}$$

There is an inductive relationship between $X(\sigma^k)$ and $X(\sigma^{k-1})$, we explore this concept further by first establishing general formulas for the braid matrices then extending these results to the polynomials defining the braid varieties. Moreover, with these results we show that the braid variety $X(\sigma^k)$ is smooth.

LEMMA 2.6 (Hughes[**15**], Chantraine-Ng-Sivek[**5**]). *One can express the braid matrix for* $\beta = \sigma^k$ *as*

$$B_\beta(z_1, \ldots, z_k) = \begin{pmatrix} F_k(z_1, \ldots, z_k) & -F_{k-1}(z_1, \ldots, z_{k-1}) \\ F_{k-1}(z_2, \ldots, z_k) & -F_{k-2}(z_2, \ldots, z_{k-1}) \end{pmatrix}$$

where

$$F_k(z_i, \ldots, z_{i+k}) = z_{i+k} F_{k-1}(z_i, \ldots, z_{i+k-1}) - F_{k-2}(z_i, \ldots, z_{i+k-2}) \tag{3}$$

with initial values $F_1(z_i) = z_i$, $F_0 \equiv 1$ *and* $F_{-1} \equiv 0$.

PROOF. We proceed with induction on k. Clearly,

$$B_{\sigma^1}(z_1) = \begin{pmatrix} z_1 & -1 \\ 1 & 0 \end{pmatrix} = \begin{pmatrix} F_1(z_1) & -F_0(\emptyset) \\ F_0(\emptyset) & -F_{-1}(\emptyset) \end{pmatrix}$$

Suppose

$$B_{\sigma^k}(z_1, \ldots, z_k) = \begin{pmatrix} F_k(z_1, \ldots, z_k) & -F_{k-1}(z_1, \ldots, z_{k-1}) \\ F_{k-1}(z_2, \ldots, z_k) & -F_{k-2}(z_2, \ldots, z_{k-1}) \end{pmatrix}$$

Then

$$\begin{aligned}B_{\sigma^{k+1}}(z_1,\dots,z_k,z_{k+1}) &= B_{\sigma^k}(z_1,\dots,z_k)B_\sigma(z_{k+1})\\ &= \begin{pmatrix} F_k(z_1,\dots,z_k) & -F_{k-1}(z_1,\dots,z_{k-1})\\ F_{k-1}(z_2,\dots,z_k) & -F_{k-2}(z_2,\dots,z_{k-1})\end{pmatrix}\begin{pmatrix} z_{k+1} & -1\\ 1 & 0\end{pmatrix}\\ &= \begin{pmatrix} z_{k+1}F_k(z_1,\dots,z_k) - F_{k-1}(z_1,\dots,z_{k-1} & -F_{k-1}(z_1,\dots,z_{k-1})\\ z_{k+1}F_{k-1}(z_2,\dots,z_k) - F_{k-2}(z_2,\dots,z_k) & -F_{k-1}(z_2,\dots,z_k)\end{pmatrix}\\ &= \begin{pmatrix} F_{k+1}(z_1,\dots,z_{k+1}) & -F_k(z_1,\dots,z_k)\\ F_k(z_2,\dots,z_k) & -F_{k-1}(z_2,\dots,z_k)\end{pmatrix}\end{aligned}$$

□

THEOREM 2.7 (Hughes [**15**]). *The braid variety $X(\sigma^k)$ is defined in $\mathbb{C}^k$ by the equation $F_k(z_1,\dots,z_k)=0$ where F_k is given by the recursion (3). Moreover, if $F_k(z_1,\dots,z_k)=0$, then $F_{k-1}(z_1,\dots,z_{k-1})\neq 0$ and $z_k = \dfrac{F_{k-2}(z_1,\dots,z_{k-2})}{F_{k-1}(z_1,\dots,z_k)}$.*

PROOF. By Lemma 2.6, we express the braid matrix as

$$B_\beta(z_1,\dots,z_k) = \begin{pmatrix} F_k(z_1,\dots,z_k) & -F_{k-1}(z_1,\dots,z_{k-1})\\ F_{k-1}(z_2,\dots,z_k) & -F_{k-2}(z_2,\dots,z_{k-1})\end{pmatrix}$$

Using the definition for a braid variety, we find that

$$\begin{aligned}X(\sigma^k) &= \left\{(z_1,\dots,z_k) : \begin{pmatrix}0 & -1\\ 1 & 0\end{pmatrix} B_\beta(z_1,\dots,z_k) \text{ is upper-triangular}\right\}\\ &= \{(z_1,\dots,z_k)\in\mathbb{C}^k : F_k(z_1,\dots,z_k)=0\}\end{aligned}$$

Given that $F_n(z_1,\dots,z_k)=0$ and $F_k = z_kF_{k-1} - F_{k-2}$. If $F_{k-1}\neq 0$, then we can solve the equation $F_k=0$ for z_k:

$$F_k = z_kF_{k-1} - F_{k-2} = 0,\ z_k = \frac{F_{k-2}}{F_{k-1}}.$$

Suppose instead that $F_{k-1}(z_1,\dots,z_{k-1})=0$ and given that $F_k(z_1,\dots,z_k)=0$ by the definition of $X(\sigma^k)$, then $F_{k-2}(z_1,\dots,z_{k-2})=0$. By proceeding with downward induction on k, we conclude that $F_k(z_1,\dots,z_k)=0$ for all k, contradicting $F_0=1$. Therefore, $F_{k-1}(z_1,\dots,z_{k-1})\neq 0$. □

COROLLARY 2.8. *We have $X(\sigma^k)\simeq\{(z_1,\dots,z_{k-1}) : F_{k-1}(z_1,\dots,z_{k-1})\neq 0\}$.*

COROLLARY 2.9. *The braid variety $X(\sigma^k)$ is smooth of complex dimension $k-1$.*

PROOF. By Corollary 2.8, $X(\sigma^k) = \{(z_1,\dots,z_{k-1}) : F_{k-1}(z_1,\dots,z_{k-1})\neq 0\}$. Since $\{(z_1,\dots,z_{k-1}) : F_{k-1}(z_1,\dots,z_{k-1})\neq 0\}$ is an open subset in $\mathbb{C}^{k-1}$, then $X(\sigma^k)$ is a smooth manifold. □

2.2. Cohomology using Alexander and Poincaré duality. Given the inductive definition of the two strand braid variety $X(\beta)$ we may determine the homology in terms of the vector space with Alexander and Poincaré duality. Our varieties are non-compact, so we have to be careful and sometimes use cohomology with compact support.

THEOREM 2.10 (Poincaré Duality). *If M is an orientable n-manifold then we have an isomorphism $\widetilde{H}^k_c(M;\mathbb{C})\simeq\widetilde{H}_{n-k}(M;\mathbb{C})$ for all k.*

THEOREM 2.11 (Alexander Duality). *If K is a locally contractible, nonempty, proper subspace of $\mathbb{R}^n$, then $\widetilde{H}_i(\mathbb{R}^n - K;\mathbb{C}) \simeq \widetilde{H}_c^{n-i-1}(K;\mathbb{C})$ for all i.*

The cohomology of two-strand braid varieties was computed in [**17**, Section 6.2, Proposition 9.13] using cluster algebra methods (compare with Theorem 3.5 below). Here we give a simpler inductive proof using Poincaré and Alexander dualities.

THEOREM 2.12. *Let $\beta = \sigma^n$, then the homology of the two-strand braid variety is given by:*

$$H^i(X(\beta)) = \begin{cases} \mathbb{C} & \text{for } 0 \leq i \leq n-1 \\ 0 & \text{otherwise.} \end{cases}$$

PROOF. We proceed by induction on n. Given Corollary 2.8, then

$$H^i(X(\sigma^2)) = H^i(\{z_1z_2 - 1 = 0\}) = H^i(\{z_1 \neq 0\}) = H^i(\mathbb{C}^*)$$

Since $H^i(\mathbb{C}^*) = \mathbb{C}$ for $i = 0, 1$, then the theorem is true for $n = 2$. Supposing the statement holds for $n = k$ we determine that

$$\begin{aligned} \widetilde{H}_i(X(\sigma^{k+1})) &= \widetilde{H}_i(\{F_{k+1} = 0\}) = \widetilde{H}_i(\{F_k \neq 0\}) \text{ (by Corollary 2.8)} \\ &= \widetilde{H}_c^{2k-1-i}(\{F_k = 0\}) \quad \text{(by Theorem 2.11)} = \widetilde{H}_c^{2k-1-i}(X(\sigma^k)) \\ &= \widetilde{H}_{2k-2-(2k-1-i)}(X(\sigma^k)) \text{ (by Theorem 2.10)} = \widetilde{H}_{i-1}(X(\sigma^k)). \end{aligned}$$

Since

$$\widetilde{H}_i(X(\sigma^{k+1})) = \begin{cases} \mathbb{C} & 1 \leq i \leq k+1 \\ 0 & \text{otherwise,} \end{cases}$$

we obtain

$$H_i(X(\sigma^{k+1})) = \begin{cases} \mathbb{C} & 0 \leq i \leq k+1 \\ 0 & \text{otherwise.} \end{cases}$$

□

2.3. The Grassmannian and the open positroid variety. The *Grassmannian* $\mathrm{Gr}(k,n)$ parametrizes all k-dimensional linear subspaces of the n-dimensional space, presented as the row span of a $k \times n$ matrix of maximal rank. Let $v_1, \ldots, v_n$ be the columns of the matrix where v_i are k-dimensional vectors. Given $I \in \binom{[n]}{k}$ the *Plücker coordinate* $\Delta_I(A)$ is the minor of $k \times k$ submatrix of A in column set I.

DEFINITION 2.13 ([**16**]). The open positroid variety $\Pi_{k,n}$ is defined as the set of elements in the Grassmannian such that there is a representative $k \times n$-matrix such that all cyclically consecutive $k \times k$ minors don't vanish

$$\Delta_{i,\ldots,i+k-1} = \det(v_i, \ldots, v_{i+k-1}) \neq 0$$

This condition does not depend on the representative, so the positroid is well-defined.

DEFINITION 2.14. Let $\Pi_{2,n}^{\circ,1}$ be the subset of the the open positroid variety such that each $\Delta_{i,i+1} = 1$ for all $1 \leq i \leq n-1$ and $\Delta_{1,n} \neq 0$.

LEMMA 2.15. *Suppose that $v_1, \ldots, v_{k+1}$ is a collection of vectors in $\mathbb{C}^2$ such that $v_1 = (1,0)$ and $\det(v_i, v_{i+1}) = 1$. Then there exists a unique collection of parameters $z_1, \ldots, z_k$ such that $B(z_1)\cdots B(z_i) = (v_{i+1} \quad -v_i)$ for all i.*

PROOF. Let $v_i = (v_i^1, v_i^2)$, we prove the statement by induction in i. For $i = 1$ we have $v_1 = (1,0)$ and $v_2 = (z,1)$ since $\det(v_i, v_{i+1}) = 1$. For $i > 1$ the vectors v_{i-1}, v_i form a basis of $\mathbb{C}^2$, so we can write $v_{i+1} = \alpha v_{i-1} + \beta v_i$. Now

$$\det(v_i, v_{i+1}) = \alpha \det(v_i, v_{i-1}) + \beta \det(v_i, v_i) = -\alpha \det(v_{i-1}, v_i) + 0 = -\alpha$$

so $\alpha = -1$ and we can denote $z_i = \beta$ and write

$$v_{i+1} = -v_{i-1} + z_i v_i. \tag{4}$$

Now

$$\begin{pmatrix} v_{i+1}^1 & -v_i^1 \\ v_{i+1}^2 & -v_i^2 \end{pmatrix} = \begin{pmatrix} v_i^1 & -v_{i-1}^1 \\ v_i^2 & -v_{i-1}^2 \end{pmatrix} \begin{pmatrix} z_i & -1 \\ 1 & 0 \end{pmatrix}$$

and by assumption of induction we have

$$B(z_1) \cdots B(z_{i-1}) = \begin{pmatrix} v_i^1 & -v_{i-1}^1 \\ v_i^2 & -v_{i-1}^2 \end{pmatrix}.$$

□

REMARK 2.16. Note that $B(z_1) \cdots B(z_i) \begin{pmatrix} 1 \\ 0 \end{pmatrix} = v_{i+1}$.

LEMMA 2.17. *Let $\Pi^{\circ}_{2,k+1}$ and $\Pi^{\circ,1}_{2,k+1}$ be as described in 2.14 and 2.13, then*

a) $\Pi^{\circ,1}_{2,k+1}$ *is isomorphic to* $X(\sigma^k)$.

b) $\Pi^{\circ}_{2,k+1}$ *is isomorphic to* $X(\sigma^k) \times (\mathbb{C}^*)^k$.

PROOF. a) We package the vectors v_k in a $2 \times (k+1)$ matrix V. Since $\Delta_{1,2} = 1$, we can use row operations to make sure that the first column of V is $(1,0)$, so we get

$$V = \begin{pmatrix} 1 & v_2^1 & \cdots & v_{k+1}^1 \\ 0 & v_2^2 & \cdots & v_{k+1}^2. \end{pmatrix} \in \Pi^{\circ,1}_{2,k+1}.$$

By Lemma 2.15 we can uniquely find the variables $z_1, \ldots, z_k$ such that

$$V = \begin{pmatrix} 1 & F_1(z_1) & \cdots & F_k(z_1, \ldots, z_k) \\ 0 & F_0 & \cdots & F_{k-1}(z_2, \ldots, z_k) \end{pmatrix}$$

Note that $\det B_i(z) = 1$, so $\det B_\beta(z_1, \ldots, z_i) = 1$ for any braid β and

$$F_i(z_1, \ldots, z_i) F_i(z_2, \ldots, z_{i+1}) - F_{i+1}(z_1, \ldots, z_{i+1}) F_{i-1}(z_2, \ldots, z_i) = 1, \tag{5}$$

so the matrix V indeed satisfies $\Delta_{i,i+1}(V) = 1$. The matrix V belongs to $\Pi^{\circ,1}_{2,k}$ if and only if $F_{k-1}(z_2, \ldots, z_k) \neq 0$. In this case, we can use row operations to ensure that $F_k(z_1, \ldots, z_k) = 0$ (we subtract from the first row $F_k(z_1, \ldots, z_k)/F_{k-1}(z_2, \ldots, z_k)$ times the second row).

The braid variety $X(\sigma^k)$ is cut out by the equation $\{(z_1, \ldots, z_k) \in \mathbb{C}^k : F_k(z_1, \ldots, z_k) = 0\}$, so we get a map from $\Pi^{\circ,1}_{2,k}$ to $X(\sigma^k)$. To construct the inverse, observe that $F_k(z_1, \ldots, z_k) = 0$ and (5) implies that

$$F_{k-1}(z_1, \ldots, z_{k-1}) F_{k-1}(z_1, \ldots, z_k) = 1,$$

so $F_{k-1}(z_2, \ldots, z_k) \neq 0$. Therefore $\Pi^{\circ,1}_{2,k} \simeq X(\sigma^k)$.

b) Similarly to the above, we can use row operations to ensure any matrix in $\Pi^{\circ}_{2,k+1}$ has the first column $(1,0)$. Now we define a map $\Pi^{\circ,1}_{2,k+1} \times (\mathbb{C}^*)^k \to \Pi^{\circ}_{2,k+1}$ by rescaling all other columns:

$$\varphi : [(v_1, v_2, \ldots, v_{k+1}), (\lambda_1, \ldots, \lambda_k)] \mapsto (v_1, \lambda_1 v_2, \ldots, \lambda_k v_{k+1}).$$

The inverse map is clear, since we get

$$\det(\lambda_{i-1}v_i, \lambda_i v_{i+1}) = \lambda_{i-1}\lambda_i,$$

and the scalars λ_i can recovered from the minors $\Delta_{i,i+1}$ for the image of φ. □

EXAMPLE 2.18. We have

$$B(z_1)B(z_2) = \begin{pmatrix} z_1 & -1 \\ 1 & 0 \end{pmatrix}\begin{pmatrix} z_2 & -1 \\ 1 & 0 \end{pmatrix} = \begin{pmatrix} z_1z_2 - 1 & -z_1 \\ z_2 & -1 \end{pmatrix}$$

$$B(z_1)B(z_2)B(z_3) = \begin{pmatrix} z_1z_2 - 1 & -z_1 \\ z_2 & -1 \end{pmatrix}\begin{pmatrix} z_3 & -1 \\ 1 & 0 \end{pmatrix} = \begin{pmatrix} z_1z_2z_3 - z_1 - z_3 & 1 - z_1z_2 \\ z_2z_3 - 1 & -z_2 \end{pmatrix}.$$

This means that we can package v_i in a matrix

$$\begin{pmatrix} v_1 & v_2 & v_3 & v_4 \end{pmatrix} = \begin{pmatrix} 1 & z_1 & z_1z_2 - 1 & z_1z_2z_3 - z_1 - z_3 \\ 0 & 1 & z_2 & z_2z_3 - 1. \end{pmatrix}$$

2.4. Cluster algebras. Cluster algebras are commutative rings that are not defined in the typical sense by generators and relations, instead it is defined by a seed s which consists of a quiver, or exchange matrix, and cluster variables, which is a finite collection of algebraically independent elements of the algebra. This seed along with a concept of mutation generates a subring of a field $\mathcal{F}$. We refer to [**21**] for more details on cluster algebras.

A cluster variety is an affine algebraic variety X defined by a collection of open charts $U \simeq (\mathbb{C}^*)^d$ where each chart U is parametrized by cluster coordinates $A_1, \ldots, A_d$ which are invertible on U and extend to regular functions on X. These coordinates can be either mutable or frozen where the coordinate is frozen if it extends to a non-vanishing regular function on X.

For each chart we assign a skew-symmetric integer matrix ε_{ij} called the exchange matrix to a quiver Q defined by

$$(\varepsilon_{ij}) = \begin{cases} a & \text{if there are } a \text{ arrows from vertex } i \text{ to vertex } j; \\ -a & \text{if there are } a \text{ arrows from vertex } j \text{ to vertex } i; \\ 0 & \text{otherwise} \end{cases}$$

For each chart U and each mutable variable A_k, there is another chart U' with cluster coordinates $A_1, \ldots, A'_k, \ldots, A_d$ and a skew-symmetric matrix ε'_{ij} related by mutation μ_k, where the mutation is defined by

$$A'_kA_k = \left(\prod_{\varepsilon_{ki}\geq 0} A_i^{\varepsilon_{ki}} + \prod_{\varepsilon_{ki}\leq 0} A_i^{-\varepsilon_{ki}}\right) \tag{6}$$

If $i \neq k$ then the cluster variables A_i remain unchanged.

When performing a mutation, we modify the quiver using the following rules:

(1) If there is a path of the vertices $i \to k \to j$, then we add an arrow from i to j.
(2) Any arrows incident to k change orientation.
(3) Remove a maximal disjoint collection of 2-cycles produced in Steps (1) and (2).

Any two charts in the cluster algebra are related by a sequence of mutations μ, and μ_k is an involution. Given these conditions the ring of functions on X is generated by all cluster variables in all charts.

We will need the notion of exchange ratios defined as the ratio of two terms in (6):

$$\widehat{y_i} = \frac{\prod_{\varepsilon_{ki}\geq 0} A_i^{\varepsilon_{ki}}}{\prod_{\varepsilon_{ki}\leq 0} A_i^{-\varepsilon_{ki}}}.$$

Let V be a rational affine algebraic variety with algebra of regular functions $\mathbb{C}[V]$ and field of rational functions $\mathbb{C}(V)$.

DEFINITION 2.19 ([**6**,**7**]). Let Σ and Σ_0 be seeds of rank r in $\mathbb{C}(V)$. Let $Q, A_i, \hat{y}_i$ denote the quiver, cluster variables and exchange ratios in Σ and use primes to denote these quantities in Σ_0. We assume that $A_{r+1}, \ldots, A_d$ are frozen. Then Σ and Σ_0 are quasi-equivalent, denoted $\Sigma \sim \Sigma_0$, if the following hold:

- The groups $\mathbf{P}, \mathbf{P}_0 \subset \mathbb{C}[V]$ of Laurent monomials in frozen variables coincide. That is, each frozen variable A_i' is a Laurent monomial in $\{A_{r+1}, \ldots, A_d\}$ and vice versa.
- Corresponding mutable variables coincide up to multiplication by an element of $\mathbf{P}$: for $i \in [r]$, there is a Laurent monomial $M_i \in \mathbf{P}$ such that $A_i = M_i A_i' \in \mathbb{C}(V)$.
- The exchange ratios (3) coincide: $\hat{y}_i = \hat{y}_i'$ for $i \in [r]$.

Quasi-equivalence is an equivalence relation on seeds. Seeds Σ, Σ_0 are related by a quasi-cluster transformation if there exists a finite sequence $\underline{\mu}$ of mutations such that $\underline{\mu}(\Sigma) \sim \Sigma_0$.

By the main result of [**6**], it is sufficient to check the conditions of quasi-equivalence in one cluster, and they will automatically hold in every other cluster.

THEOREM 2.20 (Scott [**18**], Galashin–Lam [**9**], Serhiyenko–Sherman-Bennett–Williams [**19**]). *Any open positroid variety has a cluster structure.*

For positroid varieties $\Pi^\circ_{2,k+1}$ we obtain a cluster variety of type A_{k-1} with $k+1$ frozen variables. We asssign the vectors v_i from Lemma 2.17 to the vertices of a regular polygon $\mathcal{P}$. The cluster charts in $\Pi^\circ_{2,k+1}$ are determined by triangulations of $\mathcal{P}$. Given a triangulation, the edges between the vertices i and j correspond to cluster variables determined by the Plücker coordinates $\Delta_{i,j} = \det(v_i, v_j)$.

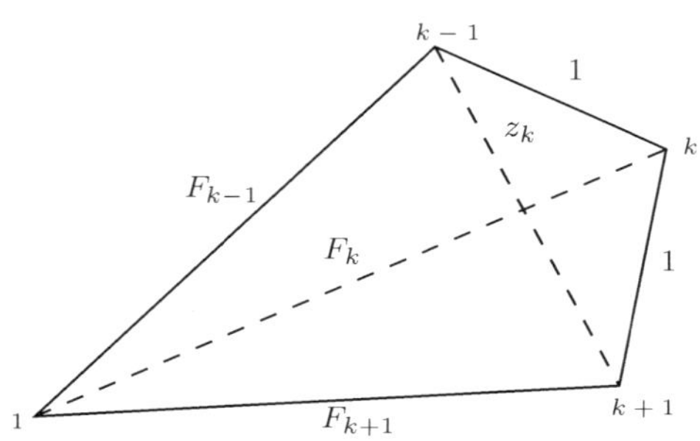

FIGURE 3. Section of the triangulation of U_{fan}, see Figure 4, between the vertices $1, k-1, k$ and $k+1$

LEMMA 2.21 (Hughes[**15**]). *In* $\Pi^{\circ,1}_{2,k+1}$ *for all* $i<j$ *we have*

$$\Delta_{i,j} = F_{j-i-1}(z_{i+1},\dots,z_{j-1}).$$

In particular, $\Delta_{i,i+2} = z_{i+1}$.

PROOF. Using the results from Lemma 2.15, we have the following relations

$$B(z_1)\dots B(z_i) = \begin{pmatrix} v_{i+1} & -v_i \end{pmatrix}$$
$$B(z_1)\dots B(z_j) = \begin{pmatrix} v_{j+1} & -v_j \end{pmatrix}$$

Given that $i<j$, we then rewrite

$$B(z_1)\dots B(z_i)B(z_{i+1})\dots B(z_j) = \begin{pmatrix} v_{j+1} & -v_j \end{pmatrix}$$
$$\begin{pmatrix} v_{i+1} & -v_i \end{pmatrix} B(z_{i+1})\dots B(z_j) = \begin{pmatrix} v_{j+1} & -v_j \end{pmatrix}$$

From Theorem 2.6, the product of the braid matrices from $i+1$ to j can be expressed as

$$B(z_{i+1})\dots B(z_j) = \begin{pmatrix} F_{j-i}(z_{i+1},\dots,z_j) & -F_{j-i-1}(z_{i+1},\dots,z_{j-1}) \\ F_{j-i-1}(z_{i+2},\dots,z_j) & -F_{j-i-2}(z_{i+2},\dots,z_{j-1}) \end{pmatrix}$$

Which allows us to rewrite the previous equation as

$$\begin{pmatrix} v_{i+1} & -v_i \end{pmatrix}\begin{pmatrix} F_{j-i}(z_{i+1},\dots,z_j) & -F_{j-i-1}(z_{i+1},\dots,z_{j-1}) \\ F_{j-i-1}(z_{i+2},\dots,z_j) & -F_{j-i-2}(z_{i+2},\dots,z_{j-1}) \end{pmatrix} = \begin{pmatrix} v_{j+1} & -v_j \end{pmatrix}$$

Here we obtain the equation

$$-v_j = -F_{j-i-1}(z_{i+1},\dots,z_{j-1})v_{i+1} + F_{j-i-2}(z_{i+2},\dots,z_{j-1})v_i$$

By finding an expression for v_j, we may now determine Δ_{ij}, since determinants are linear, we find that

$$\begin{aligned}\Delta_{ij} &= \det\begin{pmatrix} v_i & v_j \end{pmatrix} \\ &= F_{j-i-1}(z_{i+1},\dots,z_{j-1})\det\begin{pmatrix} v_i & v_{i+1} \end{pmatrix} - F_{j-i-2}(z_{i+2},\dots,z_{j-1})\det\begin{pmatrix} v_i & v_i \end{pmatrix} \\ &= F_{j-i-1}(z_{i+1},\dots,z_{j-1})(1) - F_{j-i-2}(z_{i+2},\dots,z_{j-1})(0) \\ &= F_{j-i-1}(z_{i+1},\dots,z_{j-1})\end{aligned}$$

To see that $\Delta_{i,i+2} = z_{i+1}$, we see that $\Delta_{ij} = F_{(i+2)-i-1}(z_{i+1}) = F_1(z_{i+1}) = z_{i+1}$ as desired. □

For $a<b<c<d$ we have the Plücker relation

$$\Delta_{ac}\Delta_{bd} = \Delta_{ab}\Delta_{cd} + \Delta_{ad}\Delta_{bc}. \tag{7}$$

A special case of (7) is

$$\Delta_{i,k}\Delta_{k-1,k+1} = \Delta_{i,k-1}\Delta_{k,k+1} + \Delta_{i,k+1}\Delta_{k-1,k}$$

which in $\Pi^{\circ,1}_{2,k+1}$ translates to

$$\Delta_{i,k}z_k = \Delta_{i,k-1} + \Delta_{i,k+1}$$

For $i=1$ it is indeed equivalent to our recursion (3), see Figure 3

Outer edges of $\mathcal{P}$ correspond to frozen variables, while diagonals correspond to mutable variables. In particular, $\Pi^{\circ}_{2,k+1}$ has k frozen variables, whereas in $\Pi^{\circ,1}_{2,k+1}$ we specialize the frozen Plücker coordinates, $\Delta_{i,i+1} = 1$ for $1\le i\le k$ and these can be neglected. Thus $\Pi^{\circ,1}_{2,k+1}$ has one frozen variable $\Delta_{1,k+1}$ which we denote by w. To generate the quiver, in each triangle of the triangulation we connect the

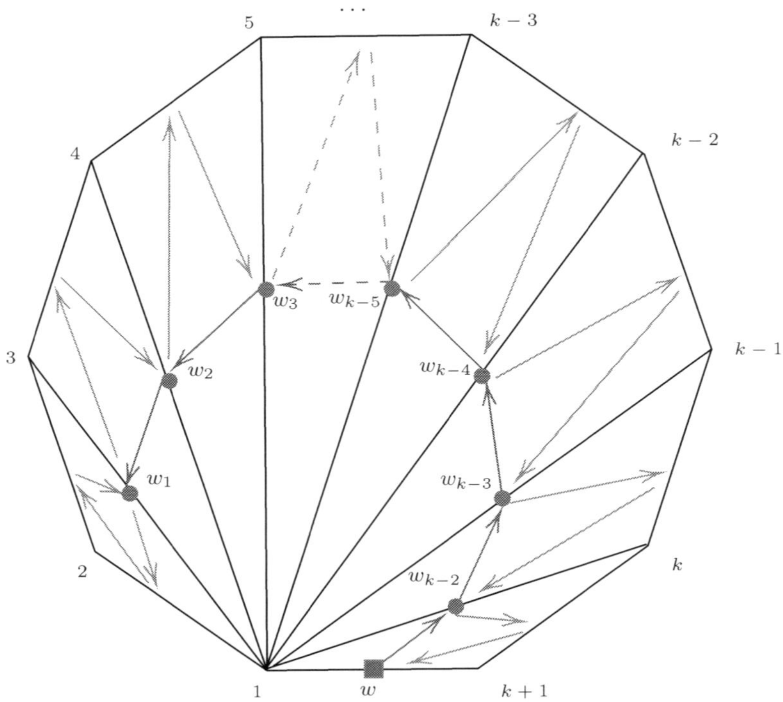

FIGURE 4. The special chart $\mathrm{U}_{\text{fan}} \in \Pi^{\circ,1}_{2,k+1}$ where each of the $k-2$ diagonals are have fixed endpoint at v_1. The Plučker coordinates, or cluster variables, correspond to the weights of the edges given by either a blue square (frozen vertices) or a green circle (mutable vertices). The quiver of the cluster chart is generated by clockwise orientation of the colored arrows in each triangle of the triangulation. This procedure produces the quiver A_{k-1}, seen in purple, with w as the singular frozen variable. In the terminology of [**3**], this chart is given by the right inductive weave.

cluster variables by arrows in a clockwise order. Mutations correspond to flips of triangulations due to the Plucker relation.

Consider the special chart U_{fan} in $\Pi^{\circ,1}_{2,k+1}$ corresponding to the "fan" triangulation where the $k-2$ diagonals are defined by $\Delta_{1,i}$ for $2 \leq i \leq k$, as seen in Figure 4. Equivalently, the chart U_{fan} is given by inequalities

$$\mathrm{U}_{\text{fan}} = \{F_{i-1}(z_2, \ldots, z_i) \neq 0, 1 \leq i \leq k\} \subset X(\sigma^k).$$

In this chart, the quiver is precisely A_{k-1} with one frozen variable w. From lemma the mutable cluster variables are precisely $w_i = F_i(z_2, \ldots, z_{i+1})$ and the frozen variable is $w = w_{k-2} = F_k(z_2, \ldots, z_{k+1})$.

3. Ring structure on cohomology using (algebraic) deRham cohomology

3.1. Constructing the forms. Define the one-form $\alpha = \frac{dw}{w}$ where $w = \Delta_{1,k+1}$ is the frozen cluster variable. Since $w \neq 0$ everywhere, α is regular everywhere.

Define the two-form as

$$\omega = \sum \varepsilon_{ij} \frac{dw_i}{w_i} \wedge \frac{dw_j}{w_j} \tag{8}$$

on some cluster chart with quiver (ε_{ij}). By [**12**, Section 2.3] (see also [**17**]) the form ω is well-defined in any other cluster chart and is given by a similar equation (8) for the mutated quiver. The cluster charts cover $X(\sigma^k)$ up to codimension 2 and $X(\sigma^k)$ is smooth, so ω extends to a regular form on $X(\sigma^k)$.

For the special chart $\mathrm{U}_{\mathrm{fan}}$ we get

$$\omega = \frac{dw}{w} \wedge \frac{dw_{k-2}}{w_{k-2}} + \sum_{i=1}^{k-3} \frac{dw_{i+1}}{w_{i+1}} \wedge \frac{dw_i}{w_i} \tag{9}$$

where $w_i = \Delta_{1,i+2}$.

We can also write the forms α and ω explicitly in the coordinates z_i. Thus far, we have expressed $X(\sigma^k)$ as an open subset in the affine space with coordinates $z_1, \ldots, z_{k-1}$ with z_k expressed as some function of these. Similarly, we may also have expressed $X(\sigma^k)$ is an open subset in the affine space with coordinates $z_2, \ldots, z_k$ with z_1 expressed as some function of these, i.e., $F_k(z_1, \ldots, z_k) = z_1 F_{k-1}(z_2, \ldots, z_k) - F_{k-2}(z_3, \ldots, z_k)$ where $F_{-1} \equiv 0$, $F_0 \equiv 1$, and $F_1(z_2) = z_2$. We will use $z_2, \ldots, z_k$ as a coordinate system on $X(\sigma^k)$ below.

LEMMA 3.1. *For all* $2 \leq i \leq k$ *and* $2 \leq n \leq k+1$ *we have*

$$\frac{\partial \Delta_{1n}}{\partial z_i} = \Delta_{1i}\Delta_{in}.$$

PROOF. We have the matrix identity

$$\begin{pmatrix} F_n(z_1, \ldots, z_n) & -F_{n-1}(z_1, \ldots, z_{n-1}) \\ F_{n-1}(z_2, \ldots, z_n) & -F_{n-2}(z_2, \ldots, z_{n-1}) \end{pmatrix} = C \begin{pmatrix} z_i & -1 \\ 1 & 0 \end{pmatrix} \widetilde{C}$$

where

$$C = \begin{pmatrix} F_{i-1}(z_1, \ldots, z_{i-1}) & -F_{i-2}(z_1, \ldots, z_{i-2}) \\ F_{i-2}(z_2, \ldots, z_{i-1}) & -F_{i-3}(z_2, \ldots, z_{i-2}), \end{pmatrix}$$

$$\widetilde{C} = \begin{pmatrix} F_{n-i}(z_{i+1}, \ldots, z_n) & -F_{n-i-1}(z_{i+2}, \ldots, z_{n-1}) \\ F_{n-i-1}(z_{i+2}, \ldots, z_n) & -F_{n-i-2}(z_{i+2}, \ldots, z_{n-1}), \end{pmatrix}$$

which implies

$$\begin{aligned} &F_{n-2}(z_2, \ldots, z_{n-1}) \\ &\quad = (F_{i-2}(z_2, \ldots, z_{i-1})z_i - F_{i-3}(z_2, \ldots, z_{i-2}))F_{n-i-1}(z_{i+2}, \ldots, z_{n-1}) \\ &\qquad - F_{i-2}(z_2, \ldots, z_{i-1})F_{n-i-2}(z_{i+2}, \ldots, z_{n-1}) \end{aligned}$$

and

$$\frac{\partial F_{n-2}(z_2, \ldots, z_{n-1})}{\partial z_i} = F_{i-2}(z_2, \ldots, z_{i-1})F_{n-i-1}(z_{i+2}, \ldots, z_{n-1}).$$

Now by Lemma 2.21 we have $\Delta_{1,n} = F_{n-2}(z_2, \ldots, z_{n-1})$ and

$$\frac{\partial \Delta_{1,n}}{\partial z_i} = F_{i-2}(z_2, \ldots, z_{i-1}) F_{n-i-1}(z_{i+2}, \ldots, z_{n-1}) = \Delta_{1,i}\Delta_{i,n}. \qquad \square$$

COROLLARY 3.2. *We have*

$$\alpha = \frac{d\Delta_{1,k+1}}{\Delta_{1,k+1}} = \frac{1}{\Delta_{1,k+1}} \sum_{i=2}^{k} \Delta_{1,i}\Delta_{i,k+1} dz_i.$$

LEMMA 3.3. *For $i \leq k$ we have*

$$\frac{1}{\Delta_{1,i}\Delta_{1,i+1}} + \ldots + \frac{1}{\Delta_{1,k}\Delta_{1,k+1}} = \frac{\Delta_{i,k+1}}{\Delta_{1,i}\Delta_{1,k+1}}.$$

PROOF. We prove it by induction in k, for $k = i$ the statement is clear since $\Delta_{i,i+1} = 1$. For the step of induction, suppose that it is true for $k-1$, then

$$\begin{aligned}\frac{1}{\Delta_{1,i}\Delta_{1,i+1}} + \ldots + \frac{1}{\Delta_{1,k-1}\Delta_{1,k}} + \frac{1}{\Delta_{1,k}\Delta_{1,k+1}} &= \frac{\Delta_{i,k}}{\Delta_{1,i}\Delta_{1,k}} + \frac{1}{\Delta_{1,k}\Delta_{1,k+1}} \\ &= \frac{\Delta_{i,k}\Delta_{1,k+1} + \Delta_{1,i}\Delta_{k,k+1}}{\Delta_{1,i}\Delta_{1,k}\Delta_{1,k+1}},\end{aligned}$$

which by Plücker relation simplifies to

$$\frac{\Delta_{1,k}\Delta_{i,k+1}}{\Delta_{1,i}\Delta_{1,k}\Delta_{1,k+1}} = \frac{\Delta_{i,k+1}}{\Delta_{1,i}\Delta_{1,k+1}}. \qquad \square$$

LEMMA 3.4. *We have*

$$\omega = \frac{1}{\Delta_{1,k+1}} \sum_{2 \leq i < j \leq k} \Delta_{1,i}\Delta_{i,j}\Delta_{j,k+1} dz_i \wedge dz_j.$$

PROOF. By Lemma 3.1 we can write

$$\begin{aligned}d\Delta_{1,s} \wedge d\Delta_{1,s+1} &= \sum_{i<j\leq s} (\Delta_{1,i}\Delta_{i,s}\Delta_{1,j}\Delta_{j,s+1} - \Delta_{1,i}\Delta_{i,s+1}\Delta_{1,j}\Delta_{j,s}) dz_i \wedge dz_j \\ &= \sum_{i<j\leq s} \Delta_{1,i}\Delta_{1,j}(\Delta_{i,s}\Delta_{j,s+1} - \Delta_{i,s+1}\Delta_{j,s}) dz_i \wedge dz_j.\end{aligned}$$

By Plücker relation we have

$$\Delta_{i,s}\Delta_{j,s+1} - \Delta_{i,s+1}\Delta_{j,s} = \Delta_{ij},$$

hence

$$d\Delta_{1,s} \wedge d\Delta_{1,s+1} = \sum_{i<j\leq s} \Delta_{1,i}\Delta_{1,j}\Delta_{i,j} dz_i \wedge dz_j.$$

The coefficient at $dz_i \wedge dz_j$ does not depend on k, so we get

$$\begin{aligned}\omega &= \sum_{s=1}^{k} \frac{d\Delta_{1,s} \wedge d\Delta_{1,s+1}}{\Delta_{1,s}\Delta_{1,s+1}} \\ &= \sum_{i<j} \Delta_{1,i}\Delta_{1,j}\Delta_{i,j} dz_i \wedge dz_j \left(\frac{1}{\Delta_{1,j}\Delta_{1,j+1}} + \ldots + \frac{1}{\Delta_{1,k}\Delta_{1,k+1}} \right).\end{aligned}$$

By Lemma 3.3 this simplifies to

$$\sum_{i<j} \frac{\Delta_{1,i}\Delta_{1,j}\Delta_{i,j}\Delta_{j,k+1} dz_i \wedge dz_j}{\Delta_{1,j}\Delta_{1,k+1}} = \frac{\sum_{i<j} \Delta_{1,i}\Delta_{i,j}\Delta_{j,k+1} dz_i \wedge dz_j}{\Delta_{1,k+1}}. \qquad \square$$

In particular, Lemma 3.4 gives a direct proof that ω is regular everywhere on $X(\sigma^k)$. See Section 3.3 for explicit examples and computations.

3.2. de Rham cohomology. By construction, $d\alpha = d\omega = 0$, so they represent some de Rham cohomology classes. The following theorem shows that these are in fact nonzero in cohomology and generate $H^*(X(\sigma^k))$ as an algebra.

THEOREM 3.5. *The forms α and ω generate $H^*(X(\sigma^k))$ as an algebra, modulo the following relations:*

1) If k is even, the only relation is $\omega^{\frac{k}{2}} = 0$. The basis in cohomology is given by:

$$1, \alpha, \omega, \alpha\omega, \ldots, \omega^{\frac{k}{2}-1}, \alpha\omega^{\frac{k}{2}-1}. \tag{10}$$

2) If k is odd, the relations are $\alpha\omega^{\frac{k-1}{2}} = \omega^{\frac{k+1}{2}} = 0$. The basis in cohomology is given by:

$$1, \alpha, \omega, \alpha\omega, \ldots, \alpha\omega^{\frac{k-3}{2}}, \omega^{\frac{k-1}{2}}. \tag{11}$$

PROOF. We work in the chart $\mathrm{U_{fan}}$, there is a natural inclusion map $i : \mathrm{U_{fan}} \to X(\sigma^k)$ and the corresponding restriction map in cohomology: $i^* : H^*(X(\sigma^k)) \to H^*(\mathrm{U_{fan}})$.

We want to first prove that the restrictions of all the forms (10) and (11) to $H^*(\mathrm{U_{fan}})$ do not vanish, this would imply that these forms do not vanish in $H^*(X(\sigma^k))$. Recall that $\mathrm{U_{fan}} \simeq (\mathbb{C}^*)^{k-1}$ with coordinates $w_1, \ldots, w_{k-2}, w = w_{k-1}$, so $H^*(\mathrm{U_{fan}})$ is isomorphic to an exterior algebra in $\frac{dw_i}{w_i}$.

Suppose k is odd then

$$\begin{aligned}\omega^{\frac{k-1}{2}} &= \left(\frac{dw_2}{w_2} \wedge \frac{dw_1}{w_1} + \ldots + \frac{dw}{w} \wedge \frac{dw_{k-2}}{w_{k-2}}\right)^{(k-1)/2} \\ &= (k-1)/2)! \frac{dw_1}{w_1} \wedge \cdots \wedge \frac{dw_{k-2}}{w_{k-2}} \wedge \frac{dw}{w}\end{aligned}$$

and

$$\begin{aligned}\alpha\omega^{\frac{k-3}{2}} &= \frac{dw}{w} \wedge \left(\frac{dw_2}{w_2} \wedge \frac{dw_1}{w_1} + \ldots + \frac{dw}{w} \wedge \frac{dw_{k-2}}{w_{k-2}}\right)^{(k-3)/2} \\ &= \frac{dw}{w} \wedge \left(\frac{dw_2}{w_2} \wedge \frac{dw_1}{w_1} + \ldots + \frac{dw_{k-2}}{w_{k-2}} \wedge \frac{dw_{k-3}}{w_{k-3}}\right)^{(k-3)/2} \\ &= \frac{dw}{w} \wedge ((k-3)/2)! \sum_{j=0}^{(k-5)/2} \frac{dw_1}{w_1} \wedge \ldots \widehat{\frac{dw_{2j+1}}{w_{2j+1}}} \cdots \wedge \frac{dw_{k-3}}{w_{k-3}}\end{aligned}$$

In particular, these are nonzero. Suppose k is even, then similarly

$$\omega^{\frac{k}{2}-1} = \sum_{j=0}^{k/2-2} \frac{dw_1}{w_1} \wedge \frac{dw_2}{w_2} \wedge \cdots \wedge \widehat{\frac{dw_{2j+1}}{w_{2j+1}}} \wedge \cdots \wedge \frac{dw_{k-2}}{w_{k-2}} \wedge \frac{dw}{w}.$$

and

$$\alpha\omega^{\frac{k-3}{2}} = ((k-1)/2)! \frac{dw_1}{w_1} \wedge \frac{dw_2}{w_2} \wedge \cdots \wedge \frac{dw_{k-3}}{w_{k-3}} \wedge \frac{dw_{k-2}}{w_{k-2}} \wedge \frac{dw}{w}$$

This implies that all the forms in (10) and (11) are nonzero in $H^*(\mathrm{U}_{\mathrm{fan}})$ and hence nonzero in $H^*(X(\sigma^k))$. On the other hand, by Theorem 2.12 the corresponding cohomology groups of $X(\sigma^k)$ are one-dimensional in each degree; therefore, we obtain a basis. □

3.3. Examples.

EXAMPLE 3.6. Braid variety associated to $\beta = \sigma^3$

$$\begin{aligned} X(\sigma^3) &= \{z_1z_2z_3 - z_3 - z_1 = 0\} \\ &= \{z_1z_2 - 1 \neq 0\} \end{aligned}$$

Using row operations and scaling the columns, we can transform any matrix in $\Pi^\circ_{2,4}$ to the form

$$V = \begin{pmatrix} 1 & z_1 & z_1z_2 - 1 & z_1z_2z_3 - z_1 - z_3 \\ 0 & 1 & z_2 & z_2z_3 - 1. \end{pmatrix} \in \Pi^{\circ,1}_{2,4}$$

Using the correspondence of cluster algebras and Grassmannians, we obtain two cluster charts, as seen in Figure 5.

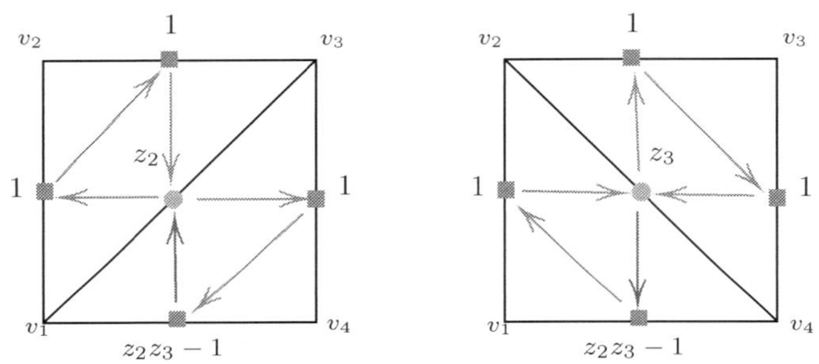

FIGURE 5. The two cluster charts for the braid variety $X(\sigma^3)$. On the left is chart U_1 where the vectors $v_i \in \Pi^{\circ,1}_{2,4}$ for $1 \leq i \leq 4$ correspond to the vertices of the polygon. The purple arrow depicts the Dynkin diagram A_1 with a frozen. On the right is chart U_2 which corresponds to the mutation of chart U_1.

Here

$$\begin{aligned} U_1 &= \{z_2 \neq 0\} \text{ with coordinates } (w_1 = z_2,\ w = z_2z_3 - 1) \\ U_2 &= \{z_3 \neq 0\} \text{ with coordinates } (w_1' = z_3,\ w = z_2z_3 - 1) \end{aligned}$$

We compute the cohomology of $X(\beta)$ using the (algebraic) de Rham cohomology on chart 1. Let $U_1 = \{w_1 = z_2 \neq 0,\ w = z_2z_3 - 1 \neq 0\}$. Then all possible forms

$$H^*(U_1) = H^*\left((\mathbb{C}^*)^2\right) = \left\langle 1, \frac{dw_1}{w_1}, \frac{dw}{w}, \frac{dw}{w} \wedge \frac{dw_1}{w_1} \right\rangle.$$

To determine the cohomology, it suffices to determine which of the above forms extend to $X(\sigma^3)$. The forms which extend are

- 1
- $\dfrac{dw}{w} = \dfrac{z_2dz_3 + z_3dz_2}{z_2z_3 - 1}$
- $\dfrac{dw}{w} \wedge \dfrac{dw_1}{w_1} = \dfrac{dz_3 \wedge dz_2}{z_2z_3 - 1}$

The 2-form can be deduced from the quiver shown in Figure 5 which agrees with [**17**]. Therefore, $H^0(X(\sigma^3)) = H^1(X(\sigma^3)) = H^2(X(\sigma^3)) = \mathbb{C}$, which agrees with Theorem 2.12.

In addition, on the chart $U_2 = \{w_1' = z_3 \neq 0,\ w = z_2z_3 - 1 \neq 0\}$, with possible forms

$$H^*(U_2) = H^*\big((\mathbb{C}^*)^2\big) = \left\langle 1, \frac{dw_1'}{w_1'}, \frac{dw}{w}, \frac{dw}{w} \wedge \frac{dw_1'}{w_1'} \right\rangle$$

the forms which extend are

- 1
- $\dfrac{dw}{w} = \dfrac{z_2dz_3 + z_3dz_2}{z_2z_3 - 1}$
- $\dfrac{dw}{w} \wedge \dfrac{dw_1'}{w_1'} = \dfrac{dz_2 \wedge dz_3}{z_2z_3 - 1}$

Indeed, the cohomology of $X(\sigma^3)$ is independent from the choice of a chart.

EXAMPLE 3.7. The braid variety associated to $\beta = \sigma^4$

$$\begin{aligned} X(\sigma^4) &= \{z_1z_2z_3z_4 - z_1z_2 - z_1z_4 - z_3z_4 + 1 = 0\} \\ &= \{z_1z_2z_3 - z_3 - z_1 \neq 0\} \end{aligned}$$

with open positroid variety of the form

$$V = \begin{pmatrix} 1 & z_1 & z_1z_2 - 1 & z_1z_2z_3 - z_1 - z_3 & z_1z_2z_3z_4 - z_1z_2 - z_1z_4 - z_3z_4 + 1 \\ 0 & 1 & z_2 & z_2z_3 - 1 & z_2z_3z_4 - z_2 - z_4 \end{pmatrix}$$
$$\in \Pi_{2,5}^{\circ,1}$$

Using the correspondence of cluster algebras and Grassmannians, we obtain one of five cluster charts, see Figure 6.

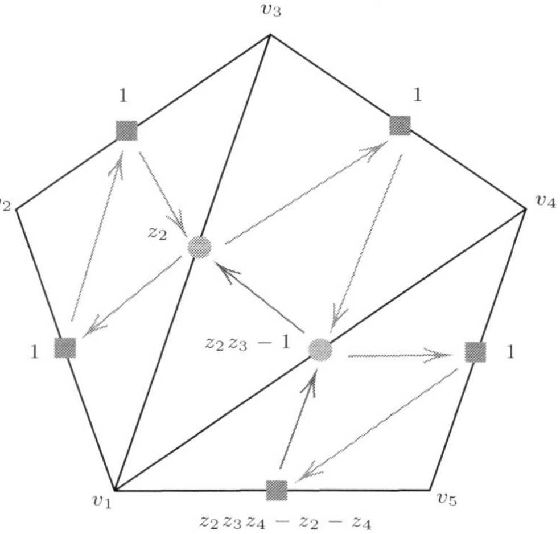

FIGURE 6. The cluster chart U_1 of $X(\sigma^4)$. One of the five possible charts given by the triangulation of the pentagon.

Here

$$U = \{w_1 := \Delta_{13} = z_2 \neq 0, w_2 := \Delta_{14} = z_2z_3 - 1 \neq 0,\\ w := \Delta_{15} = z_2z_3z_4 - z_4 - z_2 \neq 0\}$$

Using the de Rham cohomology

$$\begin{aligned} H^*(U) &= H^*((\mathbb{C}^*)^3) \\ &= \left\langle 1, \frac{dw_1}{w_1}, \frac{dw_2}{w_2}, \frac{dw}{w}, \frac{dw_1}{w_1}\wedge\frac{dw_2}{w_2}, \frac{dw_1}{w_1}\wedge\frac{dw}{w}, \frac{dw_2}{w_2}\wedge\frac{dw}{w}, \frac{dw}{w}\wedge\frac{dw_2}{w_2}\wedge\frac{dw_1}{w_1}\right\rangle \end{aligned}$$

The forms which extend to $X(\sigma^4)$ are:

- 1
- $\dfrac{dw}{w} = \dfrac{(z_3z_4 - 1)dz_2 + z_2z_4dz_3 + (z_2z_3 - 1)dz_4}{z_2z_3z_4 - z_4 - z_2}$
- $\dfrac{dw}{w}\wedge\dfrac{dw_2}{w_2} + \dfrac{dw_2}{w_2}\wedge\dfrac{dw_1}{w_1} = \dfrac{z_4dz_3\wedge dz_2 + z_3dz_4\wedge dz_2 + z_2dz_4\wedge dz_3}{z_2z_3z_4 - z_4 - z_2}$
- $\dfrac{dw}{w}\wedge\dfrac{dw_2}{w_2}\wedge\dfrac{dw_1}{w_1} = \dfrac{dz_4\wedge dz_3\wedge dz_2}{z_2z_3z_4 - z_4 - z_2}$

Therefore, $H^0(X(\sigma^4)) = H^1(X(\sigma^4)) = H^2(X(\sigma^4)) = H^3(X(\sigma^4)) = \mathbb{C}$ which agrees with Theorem 2.12.

4. Performing cuts

4.1. Cuts for braid varieties. In this section, we study various maps between braid varieties and positroid varieties. To work with such maps, it is useful to fix a specific isomorphism between $X(\sigma^k)$ and $\Pi^{\circ,1}_{2,k+1}$ which is given by lemmas below.

LEMMA 4.1. *Let* $M = \begin{pmatrix} v_1 & v_2 & \dots & v_n \end{pmatrix} \in \Pi^\circ_{2,n}$. *There is a unique matrix* $A \in GL(2,\mathbb{C})$ *such that*

$$AM = \begin{pmatrix} 1 & * & \dots & 0 \\ 0 & 1 & \dots & * \end{pmatrix} = V$$

where $\det A = \Delta_{12}^{-1}(M)$ *and* $\Delta_{ij}(V) = \Delta_{ij}(M)\cdot\det A = \dfrac{\Delta_{ij}(M)}{\Delta_{12}(M)}$.

PROOF. If $M = \begin{pmatrix} v_1 & v_2 & \dots & v_n \end{pmatrix}$, then acting on the left with the matrix $S = \begin{pmatrix} v_1 & v_n \end{pmatrix}^{-1}$, we obtain

$$S\cdot M = \frac{1}{\Delta_{1n}(M)}\begin{pmatrix} v_n^2 & -v_n^1 \\ -v_1^2 & v_1^1 \end{pmatrix}\begin{pmatrix} v_1^1 & v_2^1 & \dots & v_n^1 \\ v_1^2 & v_2^2 & \dots & v_n^2 \end{pmatrix} = \begin{pmatrix} 1 & * & \dots & 0 \\ 0 & \alpha & \dots & 1 \end{pmatrix}$$

where $\alpha = \det(S)\Delta_{12}(M) = \dfrac{\Delta_{12}(M)}{\Delta_{1n}(M)}$. Now, if we act on the left by $T = \begin{pmatrix} 1 & 0 \\ 0 & \alpha^{-1} \end{pmatrix}$, we arrive at the matrix

$$T\cdot(S\cdot M) = \begin{pmatrix} 1 & 0 \\ 0 & \alpha^{-1} \end{pmatrix}\begin{pmatrix} 1 & * & \dots & 0 \\ 0 & \alpha & \dots & 1 \end{pmatrix} = \begin{pmatrix} 1 & * & \dots & 0 \\ 0 & 1 & \dots & \alpha^{-1} \end{pmatrix}$$

Let $A = T\cdot S$, then $\det A = (\det T)(\det S) = \left(\dfrac{\Delta_{1n}(M)}{\Delta_{12}(M)}\right)\left(\dfrac{1}{\Delta_{1n}(M)}\right) = \Delta_{12}^{-1}(M)$. □

LEMMA 4.2. *Given the standard form matrix*

$$V = \begin{pmatrix} 1 & * & \dots & 0 \\ 0 & 1 & \dots & * \end{pmatrix} \tag{12}$$

where $\Delta_{i,i+1} \neq 0$, $\Delta_{1n} \neq 0$, we may rescale the vectors $(v_3, \dots, v_n)$ to $(v'_3, \dots, v'_n) = (\lambda_3 v_3, \dots, \lambda_n v_n)$ such that $\Delta'_{i,i+1} = 1$. Furthermore, such λ_i are unique.

PROOF. Let

$$v'_3 = \frac{v_3}{\Delta_{23}}, \; v'_4 = \frac{v_4 \cdot \Delta_{23}}{\Delta_{34}}, \; \dots, \; v'_k = v_k \prod_{l=2}^{k-1} \Delta_{l,l+1}^{(-1)^{k-l}}$$

Note that with the above rescaling Δ'_{1n} remains nonzero, whereas for $\Delta'_{i,i+1}$ the rescaling gives the desired result:

$$\begin{aligned} \Delta'_{i,i+1} &= \det \begin{pmatrix} v'_i & v'_{i+1} \end{pmatrix} = \det \begin{pmatrix} v_i \prod_{l=2}^{i-1} \Delta_{l,l+1}^{(-1)^{i-l}} & v_{i+1} \prod_{l=2}^{i} \Delta_{l,l+1}^{(-1)^{i+1-l}} \end{pmatrix} \\ &= \Delta_{i,i+1} \Delta_{i,i+1}^{(-1)} = 1. \end{aligned}$$

□

COROLLARY 4.3. *Given a matrix $M \in \Pi^{\circ}_{2,n}$, we can use Lemmas 4.1 and 4.2 to change M to the matrix*

$$V' = \begin{pmatrix} 1 & * & \dots & 0 \\ 0 & 1 & \dots & * \end{pmatrix}$$

such that $V' \in \Pi^{\circ,1}_{2,n}$. Furthermore, if $M \in \Pi^{\circ,1}_{2,n}$ then $\Delta_{ij}(V') = \Delta_{ij}(M)$.

PROOF. We only need to prove the last equation. If $M \in \Pi^{\circ,1}_{2,n}$ with each $\Delta_{i,i+1} = 1$, using Lemma 4.1 there exists a unique $V' = AM$, and $\Delta_{ij}(V') = \Delta_{ij}(M)/\Delta_{12}(M) = \Delta_{ij}(M)$. In particular, $\Delta_{i,i+1}(V') = 1$ for all i and we do not require the use of Lemma 4.2 to rescale the vectors. □

Let $\mathcal{P}$ be the $(k+1)$-gon corresponding to the braid variety $X(\sigma^k)$. We can choose a diagonal D_{ij} which cuts the polygon $\mathcal{P}$ in two pieces, a $(j-i+1)$-gon $\mathcal{P}_1(i,j)$ and a $(k-j+i+2)$-gon $\mathcal{P}_2(i,j)$. These correspond to braid varieties $X(\sigma^{j-i})$ and $X(\sigma^{k-j+i+1})$, respectively. We will refer to this procedure as a diagonal cut. If we denote $a = j-i$ and $b = k-j+i+1$ then $a+b = k+1$.

THEOREM 4.4. *Performing one diagonal cut on P along D_{ij} defines an injective map*

$$\Phi_{ij} : X(\sigma^a) \times X(\sigma^b) \longrightarrow X(\sigma^{a+b-1})$$

and its image is the open subset $\{\Delta_{ij} \neq 0\}$ in $X(\sigma^{a+b-1})$.

PROOF. We use the isomorphism $\Pi^{\circ,1}_{2,k+1} \simeq X(\sigma^k)$ from Theorem 2.17. We first describe the inverse map

$$\Phi_{ij}^{-1} : \{\Delta_{ij} \neq 0\} \to X(\sigma^a) \times X(\sigma^b).$$

Let $V \in \Pi^{\circ,1}_{2,k+1}$ be a $2 \times (k+1)$ matrix, choose some i, j such that $1 \leq i < j \leq k+1$ where $(i,j) \neq (1, k+1)$, to perform the diagonal cut of the $(k+1)$-gon resulting in two polygons $\mathcal{P}_a$ and $\mathcal{P}_b$ where $\mathcal{P}_a$ is a $(j-i+1)$-gon and $\mathcal{P}_b$ is a $(k-j+i+2)$-gon. Assume that $\Delta_{ij}(V) \neq 0$. Then we can decompose the matrix V into two matrices:

$$\begin{aligned} V_1 &= \begin{pmatrix} v_i & \dots & v_j \end{pmatrix} \in \mathrm{Mat}(2, a+1) \\ V_2 &= \begin{pmatrix} v_1 & \dots & v_i & v_j & \dots & v_{k+1} \end{pmatrix} \in \mathrm{Mat}(2, b+1) \end{aligned}$$

Let us prove that $V_1 \in \Pi^{\circ,1}_{2,a+1}$. As it happens $\Delta_{m,m+1}(V_1) = \Delta_{m+i-1,m+i}(V) = 1$ for $1 \leq m \leq a$, and $\Delta_{1,a+1}(V_1) = \Delta_{ij}(V) \neq 0$. We use the isomorphism $\Pi^{\circ,1}_{2,a+1} \simeq X(\sigma^a)$ from Theorem 2.17 to obtain a point in $X(\sigma^a)$ from V_1.

Next, we study the matrix V_2. We have

$$\Delta_{m,m+1}(V_2) = \begin{cases} \Delta_{m,m+1}(V) = 1 & \text{if } m < i \\ \Delta_{ij}(V) & \text{if } m = i \\ \Delta_{m+j-i-1,m+j-i}(V) = 1 & \text{if } i < m \leq k-j+i+1. \end{cases}$$

Furthermore, $\Delta_{1,b+1}(V_2) = \Delta_{1,k+1}(V) \neq 0$, so $V_2 \in \Pi^{\circ}_{2,b+1}$. We would like to use Lemmas 4.1 and 4.2 to change V_2 to a different matrix $V_2' \in \Pi^{\circ,1}_{2,b+1}$. We have two cases:

Case 1: If $i = 1$, then we first apply Lemma 4.1. Since $S = (v_1\ v_{k+1})^{-1}$ is diagonal, we simply rescale the second row of V_2 by Δ_{1j}^{-1} to get V_2 to the form (12). Next, we apply Lemma 4.2 to rescale the vectors, and get $V_2' = (v_1, v_2', \dots, v_{b+1}')$ where

$$v_m' = \begin{cases} (v^1_{m+j-2}, v^2_{m+j-2}\Delta_{1j}^{-1}) & \text{if } m \text{ is even} \\ (v^1_{m+j-2}\Delta_{1j}, v^2_{m+j-2}) & \text{if } m \text{ is odd.} \end{cases}$$

Case 2: If $i \geq 2$, then we do not need to apply Lemma 4.1, we rescale the vectors v_m for $m \geq j$. As a result, we get a matrix $V_2' = (v_1, \dots, v_i, v_j', v_{j+1}', \dots, v_{k+1}')$ where

$$v_m' = \Delta_{ij}^{(-1)^{m-j+1}} v_m.$$

Now we can describe the desired map $\Phi_{ij} : X(\sigma^a) \times X(\sigma^b) \to \{\Delta_{ij} \neq 0\}$ as follows. Given two matrices $V_1 \in \Pi^{\circ,1}_{2,a+1}, V_2' \in \Pi^{\circ,1}_{2,b+1}$, we can read off $\Delta_{ij}(V) = \Delta_{1,a+1}(V_1)$ which is nonzero by assumption. The matrix V_2' was obtained from V_2 above using multiplication by $\Delta_{ij}^{\pm 1}$, and hence is invertible, so given V_2' and Δ_{ij} we can reconstruct V_2.

Now we can reconstruct V by simply inserting V_1 into V_2. Note that if the vectors v_i and v_j from V_1 do not agree with the ones from V_2, we can always use row operations to make them agree since $\det(v_i \quad v_j) = \Delta_{ij} \neq 0$. □

THEOREM 4.5. *The map Φ_{ij} defines a quasi-equivalence of cluster varietes $\{\Delta_{ij} \neq 0\} \subset X(\sigma^{a+b-1})$ and $X(\sigma^a) \times X(\sigma^b)$. The latter has a cluster structure obtained by freezing Δ_{ij} in the cluster structure from $X(\sigma^{a+b-1})$.*

PROOF. We use the clusters in $X(\sigma^a), X(\sigma^b)$ and $X(\sigma^{a+b-1})$ defined by the triangulation in Figure 9. In particular, we get fan triangulations for $X(\sigma^a), X(\sigma^b)$.

By construction, all cluster variables corresponding to diagonals are multiplied by monomials in Δ_{ij}, but we still need to check that the exchange ratios (as in Definition 2.19) are preserved. All diagonals above Δ_{ij} are unchanged, so we need to verify that the exchange ratios do not change for diagonals $\Delta_{1,m}$. For $m < i$, this is clear. For $m = i$, the exchange ratio is

$$\frac{\Delta_{1j}}{\Delta_{ij}\Delta_{1,j-1}} = \frac{\Delta_{1j}\Delta_{ij}^{-1}}{1 \cdot \Delta_{1,j-1}}.$$

For $m = j$, the exchange ratio is

$$\frac{\Delta_{ij}\Delta_{1,j+1}}{\Delta_{1i}} = \frac{1 \cdot (\Delta_{1,j+1}\Delta_{ij})}{\Delta_{1i}}.$$

Finally, for $m > j$ we get

$$\frac{\Delta_{1,m+1}}{\Delta_{1,m-1}} = \frac{\Delta_{1,m+1}\Delta_{ij}^{(-1)^{m+1-j+1}}}{\Delta_{1,m-1}\Delta_{ij}^{(-1)^{m-1-j+1}}}$$

since $m+1-j$ and $m-1-j$ have the same parity. □

Suppose $a+b+c-2 = k$. We will study the associativity properties of our cuts along two non-intersecting diagonals D_{ij} and $D_{i'j'}$, see Figure 7. There are two general cases to consider when performing two cuts which we label as Type A or Type B. The two cuts occur at D_{ij} and $D_{i'j'}$ and will be denoted Φ_{ij} and $\Phi_{i'j'}$, respectively. Type A cuts are diagonal cuts of the form $1 \le i' \le i < j \le j' \le k+1$ given that the cuts do not degenerate to the one cut case, whereas, Type B cuts are diagonal cuts of the form $1 \le i < j \le i' < j' \le k+1$, see Figure 7.

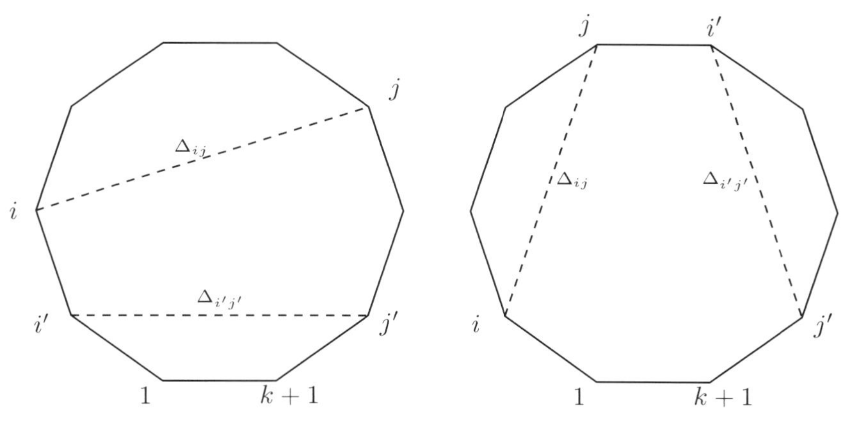

FIGURE 7. The possible cuts when performing two diagonal cuts, the dashed lines indicate these potential cuts. The polygon on the left depicts cuts of Type A and the polygon on the right depicts cuts of Type B.

THEOREM 4.6. *For Type A cuts we have a commutative diagram*

$$\begin{array}{ccc} X(\sigma^a) \times X(\sigma^b) \times X(\sigma^c) & \xrightarrow{\Phi_{ij} \times \mathrm{Id}} & X(\sigma^{a+b-1}) \times X(\sigma^c) \\ \Big\downarrow{\scriptstyle \mathrm{Id} \times \Phi_{i'j'}} & & \Big\downarrow{\scriptstyle \Phi_{i'j'}} \\ X(\sigma^a) \times X(\sigma^{b+c-1}) & \xrightarrow[\Phi_{ij}]{} & X(\sigma^{a+b+c-2}) \end{array}$$

PROOF. Let $V \in \Pi_{2,k+1}^{\circ,1}$ by Theorem 2.17 V corresponds to a point in $X(\sigma^k)$.

For Type A cuts, choose some i, j, i', j' such that $1 \leq i' \leq i < j < j' \leq k$. Similar to Theorem 4.4 involving a single diagonal cut, we describe the inverse maps then produce the desired map. Here $a = j - i, b = j' - j + i - i' + 1$ and $c = k - j' + i' + 1$. Define the matrix $V \in \Pi_{2,k+1}^{\circ,1}$ associated to $X(\sigma^k)$ as

$$V = \begin{pmatrix} v_1 & \dots & v_{i'} & \dots & v_i & \dots & v_j & \dots & v_{j'} & \dots & v_{k+1} \end{pmatrix}$$

We will be dealing with minors in several different matrices, as such we will include the matrices in the notations.

(i) First, we consider the case where we cut at along $\Delta_{ij}(V)$ then $\Delta_{i'j'}(V)$ which is described in Figure 8a by

$$X(\sigma^{a+b+c-2}) \xrightarrow{\Phi_{ij}^{-1}} X(\sigma^a) \times X(\sigma^{b+c-1}) \xrightarrow{\mathrm{Id}\times\Phi_{i'j'}^{-1}} X(\sigma^a) \times X(\sigma^b) \times X(\sigma^c)$$

By performing the initial cut $\Delta_{ij}(V)$, given by $\Phi_{ij}^{-1} : X(\sigma^{a+b+c-2}) \to X(\sigma^a) \times X(\sigma^{b+c-1})$, we decompose the matrix V into the two following matrices

$$V_1 = \begin{pmatrix} v_i & \dots & v_j \end{pmatrix} \in \mathrm{Mat}(2, a+1)$$

$$V_2 = \begin{pmatrix} v_1 & \dots & v_{i'} & \dots & v_i & v_j & \dots & v_{j'} & \dots & v_{k+1} \end{pmatrix} \in \mathrm{Mat}(2, b+c)$$

Similar to the argument in Theorem 4.4, $\Delta_{ij}(V) \neq 0$ and we find that $V_1 \in \Pi_{2,a+1}^{\circ,1} \simeq X(\sigma^a)$. Here, the rescaling of vectors v_m for $m \geq j$ in

$$V_3 = \begin{pmatrix} v_1 & \dots & v_{i'} & \dots & v_i & v'_j & \dots & v'_{j'} & \dots & v'_{k+1} \end{pmatrix}$$

is given by

$$v'_m = v_m \Delta_{ij}(V)^{(-1)^{m-j+1}} \tag{13}$$

Therefore, $V_3 \in \Pi_{2,b+c}^{\circ,1} \simeq X(\sigma^{b+c-1})$ and Φ_{ij}^{-1} is well-defined. Note that during the rescaling of the matrix V_2 into V_3 the minors of V_3 also experience rescaling by a factor of $\Delta_{ij}(V)$, hence given that $v'_{j'} = v_{j'}\Delta_{ij}(V)^{(-1)^{j'-j+1}}$

$$\Delta_{i'j'}(V_3) = \Delta_{i'j'}(V)\Delta_{ij}(V)^{(-1)^{j'-j+1}} \tag{14}$$

Now we perform the second cut $\Delta_{i'j'}(V)$ given by the map

$$X(\sigma^a) \times X(\sigma^{b+c-1}) \xrightarrow{\mathrm{Id}\times\Phi_{i'j'}} X(\sigma^a) \times X(\sigma^b) \times X(\sigma^c).$$

The matrix V_1 remains unchanged whereas V_3 decomposes into

$$V_4 = \begin{pmatrix} v_{i'} & \dots & v_i & v'_j & \dots & v'_{j'} \end{pmatrix} \in \mathrm{Mat}(2, b+1)$$

$$V_5 = \begin{pmatrix} v_1 & \dots & v_{i'} & v'_{j'} & \dots & v'_{k+1} \end{pmatrix} \in \mathrm{Mat}(2, c+1)$$

By the rescaling of matrix V_3 in the previous cutting and $\Delta_{i'j'}(V_3) \neq 0$, then $V_4 \in \Pi_{2,b+1}^{\circ,1} \simeq X(\sigma^b)$. After performing the second cut there is again a rescaling, this time of the matrix V_5 which is given by the new matrix

$$V_6 = \begin{pmatrix} v_1 & \dots & v_{i'} & v''_{j'} & \dots & v''_{k+1} \end{pmatrix}$$

where for $m \geq j'$ the vectors are

$$v''_m = v'_m \Delta_{i'j'}(V_3)^{(-1)^{m-j'+1}} = v_m \Delta_{ij}(V)^{(-1)^{m-j+1}} \Delta_{i'j'}(V_3)^{(-1)^{m-j'+1}}.$$

Given that

$$\begin{aligned}\Delta_{i'j'}(V_3)^{(-1)^{m-j'+1}} &= \Delta_{i'j'}(V)^{(-1)^{m-j'+1}}\Delta_{ij}(V)^{(-1)^{j'-j+1}(-1)^{m-j'+1}}\\ &= \Delta_{i'j'}(V)^{(-1)^{m-j'+1}}\Delta_{ij}(V)^{(-1)^{m-j}}\end{aligned}$$

and $(-1)^{m-j+1}+(-1)^{m-j}=0$ we conclude that
(15)

$$v''_m = v_m\Delta_{ij}(V)^{(-1)^{m-j+1}}\Delta_{i'j'}(V)^{(-1)^{m-j'+1}}\Delta_{ij}(V)^{(-1)^{m-j}} = v_m\Delta_{i'j'}(V)^{(-1)^{m-j'+1}}.$$

As such $V_6 \in \Pi^{\circ,1}_{2,c+1} \simeq X(\sigma^c)$. This concludes the construction of the inverse map.

To construct the desired map

$$\Phi_{ij}\circ(\mathrm{Id}\times\Phi_{i'j'}) : X(\sigma^a)\times X(\sigma^b)\times X(\sigma^c) \to X(\sigma^{a+b+c-2})$$

We reconstruct V by taking $V_1 \in \Pi^{\circ,1}_{2,a+1}$, $V_4 \in \Pi^{\circ,1}_{2,b+1}$, $V_6 \in \Pi^{\circ,1}_{2,c+1}$. We can read off $\Delta_{i'j'}(V) = \Delta_{1,b+1}(V_4)$ which is nonzero by assumption. The matrix V_5 is obtained from V_6 by multiplication of $\Delta_{i'j'}(V)^{\pm 1}$ to the vectors v_l for $l \geq i'+1$, which is well-defined since $\Delta_{i'j'}(V)$ is invertible. We reconstruct the matrix $V_3 \in \mathrm{Mat}(2, b+c)$ by inserting the matrix V_4 into V_5 in the appropriate location. Furthermore, $V_3 \in \Pi^{\circ,1}_{2,b+c} \simeq X(\sigma^{b+c-1})$ by construction. This concludes the construction of the map

$$X(\sigma^a)\times X(\sigma^b)\times X(\sigma^c) \xrightarrow{\mathrm{Id}\times\Phi_{i'j'}} X(\sigma^a)\times X(\sigma^{b+c-1})$$

Continuing the construction of the desired map, we read off

$$\Delta_{ij}(V) = \Delta_{1,a+1}(V_1)$$

which is again nonzero by assumption. The matrix V_2 is obtained from V_3 by multiplication of $\Delta_{ij}(V)^{\pm 1}$ to the vectors v_l for $l \geq i+1$. We reconstruct V by inserting V_1 into V_2 at the appropriate location, completing the construction of the map

$$X(\sigma^a)\times X(\sigma^{b+c-1}) \xrightarrow{\Phi_{ij}} X(\sigma^{a+b+c-2})$$

and producing the desired map.

(ii) Now, for the case where we cut along $\Delta_{i'j'}(V)$ then $\Delta_{ij}(V)$, described in Figure 8b by

$$X(\sigma^a)\times X(\sigma^b)\times X(\sigma^c) \xrightarrow{\Phi^{-1}_{i'j'}} X(\sigma^{a+b-1})\times X(\sigma^c) \xrightarrow{\Phi^{-1}_{ij}\times\mathrm{Id}} X(\sigma^{a+b+c-2}).$$

Perform the initial cut $\Delta_{i'j'}(V)$, to decompose V into the matrices

$$\begin{aligned}W_1 &= \begin{pmatrix} v_1 & \dots & v_{i'} & v_{j'} & \dots & v_{k+1}\end{pmatrix} \in \mathrm{Mat}(2, c+1)\\ W_2 &= \begin{pmatrix} v_{i'} & \dots & v_i & \dots & v_j & \dots & v_{j'}\end{pmatrix} \in \mathrm{Mat}(2, a+b)\end{aligned}$$

By the same argument as in Theorem 4.4 $\Delta_{i'j'} \neq 0$ and $W_2 \in \Pi^{\circ,1}_{2,a+b} \simeq X(\sigma^{a+b-1})$. Now, the matrix W_1 requires rescaling of the vectors v_m for $m \geq j'$, producing the matrix

$$W_3 = \begin{pmatrix} v_1 & \dots & v_{i'} & v'_{j'} & \dots & v'_{k+1}\end{pmatrix}$$

here

$$v'_m = v_m\Delta_{i'j'}(V)^{(-1)^{m-j'+1}}$$

which is in agreement with (15). Hence, $W_3 \in \Pi^{\circ,1}_{2,c+1} \simeq X(\sigma^c)$.

We perform the second cut $\Delta_{ij}(V)$, which separates W_2 into

$$\begin{aligned}W_4 &= \begin{pmatrix} v_{i'} & \dots & v_i & v_j & \dots & v_{j'}\end{pmatrix} \in \mathrm{Mat}(2, b+1)\\ W_5 &= \begin{pmatrix} v_i & \dots & v_j\end{pmatrix} \in \mathrm{Mat}(2, a+1)\end{aligned}$$

In this case, $W_5 \in \Pi^{\circ,1}_{2,a+1} \simeq X(\sigma^a)$, whereas the matrix $W_4 \in \Pi^\circ_{2,b+1}$ requires a rescaling for the vectors v_m for $j \leq m \leq j'$. Let

$$W_6 = \begin{pmatrix} v_{i'} & \dots & v_i & v''_j & \dots & v''_{j'} \end{pmatrix}$$

with the vectors

$$v''_m = v_m \Delta_{ij}(V)^{m-j+1}$$

which agrees with (13). Therefore, $W_6 \in \Pi^{\circ,1}_{2,b+1} \simeq X(\sigma^b)$, completing the construction of the inverse maps.

Finally, we construct the desired map

$$\Phi_{i'j'} \circ (\Phi_{ij} \times \mathrm{Id}) : X(\sigma^a) \times X(\sigma^b) \times X(\sigma^c) \to X(\sigma^{a+b+c-2})$$

We reconstruct V by taking $W_5 \in \Pi^{\circ,1}_{2,a+1}$, $W_6 \in \Pi^{\circ,1}_{2,b+1}$, $W_3 \in \Pi^{\circ,1}_{2,c+1}$. We read off $\Delta_{ij}(V) = \Delta_{1,a+1}(W_5)$ which is nonzero by assumption. The matrix W_4 is recovered from W_6 by multiplication of $\Delta^{\pm 1}_{ij}$ to the vectors v_l for $l \geq i - i' + 1$, which is well-defined since Δ_{ij} is invertible. We reconstruct $W_2 \in \Pi^{\circ,1}_{2,a+b} \simeq X(\sigma^{a+b-1})$ by inserting the matrix W_5 into W_4 in the appropriate position. Concluding the construction of the map

$$X(\sigma^a) \times X(\sigma^b) \times X(\sigma^c) \xrightarrow{\Phi_{ij} \times \mathrm{Id}} X(\sigma^{a+b-1}) \times X(\sigma^c)$$

To complete the construction, we read off $\Delta_{i'j'}(v) = \Delta_{1,c+1}(W_3)$ which is also nonzero by construction. The matrix $W_1 \in \Pi^{\circ,1}_{1,c+1} \simeq X(\sigma^c)$ is recovered from the matrix W_3 by multiplication of $\Delta_{i'j'}(V)^{\pm 1}$ to the vectors v_l for $l \geq i' + 1$. We reconstruct V by inserting W_1 into W_2 at the appropriate location, concluding the construction of the map

$$X(\sigma^{a+b-1}) \times X(\sigma^c) \xrightarrow{\Phi_{i'j'}} X(\sigma^{a+b+c-2})$$

which produces the desired map showing associativity of Type A cuts. □

LEMMA 4.7. *For $A, A' \in \Pi^\circ_{2,c+1}$ we define the map T_λ as*

$$T_\lambda : A \to A', \; a_m \to a_m \lambda^{m-j}$$

Then T_λ preserves $\Pi^{\circ,1}_{2,n}$ and defines $\mathbb{C}^*$ actions on $\Pi^\circ_{2,n}$ and $\Pi^{\circ,1}_{2,n}$.

PROOF. □

THEOREM 4.8. *For Type B cuts we have a commutative diagram*

$$\begin{array}{ccccc}
X(\sigma^a) \times X(\sigma^b) \times X(\sigma^c) & \xrightarrow{\mathrm{Id} \times \Phi_{i'j'}} & X(\sigma^a) \times X(\sigma^{b+c-1}) & \xrightarrow{\Phi_{ij}} & X(\sigma^{a+b+c-2}) \\
\big\downarrow {\scriptstyle \mathrm{Id} \times \mathrm{Id} \times T_{\Delta_{ij}}} & & & & \big\uparrow {\scriptstyle \Phi_{i'j'}} \\
X(\sigma^a) \times X(\sigma^b) \times X(\sigma^c) & & \xrightarrow{\qquad \Phi_{ij} \times \mathrm{Id} \qquad} & & X(\sigma^{a+b-1}) \times X(\sigma^c)
\end{array}$$

Here $T_{\Delta_{ij}}$ is defined as in Lemma 4.7 with $\lambda = \Delta_{ij}$. Informally, we can say that the gluing P from smaller polygons is associative only up to the additional transformation $T_{\Delta_{ij}}$.

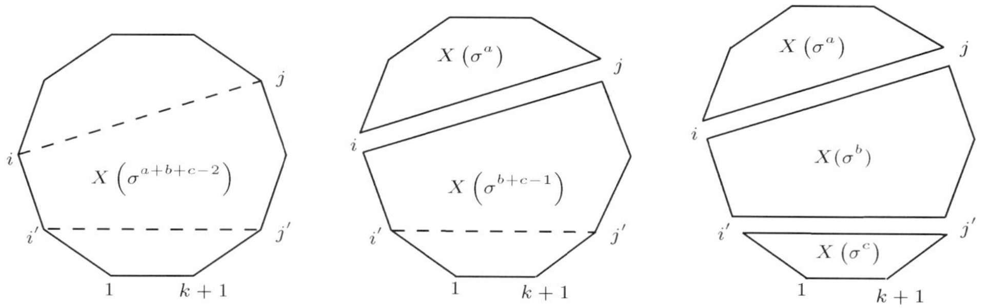

(A) Type A: Initial cut at Δ_{ij} followed by $\Delta_{i'j'}$.

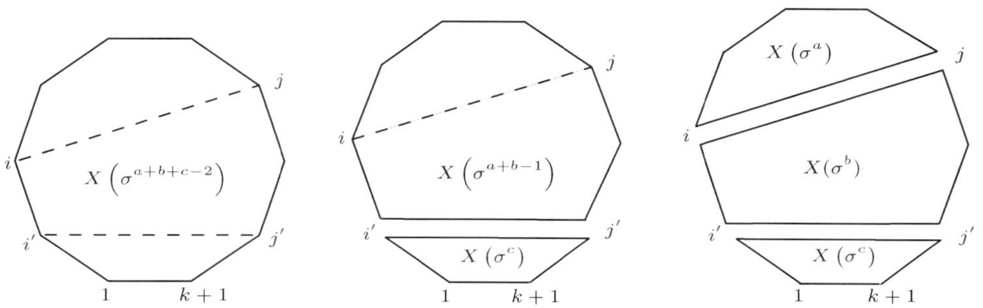

(B) Type A: Initial cut at $\Delta_{i'j'}$ followed by Δ_{ij}.

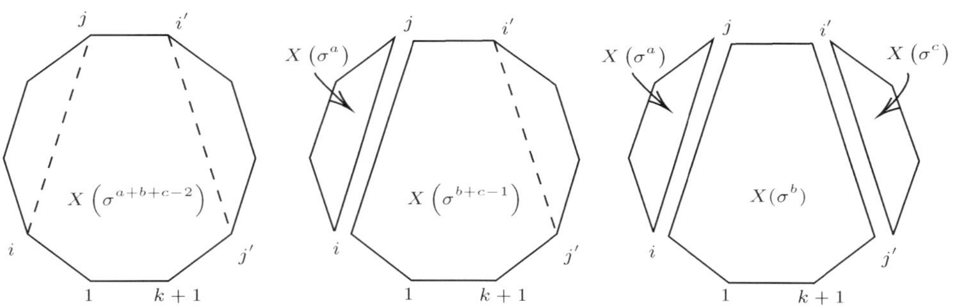

(C) Type B: Initial cut at Δ_{ij} followed by $\Delta_{i'j'}$.

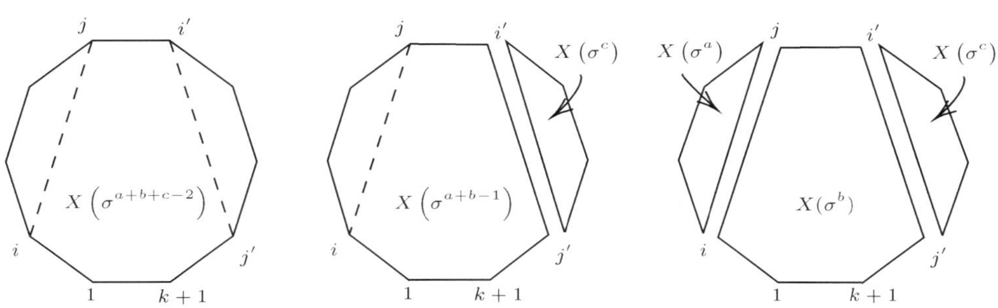

(D) Type B: Initial cut at $\Delta_{i'j'}$ followed by Δ_{ij}.

FIGURE 8. All possible variations of Type A and B cuts.

PROOF. Let $V \in \Pi^{\circ,1}_{2,k+1}$ by Theorem 2.17.

For Type B cuts, choose some i, j, i', j' such that $1 \leq i < j \leq i' < j' \leq k+1$. Similar to Theorem 4.6, we describe the inverse maps then produce the desired map. Here $a = j - i$, $b = k - j' + i' - j + i + 2$, $c = j' - i'$. Define the matrix $V \in \Pi^{\circ,1}_{2,k+1}$ associated to $X(\sigma^k)$ as

$$V = \begin{pmatrix} v_1 & \dots & v_i & \dots & v_j & \dots & v_{i'} & \dots & v_{j'} & \dots & v_{k+1} \end{pmatrix}$$

Similar to Theorem 4.6 we will be dealing with minors in several different matrices and will include the matrices in the notations.

(i) We first consider the case where we cut along $\Delta_{ij}(V)$ then $\Delta_{i'j'}(V)$, see Figure 8c, given by the map

$$X(\sigma^{a+b+c-2}) \xrightarrow{\Phi_{ij}^{-1}} X(\sigma^a) \times X(\sigma^{b+c-1}) \xrightarrow{\mathrm{Id} \times \Phi_{i'j'}^{-1}} X(\sigma^a) \times X(\sigma^b) \times X(\sigma^c)$$

Performing the initial cut $\Delta_{ij}(V)$, given by $\Phi_{ij}^{-1} : X(\sigma^{a+b+c-2}) \to X(\sigma^a) \times X(\sigma^{b+c-1})$, decomposes V into the two matrices

$$V_1 = \begin{pmatrix} v_i & \dots & v_j \end{pmatrix} \in \mathrm{Mat}(2, a+1)$$

$$V_2 = \begin{pmatrix} v_1 & \dots & v_i & v_j & \dots & v_{i'} & \dots & v_{j'} & \dots & v_{k+1} \end{pmatrix} \in \mathrm{Mat}(2, b+c)$$

By the same argument as in Theorem 4.4, $V_1 \in \Pi^{\circ,1}_{2,a+1} \simeq X(\sigma^a)$ whereas $V_2 \in \Pi^{\circ}_{2,b+c}$ and requires rescaling by $\Delta_{ij}(V)$ for the vectors v_m for $m \geq j$, resulting in the matrix

$$V_3 = \begin{pmatrix} v_1 & \dots & v_i & v'_j & \dots & v'_{i'} & \dots & v'_{j'} & \dots & v'_{k+1} \end{pmatrix}$$

where

$$v'_m = v_m \Delta_{ij}(V)^{(-1)^{m-j+1}} \tag{16}$$

Note that $\Delta_{i'j'}(V_3)$ experiences a rescaling by factor of $\Delta_{ij}(V)$, given that $v'_{i'} = V_{i'} \Delta_{ij}^{(-1)^{i'-j+1}}$ and $v'_{j'} = v_{j'} \Delta_{ij}^{(-1)^{j'-j+1}}$, the rescaled determinant is given by

$$\begin{aligned} \Delta_{i'j'}(V_3) &= \Delta_{i'j'}(V) \Delta_{ij}(V)^{(-1)^{i'-j+1}} \Delta_{ij}(V)^{(-1)^{j'-j+1}} \\ &= \Delta_{i'j'}(V) \Delta_{ij}(v)^{(-1)^{i'-j+1} + (-1)^{j'-j+1}} \end{aligned} \tag{17}$$

This completes the construction of the map $\Phi_{ij}^{-1} : X(\sigma_{a+b+c-2}) \to X(\sigma^a) \times X(\sigma^{b+c-1})$. Applying the second cut $\Delta_{i'j'}(V)$ to the matrix V_3 produces the two matrices

$$V_4 = \begin{pmatrix} v'_{i'} & \dots & v'_{j'} \end{pmatrix} \in \mathrm{Mat}(2, c+1)$$

$$V_5 = \begin{pmatrix} v_1 & \dots & v_i & v'_j & \dots & v'_{i'} & v'_{j'} & \dots & v'_{k+1} \end{pmatrix} \in \mathrm{Mat}(2, b+1)$$

Here, $V_4 \in \Pi^{\circ,1}_{2,c+1} \simeq X(\sigma^c)$. Since $V_5 \in \Pi^{\circ}_{2,b+1}$ we applying a rescaling of the vectors v'_m for $m \geq j'$ into the matrix

$$V_6 = \begin{pmatrix} v_1 & \dots & v_i & v'_j & \dots & v'_{i'} & v''_{j'} & \dots & v''_{k+1} \end{pmatrix}$$

where

$$v''_m = v'_m \Delta_{i'j'}(V_3)^{(-1)^{m-j'+1}} \tag{18}$$

Using (17) we find that

$$\begin{aligned}\Delta_{i'j'}(V_3)^{(-1)^{m-j'+1}} &= (\Delta_{i'j'}(V)\Delta_{ij}(V)^{(-1)^{i'-j+1}+(-1)^{j'-j+1}})^{(-1)^{m-j'+1}}\\ &= \Delta_{i'j'}(V)^{(-1)^{m-j'+1}}\Delta_{ij}(V)^{(-1)^{i'-j+1}(-1)^{m-j'+1}+(-1)^{j'-j+1}(-1)^{m-j'+1}}\end{aligned}$$

and $(-1)^{i'-j+1}(-1)^{m-j'+1}+(-1)^{j'-j+1}(-1)^{m-j'+1}=(-1)^{m-j'+i'-j}+(-1)^{m-j}$. Therefore

$$\begin{aligned}v''_m &= v_m\Delta_{ij}(V)^{(-1)^{m-j+1}}\Delta_{i'j'}(V)^{(-1)^{m-j'+1}}\Delta_{ij}(V)^{(-1)^{m-j'+i'+j}}\Delta_{ij}^{(-1)^{m-j}}\\ &= v_m\Delta_{i'j'}(V)^{(-1)^{m-j'+1}}\Delta_{ij}(V)^{(-1)^{m-j'+i'-j}}\end{aligned} \tag{19}$$

Now, $V_6\in\Pi^{\circ,1}_{2,b+1}\simeq X(\sigma^b)$. This concludes the construction of the inverse map, now we proceed to the construction of the desired map

$$X(\sigma^a)\times X(\sigma^b)\times X(\sigma^c)\xrightarrow{\mathrm{Id}\times\Phi_{i'j'}}X(\sigma^a)\times X(\sigma^{b+c-1})\xrightarrow{\Phi_{ij}}X(\sigma^{a+b+c-2})$$

Given $V_1\in\Pi^{\circ,1}_{2,a+1}$, $V_6\in\Pi^{\circ,1}_{2,b+1}$, $V_4\in\Pi^{\circ,1}_{2,c+1}$ we reconstruct the matrix V. First, we determine that $\Delta_{i'j'}(V)=\Delta_{1,c+1}(V_4)\neq 0$.The matrix V_5 is found by multiplication of $\Delta_{i'j'}(V)^{\pm1}$ to the vectors v_l for $l\geq i+i'-j+2$ in matrix V_6. We then reconstruct $V_3\in\Pi^{\circ,1}_{2,b+c}\simeq X(\sigma^{b+c-1})$ by inserting the matrix V_4 into the appropriate position in the matrix V_5. This completes the map $\mathrm{Id}\times\Phi_{i'j'}: X(\sigma^a)\times X(\sigma^b)\times X(\sigma^c)\to X(\sigma^a)\times X(\sigma^{b+c-1})$. Now we continue our construction of the matrix V by reading off $\Delta_{ij}(V)=\Delta_{1,a+1}(V_1)$ which is nonzero by assumption. We rescale the vectors v_l for $l\geq i+1$ in the matrix V_3 by multiplication of $\Delta_{ij}(V)^{\pm1}$ which is invertible, to obtain the matrix V_2. Finally, we insert the matrix V_1 into V_2 to obtain V. Therefore, giving us the desired map above.

(ii) Now, we consider the case were we first cut along $\Delta_{i'j'}(V)$ followed by the cut $\Delta_{ij}(V)$ and subsequently, a rescaling of $X(\sigma^c)$ by the torus action $T_{\Delta_{ij}}$, illustrated in Figure 8d, given by

$$\begin{aligned}X(\sigma^{a+b+c-2})&\xrightarrow{\Phi^{-1}_{i'j'}}X(\sigma^{a+b-1})\times X(\sigma^c)\xrightarrow{\Phi^{-1}_{ij}\times\mathrm{Id}}X(\sigma^a)\times X(\sigma^b)\times X(\sigma^c)\\ &\xrightarrow{\mathrm{Id}\times\mathrm{Id}\times T_{\Delta_{ij}}}X(\sigma^a)\times X(\sigma^b)\times X(\sigma^c)\end{aligned}$$

We perform the initial cut $\Delta_{i'j'}(V)$ to V resulting in the matrices

$$W_1=\begin{pmatrix}v_{i'} & \dots & v_{j'}\end{pmatrix}\in\mathrm{Mat}(2,c+1)$$
$$W_2=\begin{pmatrix}v_1 & \dots & v_i & \dots & v_j & \dots & v_{i'} & v_{j'} & \dots & v_{k+1}\end{pmatrix}\in\mathrm{Mat}(2,a+b)$$

By the same argument in Theorem 4.4, $V_1\in\Pi_{2,c+1}\simeq X(\sigma^c)$, whereas the matrix $V_2\in\Pi^{\circ}_{2,a+b}$ requires as rescaling of the vectors v_m for $m\geq j'$ to obtain the matrix

$$W_3=\begin{pmatrix}v_1 & \dots & v_i & \dots & v_j & \dots & v_{i'} & \widetilde{v}_{j'} & \dots & \widetilde{v}_{k+1}\end{pmatrix}$$

given by

$$\widetilde{v}_m=v_m\Delta_{i'j'}^{(-1)^{m-j'+1}} \tag{20}$$

Now, $W_3\in\Pi^{\circ,1}_{2,a+b}\simeq X(\sigma^{a+b-1})$, completing the first map.

We now perform the second cut $\Delta_{ij}(V) = \Delta_{ij}(W_2)$ by decomposing the matrix W_3 into the matrices

$$W_4 = \begin{pmatrix} v_i & \dots & v_j \end{pmatrix} \in \mathrm{Mat}(2, a+1)$$
$$W_5 = \begin{pmatrix} v_1 & \dots & v_i & v_j & \dots & v_{i'} & \widetilde{v}_{j'} & \dots & \widetilde{v}_{k+1} \end{pmatrix} \in \mathrm{Mat}(2, b+1)$$

Given that $W_4 \in \Pi^{\circ,1}_{2,a+1} \simeq X(\sigma^a)$ and $W_5 \in \Pi^{\circ}_{2,b+1}$, the matrix vectors v_m for $m \geq j$ in W_5 are rescaled into the matrix

$$W_6 = \begin{pmatrix} v_1 & \dots & v_i & v'_j & \dots & v'_{i'} & \widetilde{v}'_{j'} & \dots & \widetilde{v}'_{k+1} \end{pmatrix}$$

where for $j \leq m \leq i'$

$$v'_m = v_m \Delta_{ij}^{(-1)^{m-j+1}} \tag{21}$$

and for $m \geq j'$

$$\begin{aligned} \widetilde{v}'_m &= \widetilde{v}_m \Delta_{ij}(V)^{(-1)^{m-j'+i'-j}} \\ &= v_m \Delta_{i'j'}(V)^{(-1)^{m-j'+1}} \Delta_{ij}(V)^{(-1)^{m-j'+i'-j}} \end{aligned} \tag{22}$$

Now, $W_6 \in \Pi^{\circ,1}_{2,b+1} \simeq X(\sigma^b)$. Note that for the vectors v'_m for $j \leq m \leq i'$ (16) agrees with (21), for $j' \leq m$ (19) agrees with (22). However, the vectors v_m found in W_1 for $i' < m < j'$ do not agree with (16) and differ by a factor of $\Delta_{ij}(V)^{(-1)^{m-j+1}}$. Since $\Delta_{ij} \neq 0$, we can then apply a torus action to the matrix $W_1 \in \Pi^{\circ,1}_{2,c+1} \simeq X(\sigma^c)$ using Lemma 4.2. Let $W_1, W_7 \in \Pi^{\circ,1}_{2,c+1}$, define the torus action by the map

$$\begin{aligned} T^{-1}_{\Delta_{ij}} : W_1 &\longrightarrow W_7 \\ v_m &\longmapsto v_m \Delta_{ij}^{(-1)^{m-j-1}} \end{aligned}$$

Thus concluding the construction of the inverse maps.

Now, we construct the suitable map to establish associativity up to an additional transformation $T_{\Delta_{ij}}$, given by

$$\begin{aligned} X(\sigma^a) \times X(\sigma^b) \times X(\sigma^c) &\xrightarrow{\mathrm{Id} \times \mathrm{Id} \times T_{\Delta_{ij}}} X(\sigma^a) \times X(\sigma^b) \times X(\sigma^c) \\ &\xrightarrow{\Phi_{ij} \times \mathrm{Id}} X(\sigma^{a+b-1}) \times X(\sigma^c) \xrightarrow{\Phi_{i'j'}} X(\sigma^{a+b+c-2}) \end{aligned}$$

We reconstruct the matrix V using $W_4 \in \Pi^{\circ,1}_{2,a+1}$, $W_6 \in \Pi^{\circ,1}_{2,b+1}$, $W_7 \in \Pi^{\circ,1}_{2,c+1}$. First, we read off $\Delta_{ij}(V) = \Delta_{1,a+1}(W_4) \neq 0$ by assumption. We apply the toric action $T_{\Delta_{ij}}(W_7) = W_1 \in \Pi^{\circ,1}_{2,c+1}$. Now, we rescale the matrix W_6 by multiplication of $\Delta_{ij}^{\pm 1}$ to the vectors v_l for $l \geq j$, producing the matrix W_5. We then reinsert the matrix W_4 into W_5 at the appropriate location, arriving at the matrix $W_3 \in \Pi^{\circ,1}_{2,a+b} \simeq X(\sigma^{a+b-1})$. We then read of $\Delta_{i'j'}(V) = \Delta_{1,c+1}(W_1) \neq 0$ and multiply W_3 by a factor of $\Delta_{i'j'}(V)^{\pm 1}$ for vectors v_l for $l \geq j'$ to produce W_2. We then reinsert the matrix W_1 into W_2 arriving at the desired matrix V. Thereby completing the construction of the desired map. □

4.2. Cuts, forms and cohomology. Now we can study the effect of the cuts on the forms α and ω. More precisely, we use the map $\Phi_{ij} : X(\sigma^{j-i}) \times X(\sigma^{k-j+i+1}) \longrightarrow X(\sigma^k)$ to compute the pullbacks $\Phi^*_{ij}\alpha$ and $\Phi^*_{ij}\omega$. The forms α and ω are equivalent under cluster mutation by [**17**]; hence, we choose an arbitrary cluster chart, see Figure 9, and determine the how the forms interact with cuts.

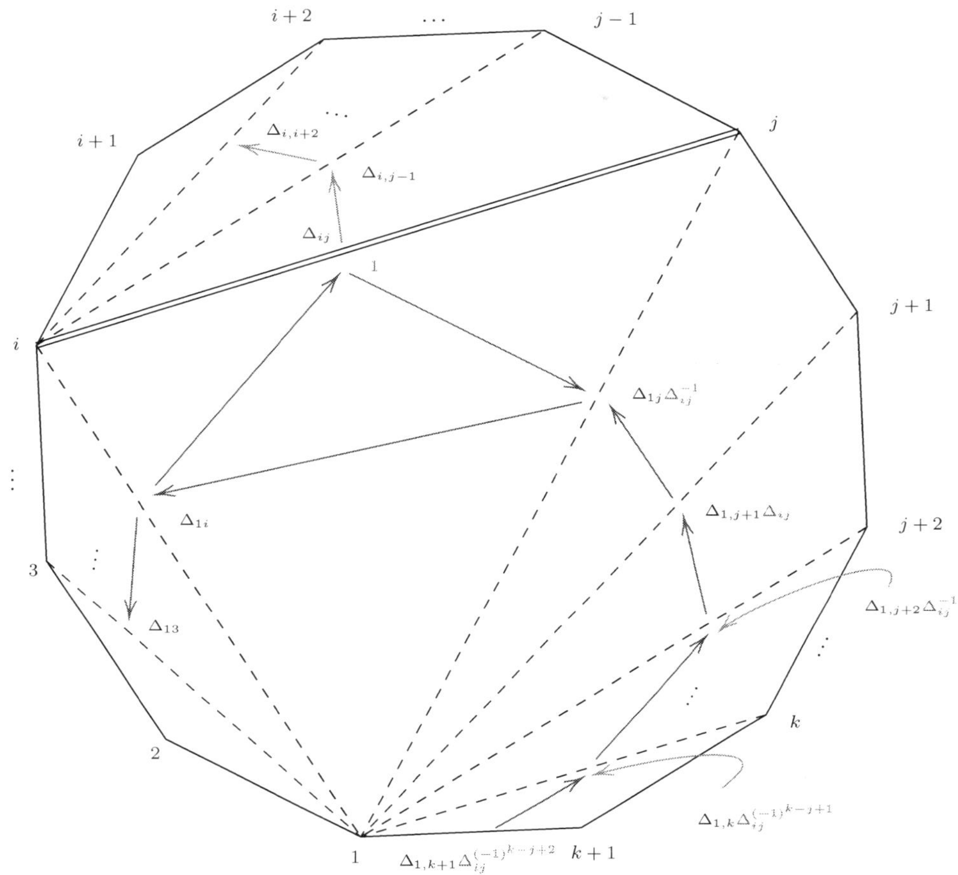

FIGURE 9. Triangulation of $(k+1)$-gon corresponding to the braid variety $X(\sigma^k)$ with its associated quiver. A cut Δ_{ij} is depicted between vertices i and j. The cluster variables from the particular triangulation are the written in black and the rescaling factor of the cluster variables from the cut Δ_{ij} are written in red.

We will denote the forms from $X(\sigma^{j-i})$ by α_1 and ω_1, and the forms from $X(\sigma^{k-j+i+1})$ by α_2 and ω_2. As an abuse of notation we use the labeling from the larger positroid $\Pi^{o,1}_{2,a+b-1}$ identified with $X(\sigma^{a+b-1})$. Technically, under the isomorphism $\Delta_{1,j-i+1} = \Phi^*_{ij}(\Delta_{ij})$, therefore, $\alpha_1 = \Phi^*_{ij}(d\log\Delta_{ij})$, similarly, $\alpha_2 = \Phi^*_{ij}(d\log\Delta_{ij}^{(-1)^{k-j+1}}\Delta_{1,k+1})$ with similar considerations made to ω_1 and ω_2.

LEMMA 4.9. *We have*

$$\Phi^*_{ij}\alpha = \alpha_2 + (-1)^{k-j}\alpha_1.$$

PROOF. Recall that $\alpha = d\log(\Delta_{1,k+1})$. By [**17**] let $\alpha_1 = d\log(\Delta_{ij})$ be the 1-form associated to $X(\sigma^{j-i})$ and $\alpha_2 = d\log(\Delta_{ij}^{(-1)^{k-j+1}}w) = d\log(\Delta_{ij}^{(-1)^{k-j+1}}\Delta_{1,k+1})$

be the 1-form associated to $X(\sigma^{k-j+i+1})$. Given these conditions we find that

$$\begin{aligned}\alpha_2 + (-1)^{k-j}\alpha_1 &= d\log(\Delta_{ij}^{(-1)^{k-j+1}}\Delta_{1,k+1}) + (-1)^{k-j} d\log(\Delta_{ij}) \\ &= d\log(\Delta_{ij}^{(-1)^{k-j+1}}) + d\log(\Delta_{1,k+1}) + (-1)^{k-j} d\log(\Delta_{ij}) \\ &= (-1)^{k-j+1} d\log(\Delta_{ij}) + d\log(\Delta_{1,k+1}) + (-1)^{k-j} d\log(\Delta_{ij}) \\ &= d\log(\Delta_{1,k+1}) = \alpha\end{aligned}$$ □

LEMMA 4.10. *We have*

$$\Phi_{ij}^*\omega = \omega_1 + \omega_2 + (-1)^{k-j}\alpha_1 \wedge \alpha_2.$$

PROOF. Consider the quiver associated to the triangulation of $X(\sigma^k)$ in Figure 9 prior to the rescaling given by the cut Δ_{ij}, by (8), the two-form ω is described as

$$\begin{aligned}\omega = {} & d\log\Delta_{1,k+1} \wedge d\log\Delta_{1,k} + d\log\Delta_{1,k} \wedge d\log\Delta_{1,k-1} \\ & + \cdots + d\log\Delta_{1,j+1} \wedge d\log\Delta_{1,j} + d\log\Delta_{1,j} \wedge d\log\Delta_{1,i} \\ & + d\log\Delta_{1,i} \wedge d\log\Delta_{1,i-1} + d\log\Delta_{1,i-1} \wedge d\log\Delta_{1,i-2} \\ & + \cdots + d\log\Delta_{14} \wedge d\log\Delta_{13} + d\log\Delta_{1,i} \wedge d\log\Delta_{ij} \\ & + d\log\Delta_{ij} \wedge d\log\Delta_{1,j} + d\log\Delta_{i,j-1} \wedge d\log\Delta_{i,j-2} \\ & + d\log\Delta_{i,j-2} \wedge d\log\Delta_{i,j-3} + \cdots + d\log\Delta_{i,i+3} \wedge d\log\Delta_{i,i+2}\end{aligned}$$

Let α_1, α_2 be the 1-form and ω_1, ω_2 be the 2-form associated to $X(\sigma^{j-i})$ and $X(\sigma^{k-j+i+1})$, respectively. By Figure 9, we define the forms associated to $X(\sigma^{j-i})$ and $X(\sigma^{k-j+i+1})$ directly from quivers as follows:

$$\alpha_1 = d\log\Delta_{ij} \tag{23}$$

$$\alpha_2 = d\log(\Delta_{1,k+1}\Delta_{ij}^{(-1)^{k-j+2}}) = d\log\Delta_{1,k+1} + (-1)^{k-j+2} d\log\Delta_{ij} \tag{24}$$

$$\begin{aligned}\omega_1 = {} & d\log\Delta_{i,j-1} \wedge d\log\Delta_{i,j-2} + d\log\Delta_{i,j-2} \wedge d\log\Delta_{i,j-3} \\ & + \cdots + d\log\Delta_{i,i+3} \wedge d\log\Delta_{i,i+2}\end{aligned}$$

While α_1, α_2, ω_1 can be easily read from the cluster chart seen in Figure 9, the 2-form ω_2 requires a bit more finesse. We notice that there is a triangle formed between the vertices $1, i, j$, to simplify the computation of ω_2, which agrees with (8), we decompose the form into parts and call them pre-triangle $\omega_{2,pre}$ for vertices between 1 and i, triangle $\omega_{2,tri}$ for the special vertices $1, i, j$ and post-triangle $\omega_{2,post}$ for vertices between j and $k+1$. By Theorem 4.4 in the rescaled braid variety $X(\sigma^{k-j+i+1})$ the Plučker coordinate $\Delta'_{ij} = \Delta_{ij}\Delta_{ij}^{-1} = 1$ resulting in $d\log\Delta'_{ij} = d\log 1 = 0$, whereas Δ_{ij} shall remain the nonzero polynomial w describing $X(\sigma^{j-i})$.

Using this decomposition, $\omega_2 = \omega_{2,pre} + \omega_{2,tri} + \omega_{2,post}$ is defined by

$$\begin{aligned}\omega_{2,pre} &= d\log\Delta_{1i}\wedge d\log\Delta_{1,i-1} + d\log\Delta_{1,i-1}\\ &\quad\wedge d\log\Delta_{1,i-2} + \cdots + d\log\Delta_{14}\wedge d\log\Delta_{13}\end{aligned}$$

$$\begin{aligned}\omega_{2,tri} &= d\log\left(\Delta_{1j}\Delta_{ij}^{-1}\right)\wedge d\log\Delta_{1i} + d\log\Delta_{1i}\\ &\quad\wedge d\log\Delta'_{ij} + d\log\Delta'_{ij}\wedge d\log\left(\Delta_{1j}\Delta_{ij}^{-1}\right)\\ &= (d\log\Delta_{1j} - d\log\Delta_{ij})\wedge d\log\Delta_{1i}\\ &= d\log\Delta_{1j}\wedge d\log\Delta_{1i} - d\log\Delta_{ij}\wedge d\log\Delta_{1i}\end{aligned}$$

$$\begin{aligned}\omega_{2,post} &= d\log\Delta_{1,k+1}\Delta_{ij}^{(-1)^{k-j+2}}\wedge d\log\Delta_{1,k}\Delta_{ij}^{(-1)^{k-j+1}}\\ &\quad+ d\log\Delta_{1,k}\Delta_{ij}^{(-1)^{k-j+1}}\wedge d\log\Delta_{1,k-1}\Delta_{ij}^{(-1)^{k-j}}\\ &\quad+\cdots+ d\log\Delta_{1,j+2}\Delta_{ij}^{-1}\wedge d\log\Delta_{1,j+1}\Delta_{ij}\\ &\quad+ d\log\Delta_{1,j+1}\Delta_{ij}\wedge d\log\Delta_{1,j}\Delta_{ij}^{-1}\\ &= (d\log\Delta_{1,k+1} + (-1)^{k-j+2}d\log\Delta_{ij})\wedge(d\log\Delta_{1,k} + (-1)^{k-j+1}d\log\Delta_{ij})\\ &\quad+ (d\log\Delta_{1,k} + (-1)^{k-j+1}d\log\Delta_{ij})\wedge(d\log\Delta_{1,k-1} + (-1)^{k-j}d\log\Delta_{ij})\\ &\quad+\cdots+(d\log\Delta_{1,j+2} - d\log\Delta_{ij})\wedge(d\log\Delta_{1,j+1} + d\log\Delta_{ij})\\ &\quad+(d\log\Delta_{1,j+1} + d\log\Delta_{ij})\wedge(d\log\Delta_{1,j} - d\log\Delta_{ij})\\ &= d\log\Delta_{1,k+1}\wedge d\log\Delta_{1,k} + (-1)^{k-j+1}d\log\Delta_{1,k+1}\wedge d\log\Delta_{ij}\\ &\quad+(-1)^{k-j+2}d\log\Delta_{ij}\wedge d\log\Delta_{1,k} + d\log\Delta_{1,k}\wedge d\log\Delta_{1,k-1}\\ &\quad+(-1)^{k-j}d\log\Delta_{1,k}\wedge d\log\Delta_{ij} + (-1)^{k-j+1}d\log\Delta_{ij}\wedge d\log\Delta_{1,k-1}\\ &\quad+\cdots+d\log\Delta_{1,j+2}\wedge d\log\Delta_{1,j+1} + d\log\Delta_{1,j+2}\wedge d\log\Delta_{ij}\\ &\quad- d\log\Delta_{i,j}\wedge d\log\Delta_{1,j+1} + d\log\Delta_{1,j+1}\wedge d\log\Delta_{1,j}\\ &\quad- d\log\Delta_{1,j+1}\wedge d\log\Delta_{ij} + d\log\Delta_{ij}\wedge d\log\Delta_{1,j}\\ &= d\log\Delta_{1,k+1}\wedge d\log\Delta_{1,k} + d\log\Delta_{1,k}\wedge d\log\Delta_{1,k-1}\\ &\quad+\cdots+d\log\Delta_{1,j+2}\wedge d\log\Delta_{1,j+1} + d\log\Delta_{1,j+1}\wedge d\log\Delta_{1,j}\\ &\quad+(-1)^{k-j+1}d\log\Delta_{1,k+1}\wedge d\log\Delta_{ij}\end{aligned}$$

Note that from (23) and (24), $\alpha_1\wedge\alpha_2 = d\log\Delta_{ij}\wedge d\log\Delta_{1,k+1}$. Therefore, the additional term $(-1)^{k-j+1}d\log\Delta_{1,k+1}\wedge d\log\Delta_{ij}$ from $\omega_{2,post}$ may be negated by $(-1)^{k-j}\alpha_1\wedge\alpha_2$, providing the necessary adjustment to acquire $\Phi_{ij}^*\omega$ as stated. □

THEOREM 4.11. *The pullback map*

$$\Phi_{ij}^* : H^*(X(\sigma^k)) \to H^*(X(\sigma^{j-i}))\otimes H^*(X(\sigma^{k-j+i+1}))$$

is injective and can be described by Lemmas 4.9 *and* 4.10

PROOF. Similar to Theorem 3.5, we want to prove that the restrictions of all forms in (10) and (11) do not vanish in $H^*(X(\sigma^{j-i}))\otimes H^*(X(\sigma^{k-j+i+1}))$, here we use the formulas from Lemmas 4.9 and 4.10.

Suppose k is odd, then we want to show that $\Phi_{ij}^*\left[\alpha\omega^{\frac{k-3}{2}}\right]$ and $\Phi_{ij}^*\left[\omega^{\frac{k-1}{2}}\right]$ are both nonzero. Since $k = a + b - 1$ is odd, then either a, b are both even or both odd.

(i) Suppose a and b are both even. Given that $\omega_1^{\frac{a}{2}-1}$, $\alpha_1\omega_1^{\frac{a}{2}-1}$, $\omega_2^{\frac{b}{2}-1}$, $\alpha_2\omega_2^{\frac{b}{2}-1}$ are nonzero by definition, then

$$\begin{aligned}
\Phi_{ij}^*\left[\alpha\omega^{\frac{k-3}{2}}\right] &= (\alpha_2+(-1)^{k-j}\alpha_1)(\omega_1+\omega_2+(-1)^{k-j}\alpha_1\wedge\alpha_2)^{\frac{k-3}{2}} \\
&= (\alpha_2+(-1)^{k-j}\alpha_1)(\omega_1+\omega_2+(-1)^{k-j}\alpha_1\wedge\alpha_2)^{\frac{a+b-4}{2}} \\
&= (\alpha_2+(-1)^{k-j}\alpha_1) \\
&\quad\times \sum_{l_1+l_2+l_3=\frac{a+b-4}{2}} \binom{\frac{a+b-4}{2}}{l_1, l_2, l_3}\omega_1^{l_1}\omega_2^{l_2}\left((-1)^{k-j}\alpha_1\wedge\alpha_2\right)^{l_3} \\
&= \binom{\frac{a+b-4}{2}}{\frac{a}{2}-1, \frac{b}{2}-1, 0}\alpha_2\omega_1^{\frac{a}{2}-1}\omega_2^{\frac{b}{2}-1}+\dots
\end{aligned}$$

with $\alpha_2\omega_2^{\frac{b}{2}-1}$, $\omega_1^{\frac{a}{2}-1}\neq 0$, then $\Phi_{ij}^*\left[\alpha\omega^{\frac{k-3}{2}}\right]$ is nonvanishing. Furthermore,

$$\begin{aligned}
\Phi_{ij}^*\left[\omega^{\frac{k-1}{2}}\right] &= (\omega_1+\omega_2+(-1)^{k-j}\alpha_1\wedge\alpha_2)^{\frac{k-1}{2}} \\
&= (\omega_1+\omega_2+(-1)^{k-j}\alpha_1\wedge\alpha_2)^{\frac{a+b-2}{2}} \\
&= \sum_{l_1+l_2+l_3=\frac{a+b-2}{2}} \binom{\frac{a+b-2}{2}}{l_1, l_2, l_3}\omega_1^{l_1}\omega_2^{l_2}\left((-1)^{k-j}\alpha_1\wedge\alpha_2\right)^{l_3} \\
&= \binom{\frac{a+b-2}{2}}{\frac{a}{2}-1, \frac{b}{2}-1, 1}\left((-1)^{k-j}\alpha_1\wedge\alpha_2\right)\omega_1^{\frac{a}{2}-1}\omega_2^{\frac{b}{2}-1}+\dots
\end{aligned}$$

where $\alpha_1\omega_1^{\frac{a}{2}-1}$, $\alpha_2\omega_2^{\frac{b}{2}-1}\neq 0$. Then $\Phi_{ij}^*\left[\omega^{\frac{k-1}{2}}\right]$ is nonvanishing.

(ii) Suppose a and b are both odd. Given that $\alpha_1\omega_1^{\frac{a-3}{2}}$, $\omega_1^{\frac{a-1}{2}}$, $\alpha_2\omega_2^{\frac{b-3}{2}}$, $\omega_2^{\frac{b-1}{2}}$ are nonzero, then

$$\begin{aligned}
\Phi_{ij}^*\left[\alpha\omega^{\frac{k-3}{2}}\right] &= (\alpha_2+(-1)^{k-j}\alpha_1) \\
&\quad\times \sum_{l_1+l_2+l_3=\frac{a+b-4}{2}} \binom{\frac{a+b-4}{2}}{l_1, l_2, l_3}\omega_1^{l_1}\omega_2^{l_2}\left((-1)^{k-j}\alpha_1\wedge\alpha_2\right)^{l_3} \\
&= (\alpha_2+(-1)^{k-j}\alpha_1)\binom{\frac{a+b-4}{2}}{\frac{a-3}{2}, \frac{b-}{2}, 0}\omega_1^{\frac{a-3}{2}}\omega_2^{\frac{b-1}{2}}+\dots \\
&= (-1)^{k-j}\binom{\frac{a+b-4}{2}}{\frac{a-3}{2}, \frac{b-}{2}, 0}\alpha_1\omega_1^{\frac{a-3}{2}}\omega_2^{\frac{b-1}{2}}+\dots
\end{aligned}$$

Given $\alpha_1\omega_1^{\frac{a-3}{2}}$, $\omega_2^{\frac{b-1}{2}}\neq 0$, then $\Phi_{ij}^*\left[\alpha\omega^{\frac{k-3}{2}}\right]$ is nonvanishing. Furthermore,

$$\begin{aligned}
\Phi_{ij}^*\left[\omega^{\frac{k-1}{2}}\right] &= \sum_{l_1+l_2+l_3=\frac{a+b-2}{2}} \binom{\frac{a+b-2}{2}}{l_1, l_2, l_3}\omega_1^{l_1}\omega_2^{l_2}\left((-1)^{k-j}\alpha_1\wedge\alpha_2\right)^{l_3} \\
&= \binom{\frac{a+b-2}{2}}{\frac{a-1}{2}, \frac{b-1}{2}, 0}\omega_1^{\frac{a-1}{2}}\omega_2^{\frac{b-1}{2}}+\dots
\end{aligned}$$

Since $\omega_1^{\frac{a-1}{2}}$, $\omega_2^{\frac{b-1}{2}}\neq 0$, then $\Phi_{ij}^*\left[\omega^{\frac{k-1}{2}}\right]$ is nonvanishing.

Now, suppose k is even, then we want to show that $\Phi_{ij}^*\left[\omega^{\frac{k}{2}-1}\right]$ and $\Phi_{ij}^*\left[\alpha\omega^{\frac{k}{2}-1}\right]$ are both nonzero. Since $k = a + b - 1$ is even, without loss of generality a is even and b is odd. Since a is even and b is odd, then $\omega_1^{\frac{a}{2}-1}$, $\alpha_1\omega_1^{\frac{a}{2}-1}$, $\alpha_2\omega_2^{\frac{b-3}{2}}$, $\omega_2^{\frac{b-1}{2}}$ are nonzero, then

$$\begin{aligned}
\Phi_{ij}^*\left[\omega^{\frac{k}{2}-1}\right] &= (\omega_1 + \omega_2 + (-1)^{k-j}\alpha_1 \wedge \alpha_2)^{\frac{k}{2}-1} \\
&= (\omega_1 + \omega_2 + (-1)^{k-j}\alpha_1 \wedge \alpha_2)^{\frac{a+b-3}{2}} \\
&= \sum_{l_1+l_2+l_3=\frac{a+b-3}{2}} \binom{\frac{a+b-3}{2}}{l_1, l_2, l_3} \omega_1^{l_1}\omega_2^{l_2}\left((-1)^{k-j}\alpha_1 \wedge \alpha_2\right)^{l_3} \\
&= \binom{\frac{a+b-3}{2}}{\frac{a}{2}-1, \frac{b-1}{2}, 0} \omega_1^{\frac{a}{2}-1}\omega_2^{\frac{b-1}{2}} + \dots
\end{aligned}$$

Since $\omega_1^{\frac{a}{2}-1}$, $\omega_2^{\frac{b-1}{2}} \neq 0$, then $\Phi_{ij}^*\left[\omega^{\frac{k}{2}-1}\right]$ is nonvanishing. Next,

$$\begin{aligned}
\Phi_{ij}^*\left[\alpha\omega^{\frac{k}{2}-1}\right] &= (\alpha_2 + (-1)^{k-j}\alpha_1)(\omega_1 + \omega_2 + (-1)^{k-j}\alpha_1 \wedge \alpha_2)^{\frac{k}{2}-1} \\
&= (\alpha_2 + (-1)^{k-j}\alpha_1)(\omega_1 + \omega_2 + (-1)^{k-j}\alpha_1 \wedge \alpha_2)^{\frac{a+b-3}{2}} \\
&= (\alpha_2 + (-1)^{k-j}\alpha_1) \\
&\quad \times \sum_{l_1+l_2+l_3=\frac{a+b-3}{2}} \binom{\frac{a+b-3}{2}}{l_1, l_2, l_3} \omega_1^{l_1}\omega_2^{l_2}\left((-1)^{k-j}\alpha_1 \wedge \alpha_2\right)^{l_3} \\
&= (\alpha_2 + (-1)^{k-j}\alpha_1)\binom{\frac{a+b-3}{2}}{\frac{a}{2}-1, \frac{b-1}{2}, 0} \omega_1^{\frac{a}{2}-1}\omega_2^{\frac{b-1}{2}} + \dots \\
&= \binom{\frac{a+b-3}{2}}{\frac{a}{2}-1, \frac{b-1}{2}, 0} \alpha_2\omega_1^{\frac{a}{2}-1}\omega_2^{\frac{b-1}{2}} + \dots
\end{aligned}$$

Since $\omega_1^{\frac{a}{2}-1}$, $\alpha_2\omega_2^{\frac{b-1}{2}} \neq 0$, then $\Phi_{ij}^*\left[\alpha\omega^{\frac{k}{2}-1}\right]$ is nonvanishing.

This implies that all the forms in (10) and (11) are nonzero in $H^*(X(\sigma^{j-i})) \otimes H^*(X(\sigma^{k-j+i+1}))$ and hence nonzero in $H^*(X(\sigma^k))$. $\square$

Acknowledgments

I wish to highly thank Eugene Gorsky, Roger Casals, Jose Simental-Rodriguez, Lauren Williams, Misha Mazin, Pavel Galashin and Catharina Stroppel for helpful conversations related to this paper. I also would like to express my sincerest gratitude to the anonymous referee of my paper, their comments and suggested edits have been highly welcome. In no particular order, I wish to thank Ian Sullivan, Trevor Oliveira-Smith, Shanon Rubin, Matthew Corbelli, Sari Ogami, Alex Black, Milo Bechthoff Weising, Alexander Simons, Greg DePaul, Brian Harvey, Laura Starkston, Becca Thomases, Tina Denena and most importantly my family. I would love to thank everyone that has greatly helped me, however I believe I would be required to write an entirely new paper simply for this purpose.

References

[1] Roger Casals, Eugene Gorsky, Mikhail Gorsky, and José Simental, *Algebraic weaves and braid varieties*, Amer. J. Math. **146** (2024), no. 6, 1469–1576. MR4855860

[2] R. Casals, E. Gorsky, M. Gorsky, and J. Simental, *Positroid links and braid varieties*, arXiv:2105.13948, 2021.

[3] Roger Casals, Eugene Gorsky, Mikhail Gorsky, Ian Le, Linhui Shen, and José Simental, *Cluster structures on braid varieties*, J. Amer. Math. Soc. **38** (2025), no. 2, 369–479, DOI 10.1090/jams/1048. MR4868947

[4] Roger Casals and Daping Weng, *Microlocal theory of Legendrian links and cluster algebras*, Geom. Topol. **28** (2024), no. 2, 901–1000, DOI 10.2140/gt.2024.28.901. MR4718130

[5] Baptiste Chantraine, Lenhard Ng, and Steven Sivek, *Representations, sheaves and Legendrian $(2,m)$ torus links*, J. Lond. Math. Soc. (2) **100** (2019), no. 1, 41–82, DOI 10.1112/jlms.12204. MR3999682

[6] Chris Fraser, *Quasi-homomorphisms of cluster algebras*, Adv. in Appl. Math. **81** (2016), 40–77, DOI 10.1016/j.aam.2016.06.005. MR3551663

[7] Chris Fraser and Melissa Sherman-Bennett, *Positroid cluster structures from relabeled plabic graphs*, Algebr. Comb. **5** (2022), no. 3, 469–513, DOI 10.5802/alco.220. MR4456866

[8] Eugene Gorsky, Oscar Kivinen, and José Simental, *Algebra and geometry of link homology: lecture notes from the IHES 2021 Summer School*, Bull. Lond. Math. Soc. **55** (2023), no. 2, 537–591, DOI 10.1112/blms.12761. MR4575924

[9] Pavel Galashin and Thomas Lam, *Positroid varieties and cluster algebras* (English, with English and French summaries), Ann. Sci. Éc. Norm. Supér. (4) **56** (2023), no. 3, 859–884, DOI 10.24033/asens.2545. MR4650160

[10] Pavel Galashin and Thomas Lam, *Positroids, knots, and q,t-Catalan numbers*, Duke Math. J. **173** (2024), no. 11, 2117–2195, DOI 10.1215/00127094-2023-0049. MR4809331

[11] P. Galashin, T. Lam, M. Sherman-Bennett, and D. Speyer, *Braid variety cluster structures, I: 3D plabic graphs*, arXiv:2210.04778, 2022.

[12] Michael Gekhtman, Michael Shapiro, and Alek Vainshtein, *Cluster algebras and Weil-Petersson forms*, Duke Math. J. **127** (2005), no. 2, 291–311, DOI 10.1215/S0012-7094-04-12723-X. MR2130414

[13] Eugene Gorsky and Matthew Hogancamp, *Hilbert schemes and y-ification of Khovanov-Rozansky homology*, Geom. Topol. **26** (2022), no. 2, 587–678, DOI 10.2140/gt.2022.26.587. MR4444266

[14] Eugene Gorsky, Matthew Hogancamp, Anton Mellit, and Keita Nakagane, *Serre duality for Khovanov-Rozansky homology*, Selecta Math. (N.S.) **25** (2019), no. 5, Paper No. 79, 33, DOI 10.1007/s00029-019-0524-5. MR4036505

[15] James Hughes, *Lagrangian fillings in A-type and their Kálmán loop orbits*, Rev. Mat. Iberoam. **39** (2023), no. 5, 1681–1723, DOI 10.4171/rmi/1436. MR4651703

[16] Allen Knutson, Thomas Lam, and David E. Speyer, *Positroid varieties: juggling and geometry*, Compos. Math. **149** (2013), no. 10, 1710–1752, DOI 10.1112/S0010437X13007240. MR3123307

[17] Thomas Lam and David E. Speyer, *Cohomology of cluster varieties I: Locally acyclic case*, Algebra Number Theory **16** (2022), no. 1, 179–230, DOI 10.2140/ant.2022.16.179. MR4384567

[18] Joshua S. Scott, *Grassmannians and cluster algebras*, Proc. London Math. Soc. (3) **92** (2006), no. 2, 345–380, DOI 10.1112/S0024611505015571. MR2205721

[19] K. Serhiyenko, M. Sherman-Bennett, and L. Williams, *Cluster structures in Schubert varieties in the Grassmannian*, Proc. Lond. Math. Soc. (3) **119** (2019), no. 6, 1694–1744, DOI 10.1112/plms.12281. MR4295518

[20] M-T.Q. Trinh, *From the Hecke category to the unipotent locus*, arXiv:2106.07444, 2021.

[21] Lauren K. Williams, *Cluster algebras: an introduction*, Bull. Amer. Math. Soc. (N.S.) **51** (2014), no. 1, 1–26, DOI 10.1090/S0273-0979-2013-01417-4. MR3119820

Department of Mathematics, University of California, Santa Barbara, California
Email address: tmscroggin@ucsb.edu

Contemporary Mathematics
Volume **827**, 2025
https://doi.org/10.1090/conm/827/16535

Symmetry of three component links

Kate Kearney

ABSTRACT. The group of admissible symmetries for a given link is called the *intrinsic symmetry group of* L. For a link of n components this is a subgroup of the group $\Gamma_n = \mathbb{Z}_2 \times (\mathbb{Z}_2^n \rtimes S_n)$. This paper examines links with three components: providing several new examples of calculations, and working to develop new tools for these calculations. In particular, we will discuss the algebraic relationship between the intrinsic symmetry groups of a link and its sublinks.

1. Introduction

A link is a collection of embedded circles in S^3. Here we consider our links to be oriented and labeled (number the components). There are a number of possible symmetries to a labeled oriented link, including the mirror image, the reversal of orientation of any of the components, and the rearrangement of the labelling of components. While some of these operations may change the link, some produce a link that is isotopic (preserving orientations) to the original. We call such transformations admissible for the link. These form a group under composition. The group of admissible symmetries for a given link, L, is called the *intrinsic symmetry group of* L, and is denoted by $\Sigma(L)$. For a link of n components this is a subgroup of the group $\Gamma_n = \mathbb{Z}_2 \times (\mathbb{Z}_2^n \rtimes S_n)$.

The intrinsic symmetry group of L was first defined by Whitten [**9**], who gave preliminary results about the group and posed three questions related to the group:

(1) What links admit a given symmetry, γ?
(2) What links have a given subgroup Σ of Γ_n as their group of symmetries?
(3) In particular, for what subgroups Σ of Γ_n does there exist a link L for which $\Sigma(L) = \Sigma$?

Hillman made progress on this problem [**5**], [**6**], examining the relationship of the Alexander polynomial and linking numbers to link symmetries, and providing examples of links with 12 of the 27 subgroups of Γ_2. Recent work by Cantarella, Cornish, Mastin and Parsley [**2**] extends this work to find examples for 21 of the the subgroups of Γ_2. Cantarella et al also examined the symmetry groups of links with 8 or fewer crossings [**1**]. These computations (primarily based on examination of linking matrices and knot polynomials as obstructions, as well as considering satellites of links) include 30 two component links, 11 three component links and 3 four component links. Further work of Livingston [**7**] uses the JSJ-decomposition

2020 *Mathematics Subject Classification.* Primary 57K10.

of the link complement to obstruct $\mathbb{A}_n$ from being the intrinsic symmetry group of a link (for $n \geq 6$).

This paper examines further examples of links with three components. Picking up where Cantarella et al left off, we will examine links with nine crossings and three components. In addition to providing several new examples of calculations, we develop new tools for these calculations to provide more insight to the general problem. In the case of two component links, the knot type of each component provides an obstruction to swapping components. Similarly, in the case that the link has three or more components, each of its sublinks can provide obstructions to certain elements of Γ_n from being admissible. We can use the algebraic relationships between $\Sigma(L)$ and the intrinsic symmetry groups of its sublinks to compute $\Sigma(L)$. At this point we have used this result simply as a computational tool for specific examples, although the hope is that it might provide insight into obstruction specific subgroups of Γ_n from being the intrinsic symmetry groups of links (for some n).

2. Definition and Properties

Given a labeled oriented link, we ask whether the rearrangement of components, or the reversal of orientation of either some of the components or the ambient S^3 produces a link which is oriented isotopic to the original. In the case of knots, this corresponds to asking whether the knot is reversible or amphichiral. For a link, we define a symmetry to be *admissible* if it produces a link which is oriented isotopic to the original. Since such symmetries can be performed consecutively we have a natural composition operation and may consider these symmetries as a group.

DEFINITION 2.1. The *intrinsic symmetry group* of a link L, denoted $\Sigma(L)$, is the group of admissible symmetries of L under composition.

We observe that if L has n components, $\Sigma(L)$ is a subgroup of $\Gamma_n = \mathbb{Z}_2 \times (\mathbb{Z}_2^n \rtimes S_n)$. The first component $\mathbb{Z}_2$ summand represents the reversal of orientation of the ambient S^3. The remaining copies of $\mathbb{Z}_2$ correspond to the reversal of orientation of any of the n components of the link (in order as labeled), with the semi-direct product with S_n representing the action of relabeling the components. The element of S_n acts on $\mathbb{Z}_2^n$ by permuting the coordinates. To define the product precisely, we may write elements of Γ_n as $(\eta, \epsilon_1, \ldots, \epsilon_n, s)$ where η and ϵ_k are each ± 1 and $s \in S_n$. Then

$$\begin{aligned}\gamma * \gamma' =&(\eta, \epsilon_1, \ldots, \epsilon_n, s) * (\eta', \epsilon_1', \ldots, \epsilon_n', s') \\ =&((\eta\eta', \epsilon_1\epsilon_{s(1)},' \ldots, \epsilon_n\epsilon_{s(n)}', s's).\end{aligned}$$

We think of γ as acting on L by

(1) Permute the order of components by action of s.
(2) Reverse the orientation of S^3 if $\eta = -1$.
(3) Reverse the orientation of components $K_s(i)$ corresponding to entries $\epsilon_i = -1$.

We follow the convention of Whitten [**9**] that the permutation of components happens before reversal of orientations. Notice that the product is defined so that $\gamma * \gamma'(L) = \gamma(\gamma'(L))$. In examples we will use the more concise notation, $m = (-1, 1, \ldots, 1, id)$, $i_k = (1, 1, \ldots, -1, \ldots, 1, id)$ (the element reversing the kth component), and $f_{mn} = (1, 1, \ldots, 1, (mn))$ (the transposition of the mth and n^{th} component labelings). Products of component reversals will be shortened, e.g.

$i_{12} = i_1 i_2$. We remind the reader that not all of these elements commute. In particular, $i_1 f_{12} = f_{12} i_2$.

In attempting to compute the intrinsic symmetry group for a given link, we look for obstructions to certain elements of Γ_n being admissible for the given link. We begin with the observation that if γ interchanges components, it must be that they have the same knot type. This is extended by a results from [**2**] and [**1**]:

PROPOSITION 2.1 (Lemma 3.2 of [**2**]). *Suppose that $L(K,i)$ is a satellite of L constructed by replacing component i with a knot or link K. Then L cannot have a pure exchange symmetry exchanging components i and j unless $L(K,i)$ and $L(K,j)$ are isotopic.*

PROPOSITION 2.2 (Proposition 5.3 of [**1**]). *The action of the Whitten group Γ_n on n-component links gives rise to the following Γ_n action on the set of $n \times n$ linking matrices* $\mathrm{Lk}(n)$*:*

$$\mathrm{LK}(\gamma(L))_{ij} = \eta \epsilon_i \epsilon_j \,\mathrm{Lk}(L)_{s(i)s(j)}$$

These provide some obstructions to certain elements of Γ_n being admissible for a given link, L. In each case, we are exploring how the relationship between the various components of the link and observing how they behave under certain symmetries. In the next section, we explore further how structures of sublinks can obstruct certain symmetries and apply these to an example.

3. Links and Sublinks

As discussed in section 2, previous work on intrinsic symmetry groups has found several obstructions to symmetry, but has primarily focused on calculating intrinsic symmetry groups for links of two components. Picking up this work to look at three component links, we make the observation that obstructions may arise not just from properties of individual components but also from two-component sublinks of our three-component links. Similarly, if we consider a link of n-components, we may find obstructions from any k-component sublink (with $k < n$).

In the case of knots, we know that the trefoil (3_1) is not equal to its mirror image [**4**]. In our notation, we determine that $m \notin \Sigma(3_1)$. Since the trefoil is reversible, $i_1 \in \Sigma(3_1)$, and hence $\Sigma(3_1) = \{id, i_1\}$.

Now consider the split link $L = U \sqcup 3_1$ (where U denotes the unknot). Since m is not an admissible element for 3_1, it cannot be admissible for L. If there were an isotopy of L and mL, then the same isotopy would produce an isotopy of 3_1 and $m3_1$, by just ignoring the movement of the unknot component.

In fact, by including Γ_1 into Γ_n by $(\eta, \epsilon_1, id) \to (\eta, 1, \ldots, 1, \epsilon_1, 1, \ldots, 1, id)$ we see that if L_i is the ith component of L, we must have a subgroup of $\Sigma(L)$ isomorphic to a subgroup of $\Sigma(L_i)$.

PROPOSITION 3.1. *If L_i is a sublink of L, let $\Sigma_i(L)$ be the subgroup of $\Sigma(L)$ of elements satisfying $s(i) = i$. There is a homomorphism from the subgroup $\Sigma_i(L)$ to $\Sigma(L_i)$.*

PROOF. Suppose γ is an element of $\Sigma_i(L)$. Then γ is an admissible symmetry of L which sends L_i to itself (but may reverse its orientation). Then we must have an isotopy of $\gamma(L)$ to L. Since $s(i) = i$, we may consider this isotopy as an isotopy of L_i, ignoring its behavior on the other link components. Then if $\pi_i(\gamma)$ is the element of Γ_1 given by ignoring all but the ith component $\pi_i : (\eta, \epsilon_1, \ldots, \epsilon_i, \ldots, \epsilon_n, id) \to (\eta, \epsilon_i, id)$, we see that $\pi_i(\gamma)$ must be an admissible symmetry of L_i. □

Other components of link L may obstruct some elements of $\Sigma(L_i)$ from being lifted to symmetries of L, so in general we do not expect to $\Sigma_i(L)$ to be surjective $\Sigma(L_i)$. It may also fail to be injective, since there may be symmetries of the $i^t h$ component that can pair with a variety of symmetries on other components (e.g. if the link is split). On the other hand, since products of elements fixing the ith component will also fix the ith component, we find that $\Sigma_i(L)$ is a subgroup of $\Sigma(L)$ and hence the projections, $\{\pi_i(\gamma)\}_{\gamma\in\Sigma_i(L)}$ form a subgroup of $\Sigma(L_i)$.

Now, considering our link, L, we observe that we can use this to obstruct certain symmetries of L when we know the symmetries of its components.

COROLLARY 3.1. *Consider a link, L, of n components, and let L_i be the ith component of L. Let $\gamma \in \Gamma_n$ be a transformation with $s(i) = i$. Let $\pi_i(\gamma)$ be the projection to the ith component. If $\pi_i(\gamma) \notin \Sigma(L_i)$ then $\gamma \notin \Sigma(L)$.*

In essence, this tells us that a link's symmetries are obstructed by the symmetries of the knots that make up the components. While this alone is a straightforward observation, we now remark that this extends to sublinks of multiple components. In fact, the proof above applies to this more general case with little modification and allows us to use known intrinsic symmetry groups to extend to obstruct particular symmetries of links with more components.

PROPOSITION 3.2. *If L' is a sublink of L, let $\Sigma'(L)$ be the subgroup of $\Sigma(L)$ of elements that permute the sublink components. There is a homomorphism from the subgroup $\Sigma'(L)$ to $\Sigma(L')$.*

PROOF. Suppose γ is an element of $\Sigma'(L)$. Then γ is an admissible symmetry of L which sends L' to itself (but may permute or reverse its orientations of components of L'). Then we must have an isotopy of $\gamma(L)$ to L. Since γ sends L' to itself, we may consider this isotopy as an isotopy of L_i, ignoring its behavior on the other link components. Then if $\pi(\gamma)$ is the element of Γ_k given by ignoring all but the components of L', we see that $\pi(\gamma)$ must be an admissible symmetry of L_i. □

Now we have a result similar to Corollary 3.1

COROLLARY 3.2. *Consider a link, L, of n components, and let L' be a sublink of L. Let $\gamma \in \Gamma_n$ be a transformation with $s(L') = L'$. Let $\pi(\gamma)$ be the projection to the components corresponding to L'. If $\pi(\gamma) \notin \Sigma(L')$ then $\gamma \notin \Sigma(L)$.*

Next we will illustrate the use of Corollary 3.2 in evaluating the intrinsic symmetry group of 3 component links. The knot names referenced follow the Rolfsen's Link Table [**8**], and diagrams come from knotinfo [**3**]. We should observe here the need to take care to follow consistent choices of orientation and component numbering. A reversal of orientation choices or re-ordering of components will change our intrinsic symmetry group to a conjugate group in Γ_n. As we apply Corollary 3.2 we should be sure to work with the numberings and orientations induced by the numbering and orientation of our original link.

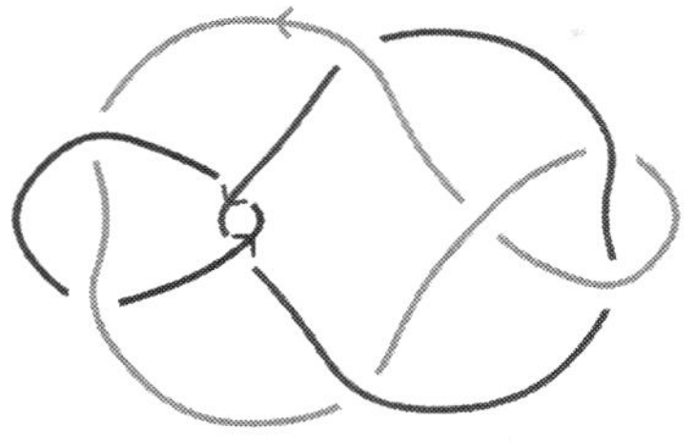

FIGURE 1. 9^3_1

EXAMPLE 3.1. Consider the link 9^3_1. We will number the components right to left across the top of the diagram (purple is 1, green is 2, blue is 3) and orient as indicated in Figure 1. All components of the link are unknots. This provides no restriction on possible symmetries. The linking matrix is $\begin{bmatrix} 0 & 1 & -1 \\ 0 & 0 & 0 \\ 0 & 0 & 0 \end{bmatrix}$.

Results from [**1**] allow us to use this to restrict to the possible elements to the subgroup generated by $i_3 f_{12}$, mi_{12}, and i_{123}.

Next we consider Proposition 3.2. The three two component sublinks are $L_{1,2} = 2^2_1$, $L_{1,3} = 2^2_1$, $L_{2,3} = 5^2_1$. From [**2**] we know that $\Sigma(2^2_1) = \Sigma_{8,2}$, which is generated by f and mi_1, and that $\Sigma(5^2_1) = \Sigma_{8,1}$, which is the subgroup of Γ^2 excluding m. This allows us to eliminate mi_{12} as possible elements of $\Sigma(9^3_1)$. A direct analysis of isotopies confirms that $i_3 f_{12}$ and i_{123} are both admissible. We conclude that $\Sigma(9^3_1) =< i_3 f_{12}, i_{123} >$.

4. Whitten's Third Question

As seen above, Corollary 3.2 can help us compute the intrinsic symmetry group for a given link, when the link has sublinks with known intrinsic symmetry groups. This addresses the first and second of Whitten's questions [**9**].

We may also make progress towards Whitten's third question (which subgroups of Γ_n can be attained as intrinsic symmetry groups), using Proposition 3.2 to extend known examples to higher component examples. Many of the subgroups of Γ_n are isomorphic to subgroups of Γ_k for some $k < n$. In these cases, if an example of a link on k components is known to have the given intrinsic symmetry group, then we may add on (as a split link) a link of $n - k$ components with an appropriate symmetry group. This produces a link with n components and the same intrinsic symmetry group.

EXAMPLE 4.1. As an example of this, consider the group $\Sigma_{4,2} =< i_1, i_2 >$. According to [**2**], the link $7a3$ (Rolfsen's 7^2_4) has intrinsic symmetry group $\Sigma_{4,2}$. Its components are a trefoil and an unknot. We select 9_{32} as a companion knot in a split link of three components, $L = 7a3 \sqcup 9_{32}$ (see Figure 2). Since 9_{32} has a different knot type from either component of $7a3$, we may rule out the possibility of any component interchanges being admissible for L. Further, 9_{32} has trivial

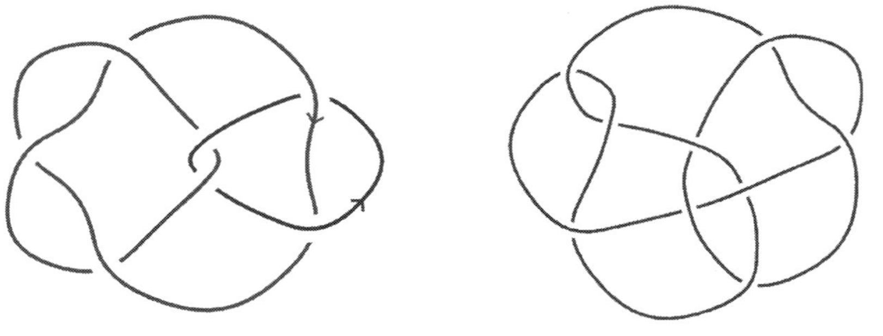

(A) $7a3 = 7^4_2$ (B) 9_{32}

FIGURE 2. Links for Example 4.1

intrinsic symmetry group, so i_3 and m cannot be admissible for L. We conclude that $\Sigma(L) = \Sigma(7a3) = \Sigma_{4,2}$.

This technique can find examples of split links for many of the possible subgroups of Γ_n, based on lifting known examples. On the other hand, we have produced split links. As observed by Livingston [**7**], it is an interesting question to ask whether non-split links exist with the same intrinsic symmetry groups. Future work might explore this as well as further applications to the computation of intrinsic symmetry groups.

References

[1] M. Berglund, J. Cantarella, M. P. Casey, E. Dannenberg, W. George, A. Johnson, A. Kelley, A. LaPointe, M. Mastin, J. Parsley, J. Rooney, and R. Whitaker, *Intrinsic symmetry groups of links with 8 and fewer crossings*, Symmetry **4** (2012), no. 1, 143–207, DOI 10.3390/sym4010143. MR2909632

[2] J. Cantarella, J. Cornish, M. Mastin, and J. Parsley, *The 27 possible intrinsic symmetry groups of two-component links*, Symmetry **4** (2012), no. 1, 129–142, DOI 10.3390/sym4010129. MR2909631

[3] J. C. Cha and C. Livingston, *KnotInfo: Table of Knot Invariants*, `http://www.indiana.edu/~knotinfo`, October 2013.

[4] M. Dehn, *Die beiden Kleeblattschlingen* (German), Math. Ann. **75** (1914), no. 3, 402–413, DOI 10.1007/BF01563732. MR1511799

[5] J. A. Hillman, *Symmetries of knots and links, and invariants of abelian coverings. I*, Kobe J. Math. **3** (1986), no. 1, 7–27. MR867798

[6] J. A. Hillman, *Symmetries of knots and links, and invariants of abelian coverings. II.*, Kobe Journal of Mathematics **3** (1987), no. 2, 149.

[7] C. Livingston, *Intrinsic symmetry groups of links*, Algebr. Geom. Topol. **23** (2023), no. 5, 2347–2368, DOI 10.2140/agt.2023.23.2347. MR4621434

[8] D. Rolfsen, *Knots and links*, Mathematics Lecture Series, vol. 7, Publish or Perish, Inc., Houston, TX, 1990. Corrected reprint of the 1976 original. MR1277811

[9] W. C. Whitten Jr., *Symmetries of links*, Trans. Amer. Math. Soc. **135** (1969), 213–222, DOI 10.2307/1995013. MR242146

GONZAGA UNIVERSITY, 502 E BOONE AVE, SPOKANE, WASHINGTON 99258
Email address: `kearney@gonzaga.edu`

Contemporary Mathematics
Volume **827**, 2025
https://doi.org/10.1090/conm/827/16536

The mixed Hilden braid group and the plat equivalence in handlebodies

Paolo Cavicchioli and Sofia Lambropoulou

ABSTRACT. Given a knot or link in the handlebody, H_g, of genus g we prove that it can always be represented as the plat closure of a braid in H_g. We further establish the Hilden braid group for the handlebody, as a subgroup of the mixed braid group, whose elements have the first g strands fixed forming the identity braid. We then formulate and prove the algebraic equivalence connecting mixed plats belonging to the same link isotopy class in H_g. The plat closure representation can be particularly suitable for computing knot invariants.

Introduction

The representation of links as plat closures of braids with an even number of strands dates back to the works of Hilden [**20**] and of Birman [**7**], who proved that any link in the 3-space, $\mathbb{R}^3$, or the 3-sphere, S^3, could be represented as the plat closure of a braid in some braid group $\mathcal{B}_{2n}$. As an example, recall that all rational knots/links (also known as 2-bridge knots/links) can be represented as the plat closure of a braid with 4 strands, leaving the last strand as the identity strand [**3**]. Further, in [**20**] Hilden described the plat isotopy of braids in the same braid group $\mathcal{B}_{2n}$, by means of a subgroup K_{2n} of $\mathcal{B}_{2n}$, the *Hilden braid group*. In [**7**] Birman extended the work of Hilden by the stabilization moves, to provide the equivalence relation for braids having isotopic plat closures. This braid equivalence is analogous to the classical Markov equivalence for braids with isotopic standard closures [**26**]. The Markov theorem together with the Alexander braiding theorem [**1, 9**] furnish the algebraic theory for the isotopy equivalence of knots and links represented as standard closures of braids.

A lot of work has been done using the plat representation. For instance, in [**6**] Bigelow studied the possibility of computing the Jones polynomial of a link, represented as plat closure of a braid, in terms of the action of that braid over a homological pairing defined on a covering of the configuration space of n points into the $2n$-punctured disc. His results were analyzed in recent years also from a computational point of view [**17, 27**] and in order to compute other link invariants,

2020 *Mathematics Subject Classification.* Primary 57K10, 57K12.
Key words and phrases. Links in the handlebody, mixed braid group, plat closure of mixed braids, mixed Markov braid group, equivalence theorem.
This research was partially funded by the National Technical University of Athens and by the University of Bologna.

such as the Conway potential function [**30**] or other polynomial invariants [**16**] as well as other invariants of links [**12**].

The above led many researchers to generalize the results obtained in $\mathbb{R}^3$ in the more general setting of closed, connected and orientable (from now on c.c.o.) 3-manifolds and handlebodies. For instance, using Heegaard surfaces, in [**14**] Doll introduced the notion of (g, b)-decomposition or generalized bridge decomposition for links in a c.c.o. 3-manifold, opening the way to the study of links in 3-manifolds via surface braid groups and their plat closures, see works of Bellingeri and Cattabriga [**5**] and Cattabriga and Gabrovšek [**11**].

In a parallel development, in [**24**], Lambropoulou and Rourke established the analogues of the Alexander and Markov theorems for knots and links in c.c.o. 3-manifolds, representing them first as mixed links in S^3 and then as standard closures of geometric mixed braids. For establishing the equivalence, they used the concept of the L-move. In [**23**] the mixed braid group $\mathcal{B}_{g,n}$ is introduced, as the subgroup of the Artin braid group $\mathcal{B}_{g+n}$ that contains the identity braid I_g on g strands as a fixed subbraid, enabling the algebraic formulation of the Markov theorem in c.c.o. 3-manifolds [**25**]. Adapting the work in [**24**] to the setting of a genus g handlebody, H_g, in [**19**] Häring and Lambropoulou established the analogues of the Alexander and Markov theorems for knots and links in H_g, by representing them as standard closures, first of geometric mixed braids and then of algebraic mixed braids, utilizing the mixed braid groups [**23**] and results from [**25**], see also [**28**].

The goal of this article is to establish the counterpart plat closure theory for handlebodies. We first introduce the notion of a braid in the handlebody H_g of genus g and then we show that it is always possible to represent a link in H_g via the plat closure of a braid in H_g (Theorem 3.5). Then we establish the *mixed Hilden braid group* for H_g as the Hilden braid subgroup, $K_{g,2n}$, of the mixed braid group $\mathcal{B}_{g,2n}$, and we provide a set of generators for $K_{g,2n}$ (Theorem 4.3). We finally formulate and prove the algebraic equivalence connecting two plats in H_g belonging to the same isotopy class via the mixed Hilden braid groups (Theorem 5.5). We are especially interested in this last part because it would be the starting point of the construction and computation of polynomial link invariants for links in the setting of handlebodies, and then extend further to the case of other 3-manifolds.

We also studied links in handlebodies in [**13**] from a different aspect. Namely, in [**13**] we proved that, given the representation of a knot or link via the plat closure of a braid in $\mathbb{R}^3$, in a handlebody or in a thickened surface, it is possible to create algorithmically another braid representing the same link type but with the standard closure, and vice versa.

With the present paper we fill in a gap in the literature, namely to study the plat closure for braids in a handlebody, by establishing first that every knot or link in H_g has indeed a plat representative and then by formulating and proving their isotopy equivalence using the mixed Hilden braid groups. It is worth adding that the plat closure representation of knots and links in thickened surfaces has been studied in [**5**] and [**11**], whilst the standard closure representation is being studied in [**10**].

In Section 1 we recall the notions of interest in the case of $\mathbb{R}^3$. Namely, the Artin braid group $\mathcal{B}_m$, the existence of standard and plat closure of braids, the braidings of knots and links to plats and to standardly closed braids, as well as the corresponding equivalence relations reflecting isotopy. In particular, in Section 1.4

we recall the definition of the Hilden braid group used in the plat equivalence. In Section 2 we recall the representation of the handlebody H_g in $\mathbb{R}^3$ by the identity braid I_g and of links and braids in H_g by mixed links and mixed braids, and the definition and a presentation of the mixed braid group $\mathcal{B}_{g,m}$. In Section 3 we first recall the standard closure of mixed braids, the mixed braiding and the mixed braid equivalence (analogues of the Alexander and Markov theorems for H_g). We then define the plat closure of braids in H_g and of mixed braids, and we state and prove the first part of the main result, the existence of a representation of links in handlebodies via plat closure of braids in Section 3.2. In Section 4 we define the Hilden braid group for H_g and the mixed Hilden braid group, as the Hilden subgroup of the mixed braid group. Then, starting from a set of generators of the classical Hilden braid group, we obtain a set of generators for the mixed Hilden braid group, which is a crucial ingredient in proving the equivalence theorem. Finally, in Section 5 we state and prove the main theorem of the paper, namely Theorem 5.5, providing the equivalence for links in handlebodies represented via plat closed braids.

1. Preliminaries: the case of $\mathbb{R}^3$

In this section we first recall classical braids and their group structure. Then their standard closure, the Alexander theorem for braiding any oriented link, as well as the braid equivalence corresponding to classical link isotopy (the Markov theorem and its L-move analogue). We then proceed with recalling the plat closure for braids and the plat braiding. Finally, we recall the Hilden braid subgroup of the braid group and the plat closure braid equivalence corresponding to classical link isotopy, all by following the work of Birman on plats [**7**]. All of this will be our basic tools to prove the equivalence theorem in the setting of handlebodies.

1.1. The Artin braid group. The Artin braid group [**2**], $\mathcal{B}_n$, is defined as the fundamental group of the unordered configuration space of n points on the real plane, i.e. a group which elements are equivalence classes of n paths $\Psi = (\psi_1, \ldots, \psi_n), \psi_i : [0,1] \to \mathbb{R}^2$ going from a set of n points of the plane to themselves, and never intersecting $(\psi_i(t) \neq \psi_j(t), \forall t \in [0,1], i, j \in 1, \ldots, n, i < j)$.

The classical presentation of B_n is generated by elements $\sigma_i, i \in 1, \ldots, n-1$, which exchanges strands i and $i+1$ as in Fig. 1, with relations:

$$\sigma_i \sigma_j = \sigma_j \sigma_i, \qquad \text{if } |i-j| \geq 2$$
$$\sigma_i \sigma_{i+1} \sigma_i = \sigma_{i+1} \sigma_i \sigma_{i+1}.$$

For further information concerning the braid group the reader can refer, for example, to [**8, 21**].

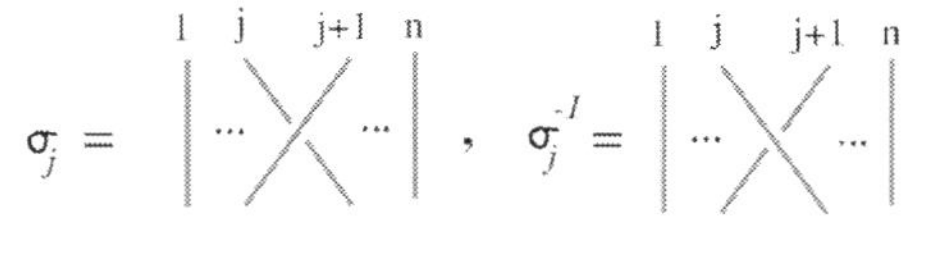

FIGURE 1. The σ_j and σ_j^{-1} generators.

1.2. The standard closure, the braiding and the braid equivalence. In order for a braid to represent a link we have to 'close' it, for example identifying the starting set of points on the plane with itself. This operation can be viewed as connecting the top endpoints to their corresponding bottom endpoints with simple unlinked arcs, like in the right part of Fig. 2. This is called *standard closure* of a braid. It is important to note that this closure naturally gives rise to an oriented link: indeed, if we consider the strands of the braid to be oriented downward, then all closing arcs will be oriented upward and the whole orientation is coherent.

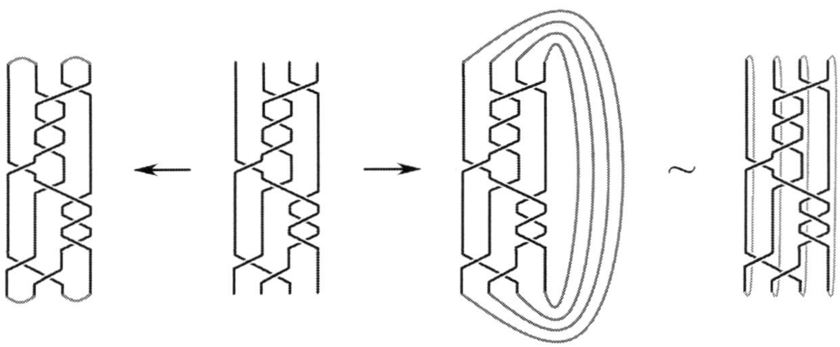

FIGURE 2. On the right, the standard closure of a braid and its isotopy to sliding the closing arcs to the back of the braid. On the left, the plat closure of the same braid.

Conversely, by a result of H. Brunn and of J.W. Alexander [**1**, **9**]:

THEOREM 1.1 (The Alexander theorem). *Every oriented link in $\mathbb{R}^3$ can be represented as the standard closure of a braid in some $\mathcal{B}_n$.*

Different braids (also with a different number of strands) can be closed to give the same (oriented) link up to isotopy. A theorem by A.A. Markov [**26**] (with an improvement by N. Weinberg [**29**], reducing the original 3 moves to 2) gives us a way of determining whether closing two braids results in isotopic links or not:

THEOREM 1.2 (The Markov theorem). *Let $\beta_n \in \mathcal{B}_n, \beta_m \in \mathcal{B}_m$ be two braids, and let L_n, L_m be the two links obtained by the standard closure of β_n, β_m respectively. Then L_n is equivalent to L_m if and only if β_n can be obtained from β_m after a finite sequence of these two moves:*

- *Conjugation:* $\beta \sim \gamma\beta\gamma^{-1}, \quad \beta, \gamma \in \mathcal{B}_n$;
- *Stabilization move:* $\beta \sim \beta\sigma_n^{\pm 1} \in \mathcal{B}_{n+1}, \quad \beta \in \mathcal{B}_n$.

Fig. 3 illustrates abstractly the 2 moves of the theorem. In the figure a red closing arc is indicated, showing the isotopy between the (standard) closures of the middle and right-hand braid.

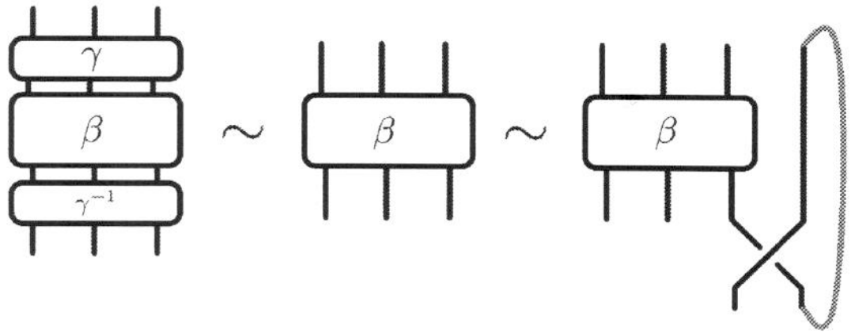

FIGURE 3. The conjugation (on the left) and the stabilization move (on the right).

Some years ago Lambropoulou and Rourke proved in [**24**] another version of the Markov theorem, with only one equivalence move: the *L-move*. This is defined as follows (see abstract illustration in Fig. 4): take a braid $\beta \in \mathcal{B}_n$ and a point P in one of its strands and not aligned to any of the crossings or endpoints of β; cut the arc at P, bend the two resulting arcs apart by a small isotopy and introduce two new vertical arcs (slightly tilted) to two new top and bottom endpoints in the vertical line of P. These two new arcs are both oriented downward and they run either both *under* all other arcs of the diagram (first instance in Fig. 4) or both *over* (third instance in Fig. 4) all other arcs of the diagram, thus performing an *under L-move* (or L_u*-move*) and a *over L-move* (or L_o*-move*) respectively. The right-hand illustration of the figure indicates an equivalent version of an L_o-move, where an in-box crossing is introduced, using braid isotopy. This last form allows to have algebraic expressions for the L-moves (see [**24**]). In the figure we have also indicated red closing arcs for the L-moves, showing that the (standard) closures of all braids are isotopic.

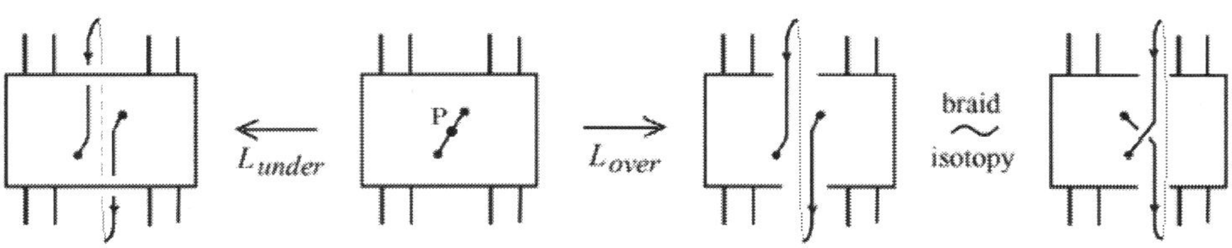

FIGURE 4. Performing an L-move. The two results are the *under* and *over* L-moves. The right-hand illustration shows the version of an L-move with an in-box crossing.

The new statement is the following:

THEOREM 1.3 (L-move braid equivalence). *Let $\beta_n \in \mathcal{B}_n, \beta_m \in \mathcal{B}_m$ be two braids, and let L_n, L_m be the two links obtained by the standard closure of β_n, β_m respectively. Then L_n is equivalent to L_m if and only if β_n can be obtained from β_m after a finite sequence of L-moves and braid isotopies.*

Note that the stabilization move of Theorem 1.2 is a special case of an L-move. Further, from Theorem 1.3 it follows that the braid conjugation of Theorem 1.2 can be achieved by the L-moves.

1.3. The plat closure and representation of links as plat closures of braids. Let B be a braid with an even number of strands, i.e. $B \in \mathcal{B}_{2n}$ some n. By [**7**] one can close the open ends by connecting adjacent top or bottom endpoints using simple arcs (see left part of Fig. 2). This procedure is called the *plat closure* of the braid and the resulting link *plat*, denoted as $\overline{B}$.

We can extend the notion of plat closure also to braids with an odd number of strands as follows: let $\beta \in \mathcal{B}_n$, we consider B' to be the image of β under the natural embedding of $\mathcal{B}_n$ into $\mathcal{B}_{n+1}$, and we define the plat closure of β as the plat closure of B' with the procedure described above. In practice, we add an extra strand in the far right of the braid β.

It is important to note that, closing in the plat way does not keep the natural orientation of the braid. In fact, when we close the braid using those short arcs, we connect two downward strands.

In [**7**, Proposition 1] Birman proved that it is always possible to describe any link as the plat closure of a braid in $\mathbb{R}^3$. This is the plat analogue of the classical Alexander theorem that uses the standard closure of braids. The idea of Birman's algorithm is simple: in a projection plane equipped with top to bottom direction, consider a knot or link diagram with no horizontal arcs. Then pull by an isotopy every local maximum to the top and every local minimum to the bottom, as abstracted in Fig. 5. The result will be a plat.

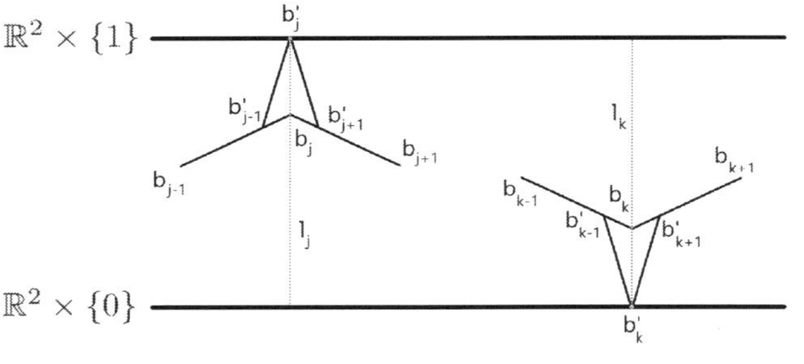

FIGURE 5. The braiding process.

In [**13**] we remarked that, if one closes a braid via the classical standard closure and push by isotopy the closing arcs right next to the endpoints, then the resulting closed braid is a plat. See rightmost part of Fig. 2. This observation provides an alternative proof of the Birman result (see [**13**, Theorem 3.2]).

1.4. The Hilden braid group. Now we want to recall a subgroup of the braid group $\mathcal{B}_{2n}$ which will be useful for the construction of the equivalence between links represented by braids closed via plat closure. Following the construction of [**20**], consider a braid as an embedding in $\mathbb{R}^3$, such that the plat closure arcs of the bottom part of the braid lie at level $\pi : \{z = 0\}$. This plane π cuts $\mathbb{R}^3$ into two halves; let the upper half be $\mathbb{R}^3_+$ and let the set of endpoints of the closure arcs in π be $A = (a_{11}, a_{12}, \ldots, a_{m1}, a_{m2})$, as in Fig. 6.

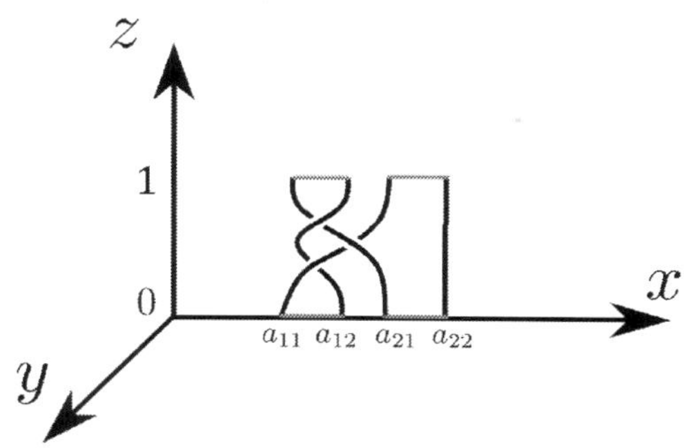

FIGURE 6. The braid embedded in $\mathbb{R}^3$ such that the closure arcs lie on the plane $\pi : \{z = 0\}$.

Then the group of isotopy classes $\Psi = [\psi]$ of orientation-preserving homeomorphisms $\psi : (\pi, A) \to (\pi, A)$ is the classical braid group $\mathcal{B}_{2n}$.

DEFINITION 1.4. The *Hilden braid group* [**20**], K_{2m}, is defined as the subgroup of $\mathcal{B}_{2m}$ generated by the equivalence classes of homeomorphisms of $\mathbb{R}^3_+$ leaving the set of closure arcs invariant on its boundary.

In [**20**] Hilden proves that K_{2m} is finitely generated and gives a set of generators in terms of elements of $\mathcal{B}_{2m}$. Moreover, he points at the following reduced set of generators, which Birman [**7**] proves is minimal.

PROPOSITION 1.5. *The Hilden braid group K_{2m} is generated by the set*

$$\{\sigma_{2i-1}, 1 \leq i \leq m;\ \mu_i = \sigma_{2i}\sigma_{2i-1}\sigma_{2i+1}\sigma_{2i}, 1 \leq i \leq m-1;\\ \rho_{ij}, 1 \leq i,j \leq m;\ \omega_{ij}, 1 \leq i,j \leq m\},$$

with:

$$\rho_{ij} = \begin{cases} (\sigma_{2j}\sigma_{2j-1})(\sigma_{2j+1}\sigma_{2j})\dots(\sigma_{2i-2}\sigma_{2i-3})\sigma_{2i-2}(\sigma_{2i-3}\sigma_{2i-2}) \\ \quad \cdot\sigma_{2i-3}^{-1}(\sigma_{2i-4}^{-1}\sigma_{2i-3}^{-1})\dots(\sigma_{2j-1}^{-1}\sigma_{2j}^{-1}), \quad \textit{if } j < i, \\ \\ (\sigma_{2j-2}^{-1}\sigma_{2j-1}^{-1})(\sigma_{2j-3}^{-1}\sigma_{2j-2}^{-1})\dots(\sigma_{2i}^{-1}\sigma_{2i+1}^{-1})\sigma_{2i}^{-1}(\sigma_{2i+1}^{-1}\sigma_{2i}^{-1}) \\ \quad \cdot\sigma_{2i+1}^{-1}(\sigma_{2i+2}\sigma_{2i+1})\dots(\sigma_{2j-1}\sigma_{2j-2}), \quad \textit{if } i < j. \end{cases}$$

$$\omega_{ij} = \begin{cases} (\sigma_{2j}\sigma_{2j-1})(\sigma_{2j+1}\sigma_{2j})\dots(\sigma_{2i-2}\sigma_{2i-3})(\sigma_{2i-1}\sigma_{2i-2}) \\ \quad \cdot\sigma_{2i-1}(\sigma_{2i-2}\sigma_{2i-1})(\sigma_{2i-3}\sigma_{2i-2})\sigma_{2i-3}^{-1} \\ (\sigma_{2i-4}^{-1}\sigma_{2i-3}^{-1})\dots(\sigma_{2j-1}^{-1}\sigma_{2j}^{-1}), \quad \textit{if } j < i, \\ \\ (\sigma_{2j-2}^{-1}\sigma_{2j-1}^{-1})(\sigma_{2j-3}^{-1}\sigma_{2j-2}^{-1})\dots(\sigma_{2i}^{-1}\sigma_{2i-1}^{-1})(\sigma_{2i+1}^{-1}\sigma_{2i}^{-1}) \\ \quad \cdot\sigma_{2i-1}(\sigma_{2i}^{-1}\sigma_{2i+1}^{-1})(\sigma_{2i-1}^{-1}\sigma_{2i}^{-1})\sigma_{2i+1}^{-1} \\ (\sigma_{2i+2}\sigma_{2i+1})\dots(\sigma_{2j-1}\sigma_{2j-2}), \quad \textit{if } i < j. \end{cases}$$

Furthermore, a minimal set of generators for K_{2m} is the following:

$$\{\sigma_1;\ \lambda_1 = \rho_{1,2} = \sigma_2\sigma_1^2\sigma_2;\ \mu_i = \sigma_{2i}\sigma_{2i-1}\sigma_{2i+1}\sigma_{2i}, 1 \leq i \leq m-1\}.$$

A visual representation of ρ_{ij} and ω_{ij} generators is depicted in Figure 7.

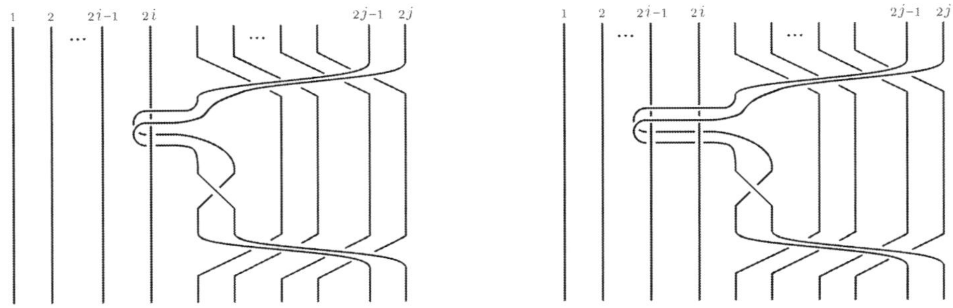

FIGURE 7. The ρ_{ij} and ω_{ij} generators in the case of $i < j$

A visual representation of the minimal set of generators is depicted in Figure 8.

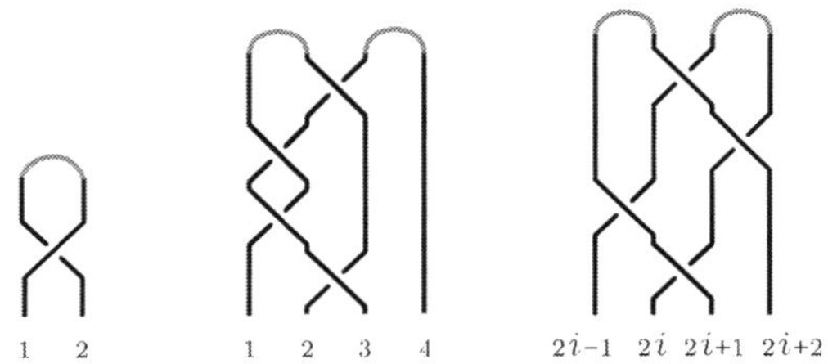

FIGURE 8. The three types of the minimal set of generators for $K_{2m} : \sigma_1, \lambda_1, \mu_i$.

1.5. Plat equivalence. In [**7**] Birman proves that every link can be represented as the plat closure of a braid in some $\mathcal{B}_{2n}$. She also proves a result concerning equivalence between plat closures of braids:

THEOREM 1.6 ([**7**]). *Let L_1, L_2 be two links, and let $B_1 \in \mathcal{B}_{2n_1}, B_2 \in \mathcal{B}_{2n_2}$ be two braids such that their plat closures define the same link types as L_1, L_2 respectively. Then L_1 is equivalent to L_2 if and only if there exists an integer $t \geq \max(n_1, n_2)$ such that, for each $m \geq t$ the elements:*

$$B_i' = B_i \sigma_{2n_i} \sigma_{2n_i+2} \dots \sigma_{2m-2} \in \mathcal{B}_{2m}, \qquad i = 1, 2$$

are in the same double coset of $\mathcal{B}_{2m}$ modulo the Hilden subgroup K_{2m}.

Fig. 9 illustrates abstractly the move in Theorem 1.6. It can be seen using Reidemeister I moves that these two braids, closed in the plat way, give rise to the same link isotopy class.

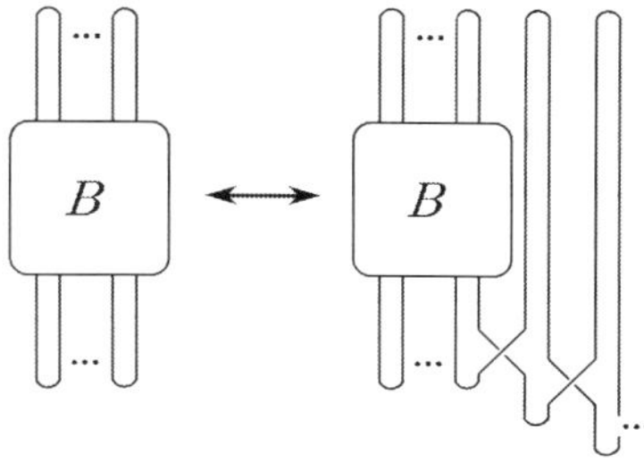

FIGURE 9. A braid $B \in \mathcal{B}_{2n}$ and the braid $B\sigma_{2n_i}\sigma_{2n_i+2}\cdots\sigma_{2m-2} \in \mathcal{B}_{2m}$.

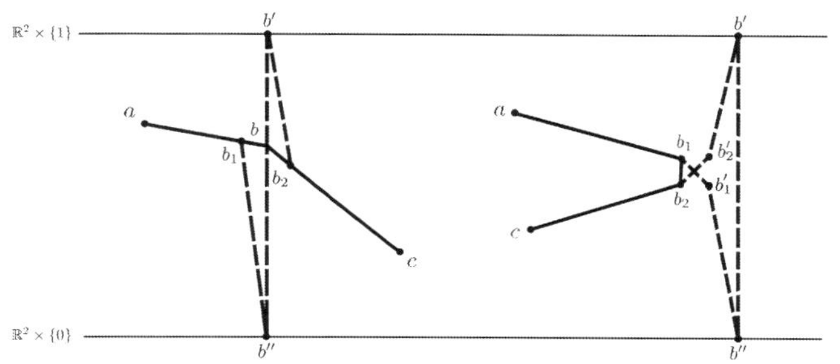

FIGURE 10. Two equivalent forms of the stabilization move.

The equivalence move in Theorem 1.6, as illustrated in Fig. 9, is a multiple application of the *stabilization move* for plats (for a proper definition see Section 5, Definition 5.1). The stabilization move is depicted in Fig. 10 in two equivalent forms up to plat equivalence: the right-hand form is created from the one on the left by performing a Reidemeister II move at the top part of the vertical arc and then suppressing one of the crossings by the equivalence; and vice-versa. In the Figure we, deliberately, have not indicated an over-arc for the crossing, as both types apply.

A crucial observation is now due.

REMARK 1.7. Both instances of the stabilization move utilized in the equivalence relation between plats are identical to the standardly closed versions of the L-moves (recall Fig. 4 in Section 1.2), forgetting orientation.

Throughout, we will usually employ capital letters for braids closed in plat form.

2. Braids and plats in handlebodies

In this section we first discuss links and braids in a handlebody and recall their representations in $\mathbb{R}^3$, the mixed links and mixed braids [**19**, **24**]. We also recall from [**19**] the standard closure of mixed braids, as adapted to handlebodies, and the analogues of braiding and braid equivalence for the standard closure. We proceed with defining the notion of plat closure for mixed braids and provide an algorithm for turning a link in a handlebody into a plat, directly analogous to that of [**7**] for $\mathbb{R}^3$, and also to [**11**] for thickened surfaces and for closed, connected and orientable 3-manifolds. We proceed with defining the appropriate Hilden braid subgroup for the mixed braid group and formulate and prove the plat closure mixed braid equivalence in the setting of handlebodies.

2.1. Handlebody representation in $\mathbb{R}^3$. Let H_g be the abstract handlebody of genus g. H_g can be constructed from a 3-ball with g 1-handles attached (see left part of Fig. 11). Note that, for $g = 0$ then H_0 is just the 3-ball, while for $g = 1$ then H_1 is the solid torus. Equivalently, H_g can be constructed as the product space of the g-punctured disc, D_g, cross the interval (middle illustration of Fig. 11): $H_g = D_g \times [0, 1]$.

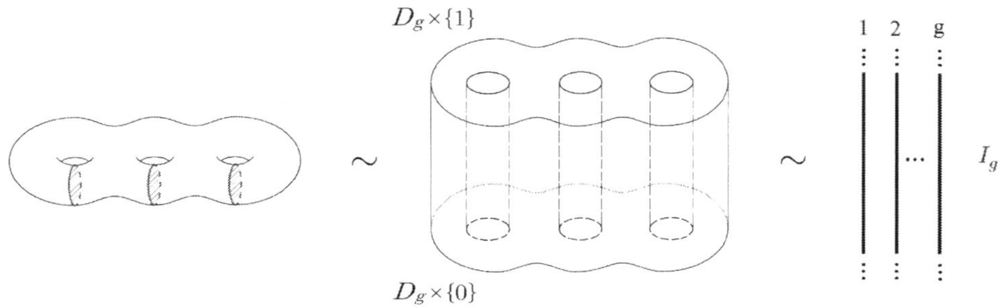

FIGURE 11. Abstract handlebody of genus 3 and its representation in $\mathbb{R}^3$.

Hence H_g, as embedded in $\mathbb{R}^3$, may be viewed as the complement of the identity braid I_g on g indefinitely extended strands (see right illustration of Fig. 11). For viewing H_g in S^3 we consider the "compactification", namely all strands of I_g meeting at the point at infinity. (Then a thickening of the resulting graph is the complementary handlebody of H_g in S^3).

2.2. Links in handlebodies and mixed links. Let now L be a link in H_g. Then, using the above representation, we can consider L to lie in the interior of $H_g = D_g \times [0, 1]$, except perhaps for finitely many points lying in the boundary components $D_g \times \{0, 1\}$. Further, fixing I_g pointwise, L may be unambiguously represented by the mixed tangle (or *mixed link*) $I_g \cup L$, see left illustration of Fig. 12. I_g forms the so-called *fixed part* of the mixed link, while L forms the so-called *moving part* of the mixed link. If we remove I_g from $I_g \cup L$ we are left with a link L in $\mathbb{R}^3$, which we shall call a *template* for the initial link L in H_g. Note that, for a given template L in $\mathbb{R}^3$ there are infinitely many links in H_g having L as the moving part and I_g as the fixed part, that is, there are infinitely many links in H_g with the given template. Further, two (polygonal) links in H_g are called *isotopic* if there is a finite sequence of the well-known Δ-moves taking one to the other. An analogous definition carries through to the corresponding mixed links, see [**19, 24**].

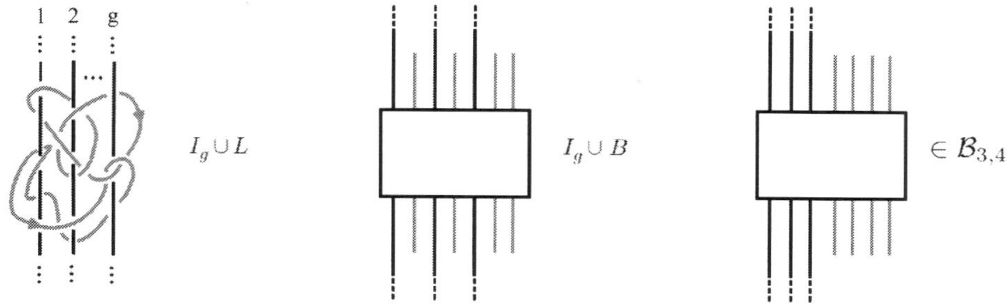

FIGURE 12. left: a mixed link, middle: an abstract geometric mixed braid, right: an abstract algebraic mixed braid.

2.3. Braids in handlebodies, mixed braids and the mixed braid group.

DEFINITION 2.1. A *braid β in H_g on n strands* is defined as an embedding ϕ of n distinct copies of $I = [0,1]$, $I_j = [0,1], j = 1, \ldots, n$ in H_g, such that each $\phi_j := \phi|_{I_j}$ is monotonous, and the top n endpoints of the braid β lie in $D_g \times \{0\}$ and the corresponding bottom n endpoints of β lie in $D_g \times \{1\}$, in the realization of H_g as $D_g \times [0,1]$.

Following [**19, 24**], the braid β in H_g is represented uniquely in S^3 by a *geometric mixed braid* on n moving strands, denoted $I_g \cup \beta$, which is an element of the Artin braid group $\mathcal{B}_{g+n}$ and which contains the identity braid I_g on g strands as a fixed subbraid (see abstraction in middle illustration of Fig. 12).

By removing I_g we are left with the *moving subbraid* β on n strands. We shall call this braid β in $\mathbb{R}^3$ a *template* for the initial braid β in H_g. Note that, for a given template β in $\mathbb{R}^3$ there are infinitely many geometric mixed braids having β as the moving subbraid and I_g as the fixed subbraid and, consequently, infinitely many braids in H_g with the given template.

A special case of a geometric mixed braid on $g + n$ strands is one for which, by removing the last n strands, we are left with the identity braid I_g. Such a mixed braid shall be called *algebraic mixed braid* (notion introduced in [**23**]). See abstraction in rightmost illustration of Fig. 12). We shall use the same notation for algebraic mixed braids as for the generality of geometric mixed braids.

It is established in [**23**] that the set of all algebraic mixed braids on g fixed strands and n moving strands form a group, the *mixed braid group*, $\mathcal{B}_{g,n}$. In other words, $\mathcal{B}_{g,n}$ is the subgroup of all elements in $\mathcal{B}_{g+n}$ for which the first g strands form the identity braid I_g. Of course, for $g = 0$, we obtain the standard definition of the Artin braid group in $\mathbb{R}^3$, $\mathcal{B}_n$, while for $g = 1$, we obtain the Artin braid group of type B, $\mathcal{B}_{1,n}$, which is the algebraic counterpart of knot theory in the solid torus and in the lens spaces (see [**22**]). It is further established in [**19**] that the groups $\mathcal{B}_{g,n}$, $n \in \mathbb{N}$, comprise the algebraic counterpart for knots and links in H_g.

In [**23**], it is proved that $\mathcal{B}_{g,n}$ has a presentation with generators: the elementary crossings among moving strands $\sigma_1, \ldots, \sigma_{n-1}$ (see Fig. 13) and the *loop generators* $\alpha_1, \ldots, \alpha_g$, where α_i represents a right-handed looping of the first moving strand around the i-th fixed strand, as depicted in Fig. 13, where also the inverse of a loop generator is depicted. Further, the generators satisfy the relations:

$$
(1)\qquad
\begin{array}{rcll}
\sigma_k\sigma_j & = & \sigma_j\sigma_k & |k-j|>1,\\
\sigma_k\sigma_{k+1}\sigma_k & = & \sigma_{k+1}\sigma_k\sigma_{k+1} & 1\le k\le n-1,\\
\alpha_r\sigma_k & = & \sigma_k\alpha_r & k\ge 2,\ 1\le r\le g,\\
\alpha_r\sigma_1\alpha_r\sigma_1 & = & \sigma_1\alpha_r\sigma_1\alpha_r & 1\le r\le g,\\
\alpha_r(\sigma_1\alpha_s\sigma_1^{-1}) & = & (\sigma_1\alpha_s\sigma_1^{-1})\alpha_r & s<r.
\end{array}
$$

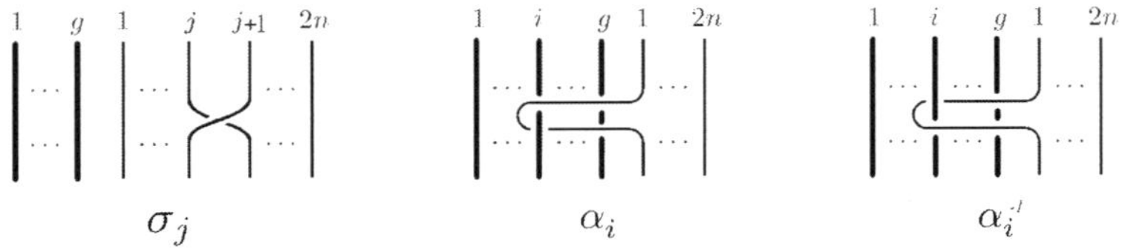

FIGURE 13. The generators of $\mathcal{B}_{g,n}$.

From the above, an algebraic mixed braid $I_g \cup \beta$ on n moving strands represents uniquely in $\mathbb{R}^3$ a braid β on n strands in $H_g = D_g \times [0,1]$, such that its endpoints lie all within a disc on the right-hand side of D_g. This leads to the following:

DEFINITION 2.2. A braid β on n strands in $H_g = D_g \times [0,1]$, such that its endpoints lie all within a disc on the right-hand side of D_g shall be called *algebraic braid in* H_g.

REMARK 2.3. Fixing a set of n points $\mathcal{C} = \{C_1, \dots, C_n\}$ in D_g we observe that the set of all braid on n strands in H_g, with endpoints in $\mathcal{C}$ forms a group by concatenation along the surface D_g. Clearly, groups related to different sets of endpoints are isomorphic. So, without loss of generality, one may assume that the set $\mathcal{C}$ lies in a disc A on the right-hand side of D_g, as in Fig. 14. By definition, these groups for different sets of endpoints coincide with the *surface braid group of* D_g *on* n *strands*, which we shall denote $\mathcal{B}_{D_g,n}$ (the surface braid groups were introduced in [**15**], see also [**4, 18**]). Furthermore, by the well-defined correspondence between mixed braids and braids in H_g, and of their operations, the surface braid group $\mathcal{B}_{D_g,n}$ is easily shown to be isomorphic to the mixed braid group $\mathcal{B}_{g,n}$.

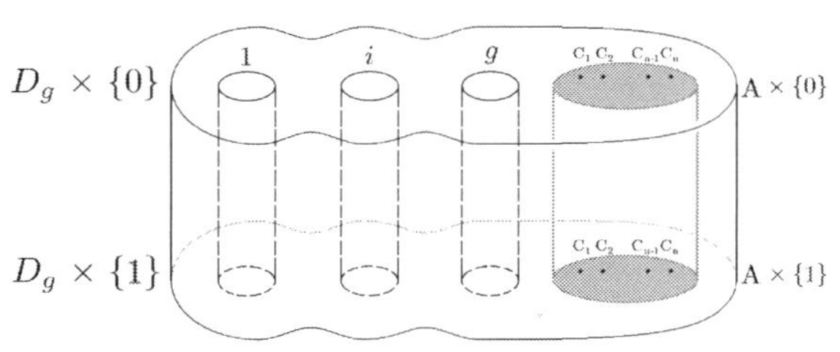

FIGURE 14. The set of points $C_1, C_2, \dots, C_n$ on the two levels $D_g \times \{0\}$ and $D_g \times \{1\}$.

3. Representing links via braids in handlebodies

3.1. The standard closure for mixed braids, the mixed braiding and the mixed braid equivalence. Let $I_g \cup \beta$ be a geometric mixed braid, representing uniquely the braid β in H_g. We recall from [**19**] that, in order to define well the (standard) closure of $I_g \cup \beta$, we first label each pair of corresponding endpoints of β with 'o' (for over) or 'u' (for under). Then, the *standard closure* of $I_g \cup \beta$ is an oriented mixed link obtained by joining with simple unlinked arcs each pair of corresponding endpoints of the moving subbraid β, according to its label: the simple arc will be lying entirely over or entirely under the rest of the mixed braid, if its label is 'o' or 'u' respectively. See Fig. 15(a) for an abstract illustration.

NOTE 3.1. The definition of standard closure for mixed braids induces the definition of *standard closure for braids in* $H_g = D_g \times [0, 1]$.

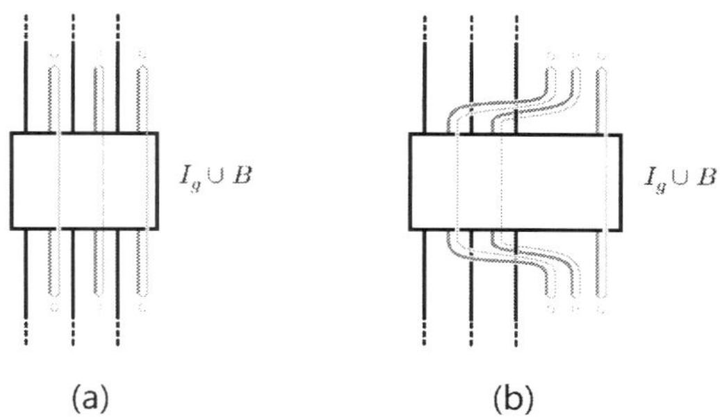

FIGURE 15. (a) the standard closure of a geometric mixed braid; (b) a parted mixed braid and its homologous closure.

As pointed out in [**19**], different choices of labels for a geometric mixed braid will, in general, result in non-isotopic mixed links. Yet, this is not the case if one restricts to algebraic mixed braids, since a closing arc can freely move through isotopy from the front to the back of the mixed braid (and vice versa), without intersecting any fixed strand.

The inverse of the closing operation is also true [**19**]: any oriented mixed link (representing uniquely an oriented link in H_g), can be braided to a geometric mixed braid with isotopic standard closure. This is an analogue of the classical Alexander theorem for knots and links in H_g. Furthermore, using the parting operation, whereby moving strands lying between fixed strands of a geometric mixed braid, along with their closing arcs, are isotoped to the right of the mixed braid away from the fixed strands, respecting their labels, a geometric mixed braid can be represented by an algebraic mixed braid with isotopic standard closure. Fig. 15(b) depicts a parted mixed braid and its homologous closure. Hence, *any oriented mixed link can be braided to an algebraic mixed braid with isotopic standard closure.* This is the algebraic analogue of the Alexander theorem for knots and links in H_g.

Moreover, *mixed link isotopy for oriented links in* H_g *is translated in* [**19**] *into mixed braid equivalence: first as L-equivalence among geometric mixed braids* (an analogue of the one-move Markov theorem), *and then as an algebraic mixed braid* *equivalence in* $\bigcup_n \mathcal{B}_{g,n}$ (an analogue of the classical Markov theorem). For details the reader may consult [**19**].

3.2. The plat closure for braids in H_g and for mixed braids.

DEFINITION 3.2. Let $I_g \cup B$ be a geometric mixed braid with an even number of moving strands, say $2n$. For defining the plat closure of $I_g \cup B$ we first number the set of top endpoints of B, from left to right, with numbers $1, 1', 2, 2', \ldots, n, n'$. To each pair (i, i') a label 'o' (for over) or 'u' (for under) is assigned. Then, we use a simple arc, say γ_i, as joining arc for the pair of endpoints (i, i'), so that γ_i runs entirely over or entirely under any fixed strand of the mixed diagram, according to the label of (i, i'). We repeat the above procedure, independently, for the set of corresponding bottom endpoints. 'Independently' in the sense that the labels of the closing arcs, say δ_i, for joining the pairs of bottom endpoints are not related to the labels of the closing arcs γ_i for joining the pairs of top endpoints. View Fig. 16 for an example. The resulting mixed link is a *plat closure* for $I_g \cup B$, called *mixed plat*, and shall be denoted $I_g \cup \overline{B}$.

In the special case where $I_g \cup B$ is an algebraic mixed braid the resulting plat shall be called *algebraic mixed plat*.

REMARK 3.3. As for the case of standard closure, the definition of plat closure for mixed braids induces the definition of *plat closure for braids in* $H_g = D_g \times [0, 1]$. Only, here we also add small local isotopies of the closing arcs of a braid B, so that the resulting plat, denoted $\overline{B}$, lies in the interior of H_g, except perhaps finitely many points.

REMARK 3.4. Different labels for closing arcs may result in non-isotopic plats. An example is illustrated in Fig. 16. Yet, note that the label of a joining arc is irrelevant if no fixed strands of I_g lie between the two endpoints. It follows, in particular, that if $I_g \cup B$ is an algebraic mixed braid, then no labels are required for forming the plat closure, as no choices are involved in this case and the plat closure is determined uniquely.

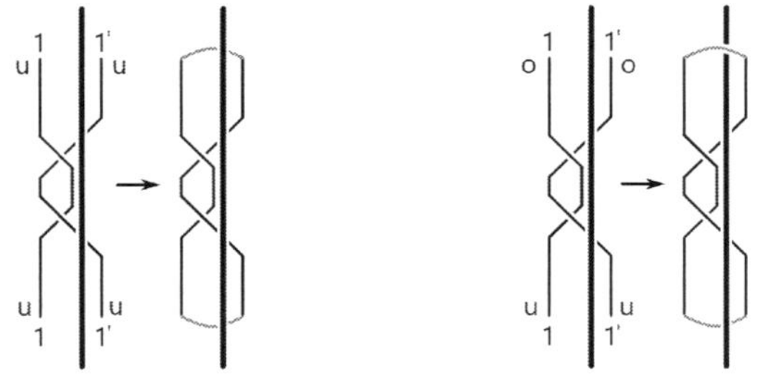

FIGURE 16. Two different examples of plat closure of a mixed braid. The two resulting links are not isotopic.

3.3. Representing mixed links as plat closures of mixed braids. Let L be a link in H_g represented by the mixed link $I_g \cup L$. We want to show that the mixed link $I_g \cup L$ may be isotoped to the (labelled) plat closure of a geometric mixed braid, providing thus an analogue of the Birman braiding theorem for knots and links in H_g. Namely, we have the following, which is also an analogue of the braiding theorem in H_g, but for plats.

THEOREM 3.5. *Let L be a link in the genus g handlebody H_g, represented by the mixed link $I_g \cup L$ in $\mathbb{R}^3$. Then L may be braided to a braid in H_g with even number of strands, whose plat closure is isotopic to L in H_g. Equivalently, $I_g \cup L$ may be braided to a geometric mixed braid with even number of moving strands, whose plat closure is isotopic to $I_g \cup L$.*

PROOF. We shall adapt the braiding algorithms of [**7**, **11**] to our setting. Indeed, let $\overrightarrow{D_g}$ be the unit vector orthogonal to D_g. It is always possible (using local isotopies, if needed) to consider $L \subset D_g \times [0.25, 0.75]$.

Consider a polygonal representation of L, with set of vertices $\{b_0, b_1, \ldots, b_n\}$ and with no edges parallel to $D_g \times \{0\}$ (which is always possible by a general position argument). Following [**7**, Lemma 2], we have that for any 3 consecutive vertices b_{j-1}, b_j, b_{j+1} of L the line l_j orthogonal to D_g through b_j admits a neighbourhood N_j which meets the link only in the edges $[b_{j-1}, b_j]$ and $[b_j, b_{j+i}]$ (see Fig. 17). We will say that L is in standard position.

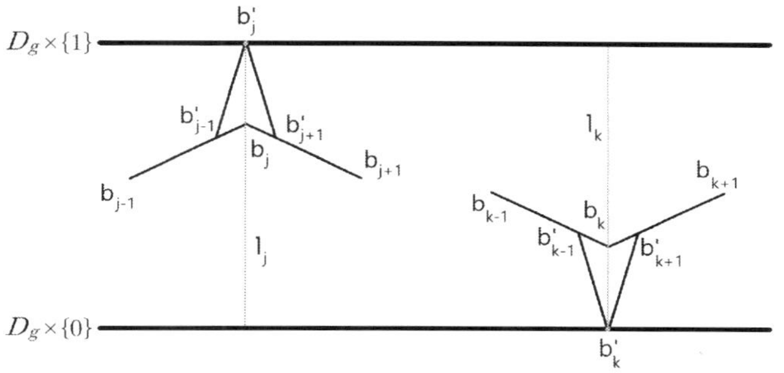

FIGURE 17. The braiding process.

Then, starting from b_0, for every three points b_{j-1}, b_j, b_{j+1} with

$$\left(\overrightarrow{[b_{j-1}, b_j]} \cdot \overrightarrow{D_g}\right) \cdot \left(\overrightarrow{[b_j, b_{j+1}]} \cdot \overrightarrow{D_g}\right) < 0$$

(meaning that b_j is either a local maximum or a local minimum) consider the intersections $[b_{j-1}, b_j] \cap N_j = b'_{j-1}$, $[b_j, b_{j+1}] \cap N_j = b'_{j+1}$. If $\overrightarrow{[b_{j-1}, b_j]} \cdot \overrightarrow{D_g} > 0$ (resp. < 0) let $b'_j = l_j \cap D_g \times \{1\}$ (resp $b'_j = l_j \cap D_g \times \{0\}$). If we replace the edge sequence b_{j-1}, b_j, b_{j+1} with $b_{j-1}, b'_{j-1}, b'_j, b'_{j+1}, b_{j+1}$ the isotopy class of the link does not change, while now the link intersects once $D_g \times \{0, 1\}$.

Repeating this operation for every three points satisfying the above condition, we will obtain an isotopic link meeting the discs $D_g \times \{0\}$ and $D_g \times \{1\}$ in exactly, say, n points, and every other disc $D_g \times \{t\}, t \in (0, 1)$, in exactly $2n$ points. It follows from Artin's definition of a braid [**2**] that such a link determines a well-defined geometric mixed braid $I_g \cup B$ on $2n$ moving strands. In turn, this mixed braid represents the link L via plat closure in $H_g = D_g \times [0, 1]$. □

Furthermore, we prove that any mixed link (resp. link in H_g) may be braided to (resp. represented by) an *algebraic mixed braid* (recall Definition 2.2) via plat closure. For this we first recall from [**7**] and adapt the definition of the spike move.

DEFINITION 3.6. Let $\overline{B}$ be a link in H_g in plat form, represented in $\mathbb{R}^3$ by the mixed plat $I_g \cup \overline{B}$. Let also b_i, b_{i+1}, b_{i+2} be three consecutive points of $\overline{B}$, such that b_{i+1} is a boundary point (i.e. $\in D_g \times \{0,1\}$) and such that the triangle $[b_i, b_{i+1}, b_{i+2}]$ is entirely contained in $D_g \times [0,1]$ and intersects $\overline{B}$ only on its sides $[b_i, b_{i+1}]$ and $[b_{i+1}, b_{i+2}]$ (see Fig. 18). Edges like $[b_i, b_{i+1}], [b_{i+1}, b_{i+2}]$ are called *boundary edges.* Let now b'_{i+1} be a new boundary point on the same boundary component as b_{i+1} and let b'_i, b'_{i+2} be interior points satisfying the conditions:

- $[b'_i, b'_{i+1}, b'_{i+2}] \cap L = \emptyset$ and $[b'_i, b'_{i+1}, b'_{i+2}]$ lies entirely in $D_g \times [0,1]$;
- for the region $[b_i, b'_i, b'_{i+2}, b_{i+2}]$ bounded by the points $b_i, b'_i, b'_{i+2}, b_{i+2}$ we have: $[b_i, b'_i, b'_{i+2}, b_{i+2}] \subset D_g \times [0,1]$ & $[b_i, b'_i, b'_{i+2}, b_{i+2}] \cap L = \{b_i, b_{i+2}\}$.

A *spike move* for the plat $\overline{B}$ consists in replacing the edges defined by the sequence b_i, b_{i+1}, b_{i+2} by the edges defined by the sequence $b_i, b'_i, b'_{i+1}, b'_{i+2}, b_{i+2}$. Then the isotopy class of the plat $\overline{B}$ does not change and the spike move preserves the plat form.

A *spike move for the mixed plat* $I_g \cup \overline{B}$ is defined analogously, except that two of the conditions above are replaced by the conditions:

$$[b'_i, b'_{i+1}, b'_{i+2}] \cap (I_g \cup \overline{B}) = \emptyset \text{ and } [b_i, b'_i, b'_{i+2}, b_{i+2}] \cap I_g = \emptyset.$$

The conditions imply, in particular, that if a fixed strand overlies (resp. underlies) the triangle $[b_i, b_{i+1}, b_{i+2}]$ then the fixed strand also overlies (resp. underlies) the replacement of the spike move. Then the isotopy class of the mixed plat $I_g \cup \overline{B}$ does not change and the spike move preserves the mixed plat form.

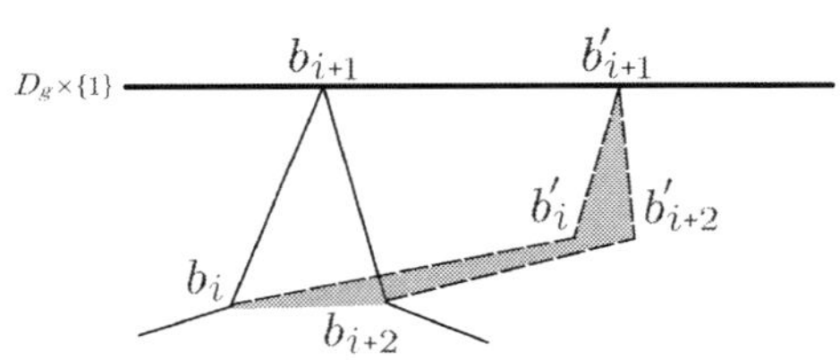

FIGURE 18. A spike move.

We can now sharpen Theorem 3.5 to representing links in H_g as plat closures of algebraic mixed braids.

THEOREM 3.7. *Let L be a link in the genus g handlebody H_g, represented by the mixed link $I_g \cup L$ in $\mathbb{R}^3$. Then L may be braided to an algebraic braid in H_g with even number of strands, whose plat closure is isotopic to L in H_g. Equivalently, $I_g \cup L$ may be braided to an algebraic mixed braid $I_g \cup B \in \mathcal{B}_{g,2n}$, for some n, representing $I_g \cup L$ via plat closure.*

PROOF. By Theorem 3.5, we first obtain from $I_g \cup L$ an isotopic geometric mixed plat $I_g \cup \overline{B}$, where $I_g \cup B \in \mathcal{B}_{g+2n}$ for some n, such that B comes with a set of $2n$ closing labels. Note that $I_g \cup B$ does not necessarily belong to the mixed braid group $\mathcal{B}_{g,2n}$. For doing this we adapt the parting technique, as in [**19**] for the standard closure.

Indeed, we perform spike moves to all boundary edges which are on the left of some strands of I_g, so that we part the mixed plat, bringing all the boundary edges to the right of all strands of I_g (see an example in Fig. 19). We shall refer to this procedure as a *plat parting* of the geometric braid. It is important to note that a spike move, by definition, respects locally the label of a closing arc. Otherwise, the closing labels do not restrict the spike moves. In the end of the parting operation we obtain an algebraic mixed plat, $I_g \cup \overline{B'}$, such that $I_g \cup B' \in \mathcal{B}_{g,2n}$. At this moment, the $2n$ closing labels do not matter anymore, as no fixed strands are in between moving strands, recall Remark 3.4.

Equivalently, on the level of H_g, using Theorem 3.5 we first obtain from L a plat in H_g and then, using appropriate spike moves, we bring the resulting plat to a new one whose closing arcs lie within a disc on the right-hand side of D_g. □

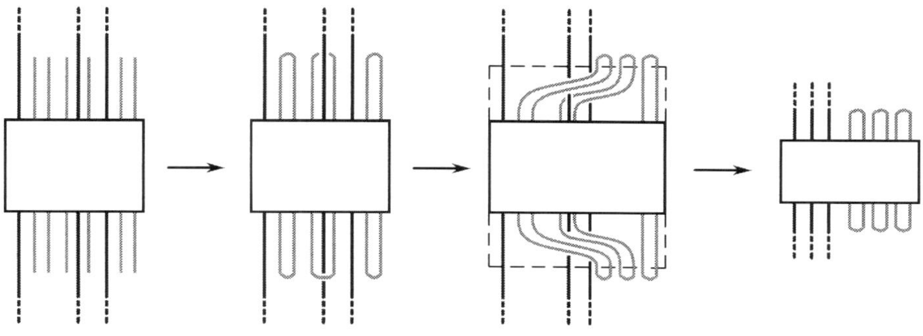

FIGURE 19. Exemplifying the process of parting a geometric mixed plat to an algebraic mixed plat.

4. The mixed Hilden braid group

As in the classical case of $\mathbb{R}^3$, we want to define the Hilden braid group also in the setting of a handlebody. We will do that adapting Definition 1.4 to the context of the surface braid group of D_g (recall Remark 2.3) resp. the mixed braid group.

DEFINITION 4.1. Let $\mathcal{C} = \{C_1, \dots, C_{2m}\}$ be a set of $2m$ points in D_g, which w.l.o.g. lie in a disc A on the right-hand side of D_g. The *Hilden group of the handlebody* H_g, denoted $K_{D_g,2m}$, is defined as the subgroup of $\mathcal{B}_{D_g,2m}$ generated by the equivalence classes of homeomorphisms of $D_g \times I$ leaving the set of closure arcs invariant on its top/bottom boundary.

Furthermore, the *mixed Hilden group* $K_{g,2m}$ is defined as the subgroup of the mixed braid group $\mathcal{B}_{g,2m}$ generated by the equivalence classes of homeomorphisms of $\mathbb{R}^3_+$ leaving the set of bottom (resp. top) plat closure arcs invariant on its boundary.

REMARK 4.2. Clearly, an isomorphism between $\mathcal{B}_{D_g,2m}$ and $\mathcal{B}_{g,2m}$ induces an isomorphism between $K_{D_g,2m}$ and $K_{g,2m}$. Hence, and following the idea of Remark 2.3, the mixed Hilden braid group can be viewed either as the Hilden subgroup of the mixed braid group of the handlebody or the Hilden subgroup of the surface braid group of D_g.

THEOREM 4.3. *A set of generators for $K_{g,2m}$ is*

$$\{\sigma_1,\ \lambda_1 = \sigma_2\sigma_1^2\sigma_2,\ \mu_i = \sigma_{2i}\sigma_{2i-1}\sigma_{2i+1}\sigma_{2i},\ 1 \leq i \leq m-1;\ \tau_{i,2j-1},\ 1 \leq i \leq g,\ 1 \leq j \leq m\},$$

where:

$$\tau_{i,1} = \alpha_i\,\sigma_1\alpha_i\sigma_1$$

$$\tau_{i,2j-1} = \sigma_{2j-2}\sigma_{2j-3}\sigma_{2j-1}\sigma_{2j-2}\ldots\sigma_2\sigma_1\sigma_3\sigma_2\alpha_i\sigma_1\,\alpha_i\,\sigma_1\sigma_2^{-1}\sigma_3^{-1}\sigma_1^{-1}\sigma_2^{-1}\ldots\sigma_{2j-2}^{-1}\sigma_{2j-1}^{-1}\sigma_{2j-3}^{-1}\sigma_{2j-2}^{-1}$$

The generators $\sigma_1, \lambda_1, \mu_i, \tau_{i,2j-1}$ shall be referred to as 'mixed Hilden generators'. View Fig. 20 for the generators σ_1, λ_1 and μ_i and Fig. 21 for the generators $\tau_{i,2j-1}$.

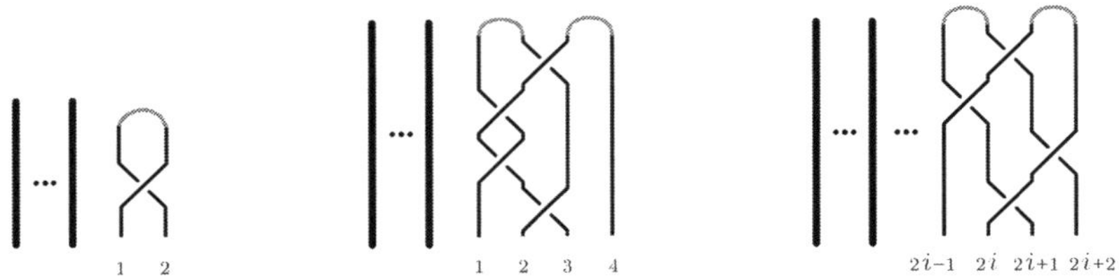

FIGURE 20. The mixed Hilden generators σ_1, λ_1 and μ_i.

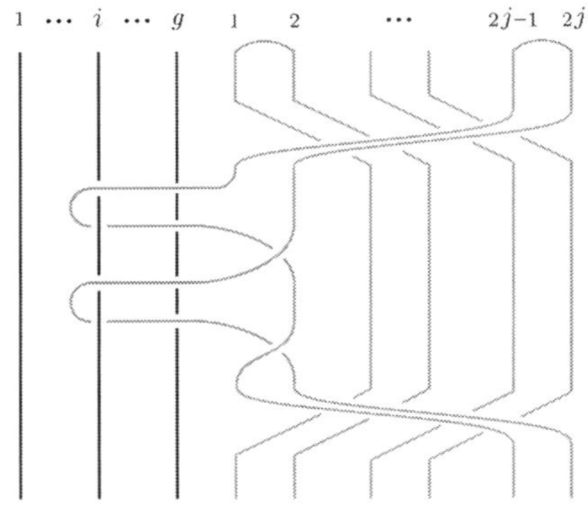

FIGURE 21. The mixed Hilden generator $\tau_{i,2j-1}$.

PROOF. Since $\mathcal{B}_{g,2m}$ is a subgroup of $\mathcal{B}_{g+2m}$, the idea is to make use of Proposition 1.5. For this, we need to also argue that $K_{g,2m}$ is a subgroup of K_{g+2m}. If g *is even* then K_{g+2m} is well-defined and, hence, $K_{g,2m}$ coincides with the subgroup of K_{g+2m} keeping fixed the first g strands. In the case of g *odd* this does not make sense, because the mixed Hilden subgroup is defined only for an even number of moving strands. For this reason we will introduce in this case a *dummy strand* with number zero on the left of the first strand of any element of K_{g+2m} as well as on the left of the first fixed strand of any element of $K_{g,2m}$. The dummy strand does not interact with any other strand of the braid. In this way, every braid will have $(1+g+2m)$ strands, and we can define then the dummied $K'_{g,2m}$ to be the Hilden subgroup of the dummied K'_{g+2m}, for g odd. With the above trick, we can always assume that K_{g+2m} is well-defined and $K_{g,2m}$ is a subgroup of K_{g+2m}.

In using Proposition 1.5, we will start by examining the full set of generators for the group K_{g+2m}. Then, the generators of $K_{g,2m}$ will be the ones inherited by K_{g+2m}, but with different enumeration and some small adjustments for both cases for g, even or odd. First of all, since the first g strands of an element of $K_{g,2m}$ are fixed, the σ_i generators involving the first g strands need to be eliminated. Likewise for the generators μ_i with indices smaller than g, since the first g strands do not participate in the plat closure. Second, we can divide ρ_{ij} and ω_{ij} generators into three families depending on their indices:

- $2i, 2j \leq g$: these generators need to be eliminated, since they move fixed strands;
- $2i \leq g,\ 2j > g$: these generators still exist in $K_{g,2m}$, but we need to write them using generators of the mixed braid group $B_{g,2m}$ and renumber the indices;
- $2i, 2j > g$: these generators remain, but a renumbering of the indices is needed.

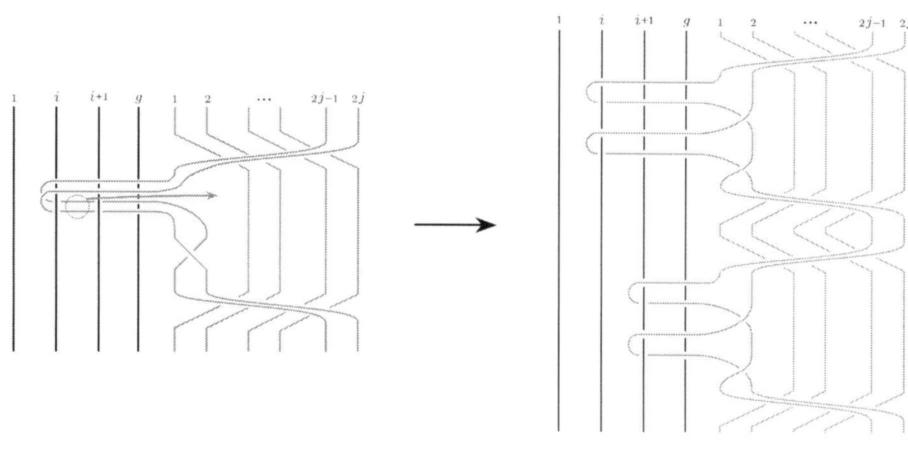

FIGURE 22. Finding the expression for ω_{ij} as a product of τ generators.

Now that the set of generators is found, we need to renumber their indices keeping in mind that the $(i+g)$-th strand in $\mathcal{B}_{g+2m}$ is numbered i in $\mathcal{B}_{g,2m}$. We rename the ρ_{ij} generators of the second family to $\tau_{i,2j-1}$ for $i = 1,\ldots,g$, $j = 1,\ldots,m$, that takes the strands numbered $2j-1, 2j$ around the i-th fixed strand. Then an ω_{ij} generator of the second family can be expressed using the $\tau_{k,l}$-type generators. Indeed, Figure 22 illustrates the expression of the ω_{ij} generator as the product $\tau_{i,2j-1} \cdot \tau_{i+1,2j-1}$.

Note that, in the case of g odd, we will have also the definition of $\tau_{0,2j-1}$, $j = 1,\ldots,m$. In this way we found a way to represent generators ρ_{ij} and ω_{ij} using the generators of $\mathcal{B}_{g,2m}$. At this point, if g is even, the construction is completed. If g is odd, we have to remove the dummy strand, numbered zero. To perform this transformation we follow the natural inclusion $K_{g+1,2m} \to K_{g,2m}$, where the leftmost strand is forgotten. This means that the generators $\tau_{0,2j-1}$ are now trivial elements of the group.

Last but not least, using the relations found by Birman in [**7**], which led to a minimal set of generators, we can introduce the generator λ_1 and omit from the set of generators the third family of ρ_{ij}'s and ω_{ij}'s, the ones with $i, j > g$, since they can be represented using the other generators, as in the classical case. □

REMARK 4.4. It is important to note that not all $\tau_{i,2j-1}$ generators are needed: in fact, for $j \neq 1$ it is possible to obtain $\tau_{i,2j-1}$ by composing μ_k generators with $\tau_{i,1}$.

5. The equivalence between plat closed mixed braids

Before going on, let us recall from [**7**] the definition of stabilization move for plat closed braids, adapted to the case of a link in the handlebody H_g, or equivalently, a mixed link.

DEFINITION 5.1. Let a, b, c be consecutive vertices of our link in H_g (resp. a mixed link $I_g \cup L$) in polygonal form. Let b' (resp. b'') be the projections of b onto $D_g \times \{1\}$ (resp. $D_g \times \{0\}$), and b_1, b_2 the intersection points of a sufficiently small neighbourhood of b with the segments $[a, b]$ and $[b, c]$. A *stabilization move* is the substitution of $[a, b]$ and $[b, c]$ with the segments $[a, b_1], [b_1, b''], [b'', b'], [b', b_2]$ and $[b_2, c]$, in its one form, and the substitution with the segments

$$[a, b_1], [b_1, b'_1], [b'_1, b''], [b'', b'], [b', b'_2], [b'_2, b_2] \text{ and } [b_2, c],$$

in its second form, whereby a crossing is also introduced at the point b. View Fig. 23.

By definition, a stabilization move is an isotopy.

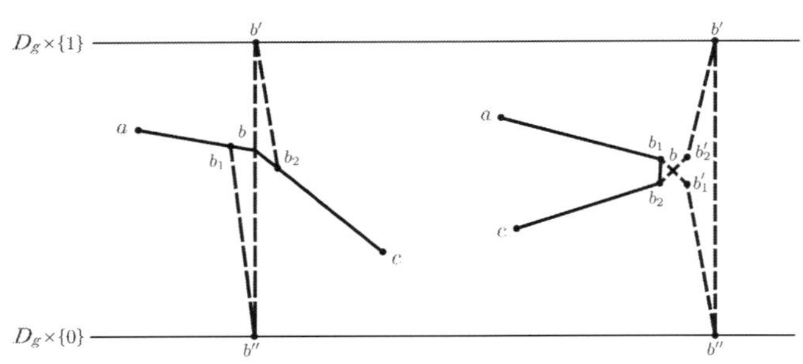

FIGURE 23. The two forms of the stabilization move for H_g.

Let us now consider two plat closed mixed braids $B, B' \in \mathcal{B}_{g,2n}$ such that one can be obtained by the other by only applying elements of the mixed Hilden group $K_{g,2n}$. Then, since all the generators of $K_{g,2n}$ represent local isotopies for a plat closed mixed braid, the two braids represent the same link isotopy class via plat closure. To formulate and prove the converse statement, a previous result of Birman helps us:

LEMMA 5.2 ([**7**, Lemma 8]). *Let L and L' be two isotopic polygonal links in general position, i.e. with no horizontal edges and such that each vertex has nothing above or underneath it in a small neighbourhood of it. Then it is possible to connect the two links by a finite sequence of spike moves and stabilization moves.*

The lemma is proved in the case of $\mathbb{R}^3$, by examining the possible Δ-moves. The only difference when adapting to H_g is that the Δ-moves miss the vertical lines of the punctures. But this fact does not present any obstacle, since the spike moves and the stabilization moves are the same moves as in $\mathbb{R}^3$ and they also miss the vertical lines (the same arguments carry through also to the setting of mixed links). The spike move for the case of H_g is defined in Definition 3.6, while the stabilization move is defined just above, in Definition 5.1. Namely we have:

LEMMA 5.3. *Let L and L' be two isotopic polygonal links in the handlebody H_g, resp. mixed links, in general position, i.e. with no horizontal edges and such that each vertex has nothing above or underneath it in a small neighbourhood of it. Then it is possible to connect the two links by a finite sequence of spike moves and stabilization moves.*

Before formulating and proving the plat equivalence for the handlebody we will introduce a special case of a stabilization move, the st_k move.

DEFINITION 5.4. An st_k *move* is a stabilization move between two algebraic mixed braids $B, B' \in \cup_{n\in\mathbb{N}}\mathcal{B}_{g,2n}$, defined as:

$$\begin{array}{rccl} st_k: & \mathcal{B}_{g,2n} & \to & \mathcal{B}_{g,2n+2} \\ & B & \mapsto & B\sigma_{2k}^{-1}\sigma_{2k+1}^{-1}\cdots\sigma_{2n-1}^{-1}\sigma_{2n}\sigma_{2n-1}\cdots\sigma_{2k+1}\sigma_{2k} \end{array}$$

This move, depicted abstractly in Fig. 24, represents a special case of a stabilization move between algebraic mixed braids, taking place at the bottom of an even numbered strand $2k$, with a positive crossing.

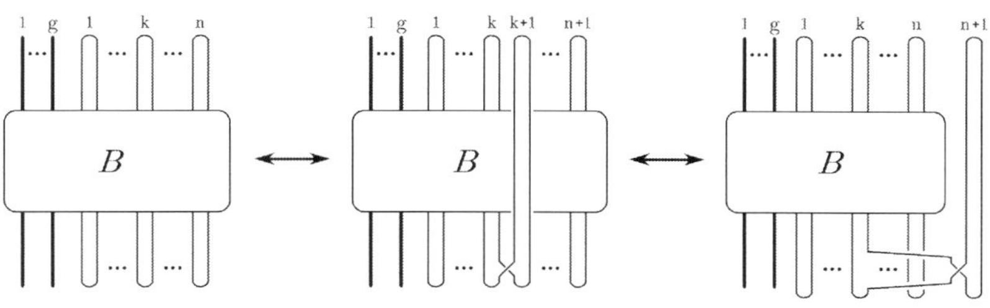

FIGURE 24. The st_k move.

THEOREM 5.5. *Two braids $B, B' \in \cup_{n\in\mathbb{N}}\mathcal{B}_{g,2n}$ determine, via plat closure, equivalent links in the handlebody H_g of genus g if and only if they are connected by braid relations and a finite sequence of the following moves:*

$$\begin{array}{lrcccl} h_\sigma: & \sigma_1 B & \longleftrightarrow & B & \longleftrightarrow & B\sigma_1, \\ h_\lambda: & \sigma_2\sigma_1^2\sigma_2 B & \longleftrightarrow & B & \longleftrightarrow & B\sigma_2\sigma_1^2\sigma_2, \\ h_\mu: & \sigma_{2i}\sigma_{2i-1}\sigma_{2i+1}\sigma_{2i} B & \longleftrightarrow & B & \longleftrightarrow & B\sigma_{2i}\sigma_{2i-1}\sigma_{2i+1}\sigma_{2i}, \quad i = 1, \ldots, n \\ h_\tau: & \alpha_j\sigma_1\alpha_j\sigma_1 B & \longleftrightarrow & B & \longleftrightarrow & B\alpha_j\sigma_1\alpha_j\sigma_1, \quad j = 1, \ldots, g \\ st_n: & & & B & \longleftrightarrow & B\sigma_{2n} \end{array}$$

The moves $h_\sigma, h_\lambda, h_\mu, h_\tau$ reflect the generators of the mixed Hilden braid group (recall Figs. 20 and 21), so do not alter the number of strands, while the move st_n, defined above, realizes the stabilization move that may take place at the bottom right of the last strand numbered $2n$.

PROOF. The one direction of the proof is obvious. Given now a link L in H_g in standard position, we obtain by Theorem 3.5 an isotopic link (a plat) with n points of intersection with $D_g \times \{0\}$, say $Y^u = \{Y_1^u, \dots, Y_n^u\}$, and n points of intersection with $D_g \times \{1\}$, say $Y^l = \{Y_1^l, \dots, Y_n^l\}$. In order to find an element $B \in \mathcal{B}_{g,2n}$ such that its plat closure is isotopic L, we have to perform plat parting to the geometric mixed plat. The plat parting comprises a sequence of specific spike moves and is described in the proof of Theorem 3.7.

For each parting move there is a choice of performing it by pulling over or under strands to the right. For proving Theorem 5.5 we will argue that these choices are included in the equivalence relation. Indeed, for a parting choose two paths p and p' in the configuration space of n points on D_g connecting, respectively, the two sets of boundary points Y^u and Y^l to a set of points $\mathcal{C} = \{C_1, \dots, C_n\}$, where C_i is an internal point of the disk A (recall Fig. 14). This choice of paths is not unique: if p and q are two choices for Y^l, the composition of p and the inverse of q is an element in the fundamental group of the configuration space of $\mathcal{C}$, that could be realized as a composition of elements of the mixed Hilden braid group. The same construction holds for the set of upper boundary points. This means that any two element of $\mathcal{B}_{g,2n}$ obtained as above for the same link L in standard position are connected by moves $h_\sigma, h_\lambda, h_\mu, h_\tau$. It would be illuminating to consider the special case of the choice of the plat parting (recall Fig. 21) by pulling a pair of consecutive strands *over* a fixed strand versus the choice of pulling them *under* the same fixed strand. Such a choice difference shows in the final algebraic braid words as a mixed Hilden generator $\tau_{i,2j-1}$ (recall Fig. 21).

By Lemma 5.2, it is now sufficient to show how stabilization and spike moves change the algebraic mixed braid representative B of a link $\overline{B}$, the plat closure of B. Recall that a stabilization move only adds a trivial loop to $\overline{B}$ at any point. Using the sequence of moves depicted in Fig. 25 and sliding the stabilization, it is always possible to consider it taking place at the bottom right of an even strand, up to spike moves. Then, by spike moves, it is always possible to consider it happening on the top of the braid. After sliding the newly generated arcs to the right of the braid, we will obtain something as depicted in the rightmost part of Fig. 24. Moreover, as demonstrated in Fig. 26 the move can be assumed to have a positive crossing only, as in move st_k in Definition 5.4. It is straightforward to check that this move is the st_k move, if the stabilization is happening on the strand numbered $2k$.

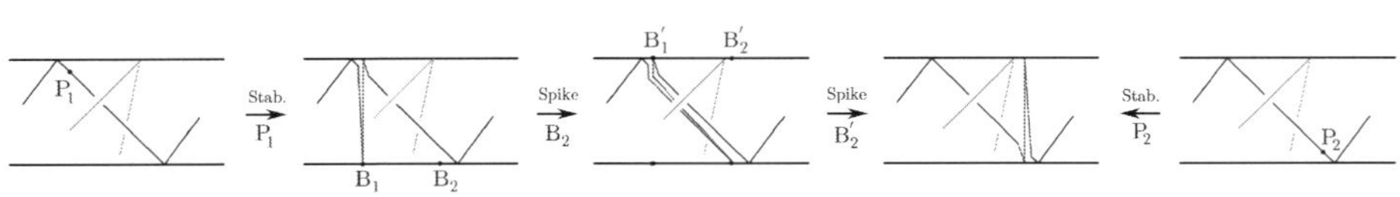

FIGURE 25. The procedure to slide a stabilization move to the bottom right part of an even numbered strand.

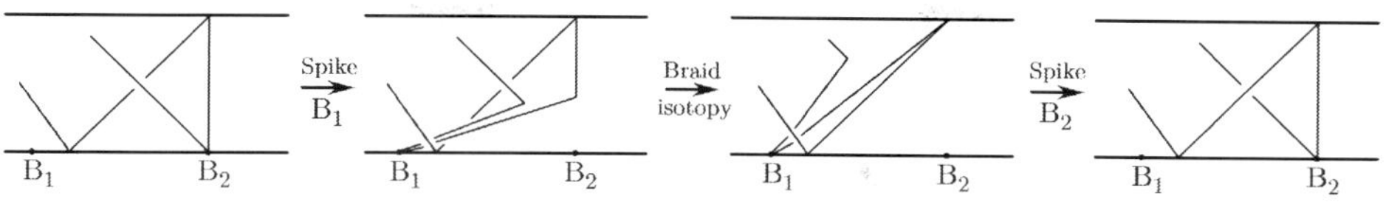

FIGURE 26. The procedure to turn a negative crossing into a positive one.

We shall now argue that in the equivalence theorem we only need the st_n move. Indeed, note that before performing an st_k move, one can bring the pair of strands $(2k-1, 2k)$ to the far right of the braid by using h_μ moves. After performing the stabilization st_n, all the four strands $2n-1, 2n, 2n+1, 2n+2$ can be brought again to the previous positions $2k-1, 2k, 2k+1$ and $2k+2$ by means of h_μ moves. The result, after braid isotopy, is equivalent to the application of the st_k move to the initial braid. For the abstraction of the whole process, the reader is referred to Fig. 27.

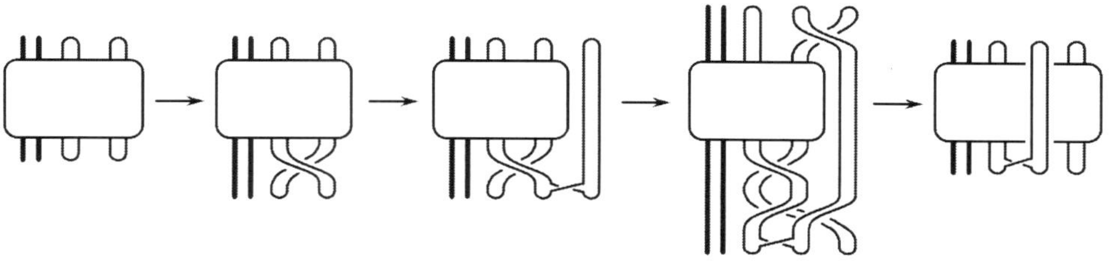

FIGURE 27. The process of performing st_k by means of h_μ, st_n and braid isotopy moves.

Suppose, now, that starting from an algebraic mixed plat we perform a stabilization move on a loop generator, obtaining a geometric mixed plat. In this case we perform the parting in a way compatible to the layering of the stabilization move, so as to obtain an algebraic stabilization move, which is analyzed above. See Fig. 28.

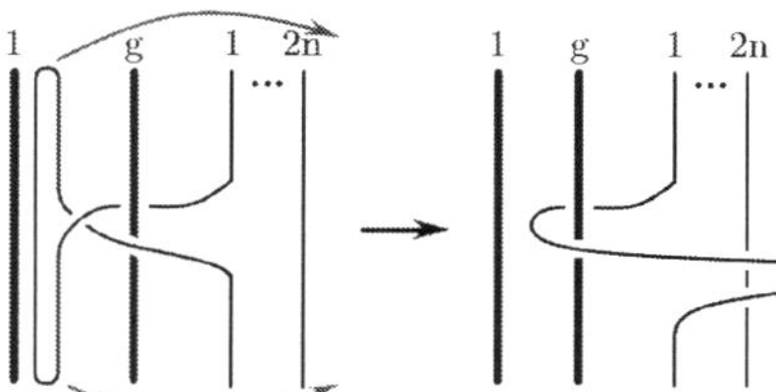

FIGURE 28. Transferring a geometric stabilization move to an algebraic one.

Let us, finally, consider a spike move taking place in $\overline{B}$, obtaining $\overline{B'}$. Suppose that the spike move brings the upper boundary point C_i into another point in the disk A, say C_i'. Following Fig. 18, where C_i coincides with b_{i+1} and C_i' with b_{i+1}', we can find $t_0 \in [0,1]$ such that $[b_i, b_{i+2}]$ lie in $D_g \times \{t_0\}$. Considering the middle point of $[b_i, b_{i+2}]$ as C_i^*, we can express the spike move as a path bringing C_i^* to C_i', following the median of the abstract triangle $[b_i, C_i', b_{i+2}]$. The plane $D_g \times \{t_0\}$ divides the braid into two parts, say B_1 the upper part and B_2 the lower part. Now consider $B_1 B_1^{-1}$, where the two spike moves connect together onto the base $[b_i, b_{i+2}]$. Since $B_1 B_1^{-1}$ is a trivial element in $B_{g,2n}$, it can be represented via elements of the Hilden braid group. In particular, the two spike moves bound a disk and can be represented as a product of the moves $h_\sigma, h_\lambda, h_\mu, h_\tau$. This means that also the initial spike move could be represented via a combination of these four moves. The symmetric construction holds in the case of a spike move on a lower boundary point.

Suppose, finally, that starting from an algebraic mixed plat we perform a spike move toward the left part of the braid, obtaining a geometric mixed plat. In this case, we just use for the parting the inverse move.

The proof of the Theorem is now complete. □

In the spirit of Birman's proof of Theorem 1.6, Theorem 5.5 can be re-stated equivalently as follows:

THEOREM 5.6. *Let $L_i, i = 1, 2$ be two links in a handlebody H_g, represented by the plat closure of $B_i \in \mathcal{B}_{g,2n_i}, i = 1, 2$. Then the two links are isotopic if and only if there exists an integer $t \geq \max n_1, n_2$ such that, for each $m \geq t$ the two elements:*

$$B_i' = B_i \sigma_{2n_i} \sigma_{2n_i+2} \cdots \sigma_{2m-2} \in \mathcal{B}_{g,2m}, \quad i = 1, 2$$

are in the same double coset of $\mathcal{B}_{g,2m}$ modulo the mixed Hilden subgroup $K_{g,2m}$.

REMARK 5.7. In **[11]** Cattabriga and Gabrovšek formulate and prove the plat equivalence for links in a thickened surface. Following their work and suppressing the b-type loop generators (looping around the handles), one would obtain the statement of Theorem 4.3.

References

[1] J. W. Alexander, *A lemma on systems of knotted curves*, Proc. Natl. Acad. Sci. USA, **9** (1923), no. 3, 93–95.

[2] E. Artin, *Theory of braids*, Ann. of Math. (2) **48** (1947), 101–126, DOI 10.2307/1969218. MR19087

[3] C. Bankwitz and H. G. Schumann, *Über viergeflechte* (German), Abh. Math. Sem. Univ. Hamburg **10** (1934), no. 1, 263–284, DOI 10.1007/BF02940679. MR3069630

[4] P. Bellingeri, *On presentations of surface braid groups*, J. Algebra **274** (2004), no. 2, 543–563, DOI 10.1016/j.jalgebra.2003.12.009. MR2043362

[5] P. Bellingeri and A. Cattabriga, *Hilden braid groups*, J. Knot Theory Ramifications **21** (2012), no. 3, 1250029, 22, DOI 10.1142/S0218216511009534. MR2887661

[6] S. Bigelow, *A homological definition of the Jones polynomial*, Invariants of knots and 3-manifolds (Kyoto, 2001), Geom. Topol. Monogr., vol. 4, Geom. Topol. Publ., Coventry, 2002, pp. 29–41, DOI 10.2140/gtm.2002.4.29. MR2002601

[7] J. S. Birman, *On the stable equivalence of plat representations of knots and links*, Canadian J. Math. **28** (1976), no. 2, 264–290, DOI 10.4153/CJM-1976-030-1. MR402715

[8] J. S. Birman and T. E. Brendle, *Braids: a survey*, Handbook of knot theory, Elsevier B. V., Amsterdam, 2005, pp. 19–103, DOI 10.1016/B978-044451452-3/50003-4. MR2179260

[9] H. Brunn, *Über verknotete kurven*, Verh. des intern. Math. Congr. **1** (1897), 256–259.

[10] A. Cattabriga, P. Cavicchioli, and S. Lambropoulou, *The study of links in thickened surfaces via the standard closure of surface braids*, work in progress.

[11] A. Cattabriga and B. Gabrovšek, *A Markov theorem for generalized plat decomposition*, Ann. Sc. Norm. Super. Pisa Cl. Sci. (5) **20** (2020), no. 4, 1273–1294. MR4201181

[12] P. Cavicchioli, *An algorithmic method to compute plat slide moves in 3-manifolds of Heegaard genus two*, Discrete Math. **346** (2023), no. 12, Paper No. 113627, 22, DOI 10.1016/j.disc.2023.113627. MR4627719

[13] P. Cavicchioli and S. Lambropoulou, *Passing from plat closure to standard closure of braids in* $\mathbb{R}^3$*, in handlebodies and in thickened surfaces*, to appear in Ars Mathematica Contemporanea, DOI 10.26493/1855-3974.3317.25b.

[14] H. Doll, *A generalized bridge number for links in 3-manifolds*, Math. Ann. **294** (1992), no. 4, 701–717, DOI 10.1007/BF01934349. MR1190452

[15] E. Fadell and L. Neuwirth, *Configuration spaces*, Math. Scand. **10** (1962), 111–118, DOI 10.7146/math.scand.a-10517. MR141126

[16] S. Garnerone, A. Marzuoli, and M. Rasetti, *Quantum computation of universal link invariants*, Open Syst. Inf. Dyn. **13** (2006), no. 4, 373–382, DOI 10.1007/s11080-006-9019-x. MR2286535

[17] S. Garnerone, A. Marzuoli, and M. Rasetti, *Quantum automata, braid group and link polynomials*, Quantum Inf. Comput. **7** (2007), no. 5-6, 479–503. MR2347217

[18] J. González-Meneses, *New presentations of surface braid groups*, J. Knot Theory Ramifications **10** (2001), no. 3, 431–451, DOI 10.1142/S0218216501000949. MR1825967

[19] R. Häring-Oldenburg and S. Lambropoulou, *Knot theory in handlebodies*, J. Knot Theory Ramifications **11** (2002), no. 6, 921–943, DOI 10.1142/S0218216502002050. Knots 2000 Korea, Vol. 3 (Yongpyong). MR1936243

[20] H. M. Hilden, *Generators for two groups related to the braid group*, Pacific J. Math. **59** (1975), no. 2, 475–486. MR388372

[21] C. Kassel and V. Turaev, *Braid groups*, Graduate Texts in Mathematics, vol. 247, Springer, New York, 2008. With the graphical assistance of Olivier Dodane, DOI 10.1007/978-0-387-68548-9. MR2435235

[22] S. S. F. Lambropoulou, *Solid torus links and Hecke algebras of* $\mathcal{B}$*-type*, Proceedings of the Conference on Quantum Topology (Manhattan, KS, 1993), World Sci. Publ., River Edge, NJ, 1994, pp. 225–245. MR1309934

[23] S. Lambropoulou, *Braid structures in knot complements, handlebodies and 3-manifolds*, Knots in Hellas '98 (Delphi), Ser. Knots Everything, vol. 24, World Sci. Publ., River Edge, NJ, 2000, pp. 274–289, DOI 10.1142/9789812792679_0017. MR1865712

[24] S. Lambropoulou and C. P. Rourke, *Markov's theorem in 3-manifolds*, Topology Appl. **78** (1997), no. 1-2, 95–122, DOI 10.1016/S0166-8641(96)00151-4. Special issue on braid groups and related topics (Jerusalem, 1995). MR1465027

[25] S. Lambropoulou and C. P. Rourke, *Algebraic Markov equivalence for links in three-manifolds*, Compos. Math. **142** (2006), no. 4, 1039–1062, DOI 10.1112/S0010437X06002144. MR2249541

[26] A. A. Markov, Über die freie äquivalenz geschlossener zöpfe (in german), Recueil mathematique Moscou, 1, 73–78; *Mat. Sb*, **43**, 1936, 1935.

[27] P. W. Shor and S. P. Jordan, *Estimating Jones polynomials is a complete problem for one clean qubit*, Quantum Inf. Comput. **8** (2008), no. 8-9, 681–714. MR2453785

[28] A. B. Sossinsky, *Preparation theorems for isotopy invariants of links in 3-manifolds*, Quantum groups (Leningrad, 1990), Lecture Notes in Math., vol. 1510, Springer, Berlin, 1992, pp. 354–362, DOI 10.1007/BFb0101202. MR1183501

[29] N. Weinberg, Sur l'equivalence libre des tresses fermée, *Comptes Rendus (Doklady) de l'Académie des Sciences de l'URSS*, **23** (1939), 215–216.

[30] K.-H. Yun, *On braid-plat relations in Conway function*, Honam Math. J. **33** (2011), no. 3, 407–418, DOI 10.5831/HMJ.2011.33.3.407. MR2866407

Dipartimento di Matematica, Università di Bologna, Piazza di Porta S. Donato, 5, 40126, Bologna, Italy

Email address: `paolo.cavicchioli@unibo.it`

URL: `https://www.unibo.it/sitoweb/paolo.cavicchioli`

School of Applied Mathematical and Physical Sciences, National Technical University of Athens, Zografou Campus, 9 Iroon Polytechneiou st., 15772 Athens, Greece

Email address: `sofia@math.ntua.gr`

URL: `http://www.math.ntua.gr/~sofia`

Contemporary Mathematics
Volume **827**, 2025
https://doi.org/10.1090/conm/827/16537

Meridional rank, welded knots, and bridge trisections

Jason Joseph and Puttipong Pongtanapaisan

ABSTRACT. This paper builds on work of the authors investigating the relatinship between meridional rank and bridge number for knotted objects that are not classical links. We compute the meridional rank of families of surfaces in the 4-sphere, virtual links, and welded links by adapting Coxeter quotients of classical links to these new situations. For many of these families we show that the meridional rank is equal to the bridge number.

1. Introduction

In the classical setting, the bridge number $\beta(L)$ is a useful complexity measure of a link L in S^3 (Definition 2.4). For suitable classes of links, the bridge number can be used to estimate other geometric invariants such as hyperbolic volume [**4**, Theorem 1.5], distortion [**5**, Theorem 1.1], ropelength [**9**, Theorem 5], and total curvature [**23**, Corollary 3.2]. Cappell and Shaneson's Meridional Rank Conjecture posits that $b(L)$ equals the meridional rank $\mu(L)$, which is an algebraic quantity defined as the minimum number of generators freely homotopic to a meridian needed to generate $\pi_1(S^3 \backslash L)$. Tools such as knot contact homology and Coxeter quotients have been used to verify the Meridional Rank Conjecture for several families of links [**1**–**3**, **6**, **8**], but no counterexamples have been discovered.

In this paper, we investigate a higher dimensional analogue of the Meridional Rank Conjecture for ribbon T^2-knots: a particular kind of knotted torus in 4-space. We first represent them by virtual knot diagrams via Satoh's *Tube map* correspondence [**20**, **28**] (Figure 6). The meridional rank and bridge number can then be conveniently calculated on these generalized knot diagrams, which look just like classical knot diagrams with a new type of crossing.

We will show that if a ribbon T^2-knot F can be represented as a diagram obtained by performing appropriate local operations to a diagram of a twisted link or an arborescent link (Definition 2.10 and Definition 2.12) associated to bipartite trees with even weights, then the meridional rank equals the bridge number. These classical links were shown by Baader et al. to satisfy the classical MRC [**1**], and we sometimes refer to them as *BBKM links*.

A naive way to try to estimate the meridional rank is to take some classical knot diagram D corresponding to some T^2-knot $Tube(D)$, whose meridonal rank is known, and turn some of the crossings into virtual crossings. This process can, however, cause the meridional rank to drop. For instance, the left image in

2020 *Mathematics Subject Classification.* Primary 57K12.

Figure 1 represents a T^2-knot F such that $\mu(F) = 2$ obtained by spinning the trefoil [**28**]. After one crossing is turned into a virtual crossing (the right image of Figure 1), we get the unknotted T^2-knot U which has $\mu(U) = 1$.

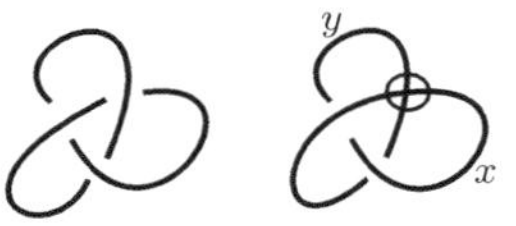

FIGURE 1. Turning some crossing virtual can cause the meridional rank to drop. The group for the T^2-knot F on the left has presentation $\langle x, y \mid xyx = yxy \rangle$. On the other hand, the T^2-knot on the right has presentation $\langle x, y \mid y = x \rangle$, and one generator can be removed (See Subsection 2.1.3 for how the group is calculated).

On the other hand, if one is careful about the location, where a crossing is changed into a virtual crossing, one can get an effective lower bound. Moves shown in Figure 16 known as *flank moves* and *flank-switch moves* have appeared in the literature before (the names of the moves are different depending on the authors [**10**]) to generate interesting virtual knots. Versions of these moves will also be useful for our goal of computing the meridional rank (Figure 16). In the statement below, a *twist region* of a knot diagram D is a sequence of bigon regions of D arranged end to end such that there are no other bigons on either end of the sequence.

THEOREM 1.1. *Let D be a twisted link diagram or an arborescent link diagram associated to bipartite trees with even weights such that $\mu(D) = n$. If D' is the result of performing some of the operations below, then $\mu(Tube(D')) \geq n$:*

(1) *Turning up to $k - 1$ classical crossings into virtual crossings in a twist region made up of k crossings that corresponds to either a single Coxeter generator or two Coxeter generators, where k is even.*
(2) *Performing any number of flank-switch moves anywhere.*
(3) *Performing a flank 1-move (Bottom row of Figure 9) on a twist region in a rational tangle corresponding to two Coxeter generators.*

FIGURE 2. At left, a classical BBKM knot D. At right, moves from Theorem 1.1 applied to D: the leftmost and the middle virtual crossings demonstrates situation (1), the rightmost two virtual crossings demonstrate situation (3) of Theorem 1.1

We refer to links after we applied the process in Theorem 1.1 as *virtual BBKM links*. Figure 2 illustrates a virtual BBKM link and the classical link it came from. We remark that the moves in Theorem 1.1 can create new T^2-knots that are not equivalent to T^2 knots that are representable by classical knot diagrams [**17**]. Epimorphisms of knot groups are interesting in its own right (see [**12, 30**], for instance). Theorem 1.1 then provides many epimorphisms from knotted surface groups to other groups.

To bound the meridional rank from above, we turn to a more geometric and combinatorial invariant. The *Wirtinger number* is a combinatorial upper bound for the bridge number of a classical knot, introduced by Blair et al. [**4**]. In that work, they prove that the Wirtinger number is in fact equal to the bridge number. The second author's result on the Wirtinger number of virtual links [**27**], and Theorem 1.1 allows us to compute the exact values of the bridge numbers and meridional ranks of infinitely many ribbon T^2-knots.

COROLLARY 1.2. *Let K be a welded knot obtained from Theorem* 1.1, *and let $F = Tube(K)$, a ribbon T^2-knot. Then, $\mu(F) = \beta(F)$.*

To arrive at the corollary above, we use the concept of *bridge trisections* of knotted surfaces in S^4, which were introduced by Meier and Zupan in [**22**]. These are the 4D-analogue of bridge splittings of classical links. As such, they are decompositions of a surface into 3 trivial disk systems, the analogues of trivial tangles. Let c_i, where $i = 1, 2, 3$ denote the number of disks in each trivial disk system. Then, our version of the bridge number is related to bridge trisections through the following proposition.

PROPOSITION 1.3. *Let S be a knotted surface. Then $\beta(S) = \min_{\mathcal{T}} (\min\{c_1, c_2, c_3\})$, where the minimum is taken over all bridge trisections $\mathcal{T}$ of S.*

Our bridge number analysis is applicable to the *bridge trisection index* $b(S)$, which can be regarded as the analogue of trisection genus [**11**] in the world of knotted surfaces. Note that this is not a bridge number in the sense of counting minima, but is related to it: $3\beta(S) - \chi(S) \leq b(S)$. We combine the results on welded knots with an adaptation of the Tube map from [**18**] to prove the following corollary about bridge trisection index of ribbon tori. The modified BBKM knots with virtual crossings and n minima naturally give rise to tori in S^4 with n minima, and in fact a bridge trisection diagrams with $3n$ arcs via Satoh's Tube map. This upper bound, together with Theorem 1.1, shows that the welded bridge number can determine the bridge trisection number of knotted surfaces in certain cases.

COROLLARY 1.4. *If F is a ribbon T^2-knot satisfying the hypotheses of Theorem* 1.1, *then $b(F) = 3n$.*

2. Preliminaries

2.1. Basics on virtual knot theory. We begin by reminding the readers of a convenient way to represent ribbon tori with generalized knot diagrams.

DEFINITION 2.1. A *virtual knot diagram* is an immersion of a circle into the plane where each double point is decorated as either a *classical crossing* or a *virtual crossing* as shown in Figure 3. A *virtual knot* is an equivalence class of virtual knot diagrams modulo planar isotopies and *extended Reidemeister moves*, which consist of the three classical Reidemeister moves, three additional moves just like the classical Reidemeister moves but with only virtual crossings involved, and one more

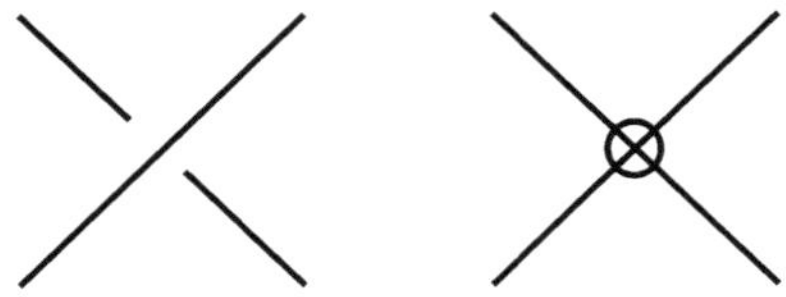

FIGURE 3. (Left) A classical crossing. (Right) A virtual crossing.

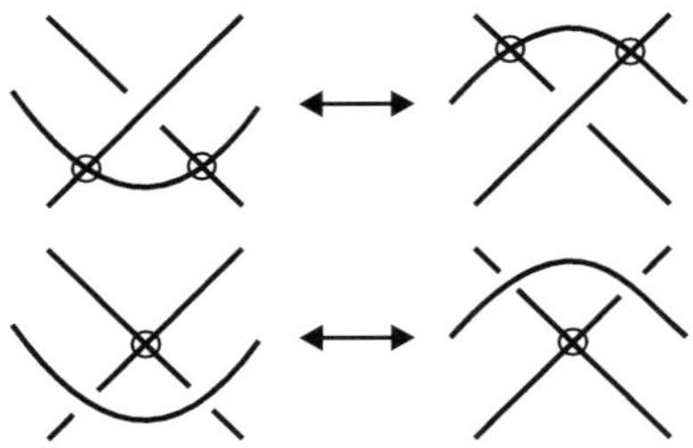

FIGURE 4. (Top) The mixed move. (Bottom) The welded move.

move called the *mixed move* depicted in the first row of Figure 4. If we additionally allow the *welded move* (bottom of Figure 4), then we call an equivalence class a *welded knot*.

To better understand the proof of Corollary 1.2, we need to introduce another way to represent virtual and welded knots and links.

DEFINITION 2.2. Given an oriented virtual link diagram $p : S^1 \sqcup S^1 \sqcup \cdots \sqcup S^1 \to \mathbb{R}^2$, its *Gauss diagram* D is a decoration of the oriented circles in the domain of p such that the preimages of the classical crossings are connected by *chords*, which are signed arrows starting from the over crossing to the under crossing. The sign of the arrow indicates the sign of a crossing using the right-hand rule. The classical Reidemeister moves can be translated to moves on the Gauss diagrams. Virtual links are then in one-to-one correspondence with Gauss diagrams modulo the Reidemeister moves [**13**]. It is well-known that a Gauss diagram does not always represent a classical link, but every Gauss diagram corresponds to some virtual link. In a sense, Gauss diagrams give simpler representations of virtual links than virtual link diagrams since virtual crossings are not present.

We remark that for this paper, the signs of the crossings near each arrow is not important for the arguments. Thus, we will omit them in the drawings. For an example of a Gauss diagram, consult Figure 5. This corresponds to the virtual knot diagram depicted in the right image of Figure 1. To be more precise, start by picking a base-point on the diagram that is furthest to the left. As we traverse clockwise, we see two over-crossings followed by two under-crossings. However, the first under-crossing encountered pairs with the first over-crossing. Thus, the two arrows in the Gauss diagram cross each other. A topological interpretation relevant to this paper is presented next.

2.1.1. *Ribbon surfaces.* Recall that a d-dimensional handle of index k is a d-ball parametrized as $D^k \times D^{d-k}$. In this paper, we use the standard notation T^n to denote the n-dimensional torus. An orientable knotted surface $S \subset S^4$ is *ribbon* if it bounds an immersed handlebody in S^4 with only ribbon intersections. A *ribbon*

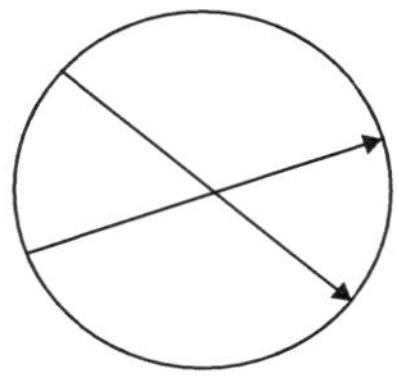

FIGURE 5. A Gauss diagram for the virtual knot diagram in the right image of Figure 1.

intersection is a disk in S^4 which is the image of a pair of disks in the handlebody, one properly embedded in the handlebody (so that its boundary lies on S) and one embedded in the interior of the handlebody. A neighborhood of the latter disk inside the handlebody can be pushed into B^5 to obtain an embedded handlebody in B^5 with boundary S and with only index 0 and 1 critical points.

2.1.2. *The Tube map.* Satoh defined the *Tube map* in [**28**], which takes a virtual knot (arc) diagram to a diagram of a ribbon torus (sphere), and proved that this respects the virtual Reidemeister moves and the welded move, i.e. if two diagrams are welded equivalent, then their tubes are isotopic. Moreover, he proved the Tube map is surjective, i.e. every ribbon torus or sphere is the tube of a welded knot or arc. The tube map gives a convenient combinatorial way to encode ribbon surfaces, which is compatible with bridge trisections. See Figure 6 for a demonstration of the tube map.

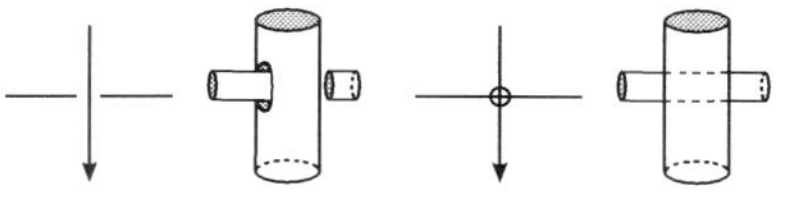

FIGURE 6. Any ribbon T^2-knot is represented by a virtual knot diagram.

2.1.3. *Fundamental group calculations.* Satoh also showed in [**28**] that the fundamental group of the complement of ribbon T^2-knots in $\mathbb{R}^4$ can be done by the following algorithm on virtual knot diagrams.

We first orient the virtual knot diagram D with n classical crossings, and choose a base-point on D. Starting at the base-point, we label the arcs $a_1, a_2, \cdots, a_n$ so that at each under-crossing, a_i is the incoming arc and a_{i+1} is the outgoing arc. For $i = 1, ..., , n$, let $\varepsilon_i = \pm 1$ be the sign of the i-th crossing. Then, the group of $Tube(D)$ has the following presentation:

$$\langle a_1, \cdots, a_n \mid a_{i+1} = a_j^{\varepsilon_i} a_i a_j^{-\varepsilon_i} \ (i = 1, \cdots, n) \rangle.$$

2.2. Definitions of the bridge number. We recall a definition from Morse theory [**24**].

DEFINITION 2.3. A smooth function from a manifold M to the real line $f : M \to \mathbb{R}$ is *Morse* if all critical points are nondegenerate.

Let K be an ambient isotopy class of a link in S^3 and let $h : S^3 \to \mathbb{R}$ be the standard height function. We will only consider smooth embeddings γ of K such that the restriction $h|_\gamma$ is Morse and all critical points of $h|_\gamma$ have distinct critical values. To discuss a particular embedding of a link type, we will write $\gamma \in K$. We will also let γ denote the image of the embedding.

DEFINITION 2.4. The *bridge number* of an embedding $\gamma \in K$, $\beta(\gamma)$, is defined to be the number of local minima of $h|_{\gamma}$. The *bridge number* of a link K, $\beta(K)$, is the minimum of $\beta(\gamma)$ over all $\gamma \in K$.

While the definition above is natural for classical links, the diagrammatic formulations and are suitable for generalizations to other objects such as virtual knots and links. In what follows, we review various definitions for the bridge numbers of virtual links. When we restrict to classical links, they are all equal.

2.2.1. *Overpass perspective.* Let D be a virtual knot diagram. An *overpass* (resp. *underpass*) is a path of D that contains at least one over-crossing (resp. under-crossing), but no under-crossing (resp. over-crossing). Any knot diagram can be bridge decomposed as an alternating sequence of overpasses and underpasses (see [**15**], for example). The *overpass bridge number* $\beta_O(K)$ of K is the minimum number of overpasses over all bridge decompositions of K. This formulation is useful for illustrating the relationship between the bridge number and the crossing number [**7**].

2.2.2. *Height function on the plane perspective.* Let D be a virtual knot diagram, and let $h : \mathbb{R}^2 \to \mathbb{R}$ be the standard projection map $h(x, y) = y$. We define the $\mathbb{R}^2$*-bridge number* $\beta_{\mathbb{R}^2}(D)$ of D to be the number of minima on D for h. We then define the *Morse bridge number* $\beta_{\mathbb{R}^2}(K)$ of a virtual knot K to be the minimum number $\beta_{\mathbb{R}^2}(D)$ over all virtual knot diagrams representing K. If one is interested in studying flat knots and links (see [**14**], for example), then this formulation is more suitable than the overpass perspective as there is no information on over-crossings or under-crossings on flat knot diagrams.

2.2.3. *Diagram coloring perspective.* For more details, see [**4, 27**]. Let D be a diagram of K.

By a strand, we mean a subarc of a diagram between two under-crossings. A stand may go over virtual crossings.

DEFINITION 2.5. A *k-partial coloring* of a diagram D is an assignment of the color blue to a subset of k strands of D.

DEFINITION 2.6. Given a k-partial coloring of D, we can perform a *coloring move* at any crossing where the over-strand and one of the under-strands is colored: this colors the other under-strand, producing a $(k + 1)$-partial coloring.

See highlighted crossings in Figure 7.

DEFINITION 2.7. If there is a k-partial coloring of a diagram D such that a sequence of coloring moves will color all of D, then we say that D is *k-colorable*.

DEFINITION 2.8. The initially colored strands of the k-partial coloring are called the *seed strands*.

DEFINITION 2.9. The minimum k such that D is k-colorable is called the *Wirtinger number* $\omega(D)$ of D. The Wirtinger number $\omega(K)$ of a knot K is the smallest value $\omega(D)$ minimized over all diagrams of K.

It is well-known that $\beta(K) = \beta_O(K) = \beta_{\mathbb{R}^2}(K)$. For the first equality, see [**29**]. The second equality is also straightforward. To see this, take an embedding

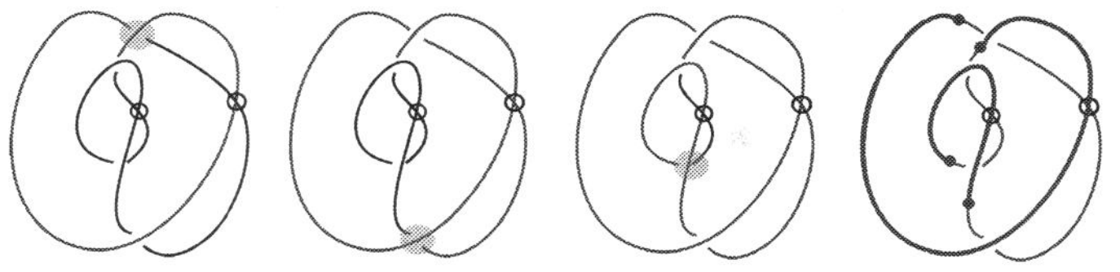

FIGURE 7. A virtual knot diagram D with $\omega(D) = 1, \beta_O(D) = 2$, and $\beta_{\mathbb{R}^2}(D) = 3$. The diagram D represents a nontrivial virtual knot, but a welded trivial knot.

realizing $\beta(K)$. Projecting to the yz-plane gives a diagram with $\beta(K)$ local maxima with respect to the height function on the plane. Conversely, given a diagram with $\beta_{\mathbb{R}^2}(K)$, we can put the diagram to be the yz-plane. This gives a 3D-realization with $\beta_{\mathbb{R}^2}(K)$ local maxima with respect to the standard height function $h : S^3 \to \mathbb{R}$ after a slight perturbation. The fact that $\beta(K) = \omega(K)$ is the main result of [**4**]. This definition is very computer friendly and the authors of [**1**] use it to verify the Meridional Rank Conjecture for many knots.

Due to Nakanishi and Satoh, $\beta_O(K) \neq \beta_{\mathbb{R}^2}(K)$ for virtual knots, but the two quantities are equivalent for welded knots [**25**].

To demonstrate the definitions on virtual knot diagrams, consider a diagram D representing K in Figure 7. Note that $\beta_{\mathbb{R}^2}(D) = 3$ since D has three minima with respect to the standard height function on the page. The overpasses of D are represented by bold segments in the rightmost picture of Figure 7 so $\beta_O(D) = 2$. Finally, observe that $\omega(D) = 1$ as D is 1-colorable, with the seed indicated in the leftmost picture of Figure 7. This demonstrates the effectiveness of the Wirtinger number. The main result of [**27**] shows that $\omega(K) = \beta_O(K)$ for virtual knots. Using a very similar argument, one can deduce that $\omega(K) = \beta_O(K)$ for welded knots as well.

2.3. Constructions of classical links. In this section, we discuss some ways to construct families of classical knots and links. For certain parameters, these knots are shown to satisfy the Meridional Rank Conjecture [**1**].

The first family of interest was introduced in [**1**]. For an example of twisted link, consult Figure 8.

DEFINITION 2.10. Consider a classical knot diagram D that is reduced so that it does not admit simplifications via Reidemeister I and II moves. There are two associated checkerboard surfaces. One can construct a graph Γ from this surface, with one vertex for each disk and one edge for each band, weighted according to the number of signed half-twists in the band. If a classical knot K admits a diagram with a checkerboard surface such that the edge weights of the induced graph Γ are all at least two in absolute value, and such that the plane dual graph Γ^* has no multiple edges, then K is called a *twisted knot*.

The next family of knots comes from piecing together trivial tangles in an organized way. Before we describe the construction, we give a precise definition of what we mean by piecing together.

DEFINITION 2.11. Consider two Seifert surfaces S_1 and S_2. A surface S is a *plumbing* of S_1 and S_2 if

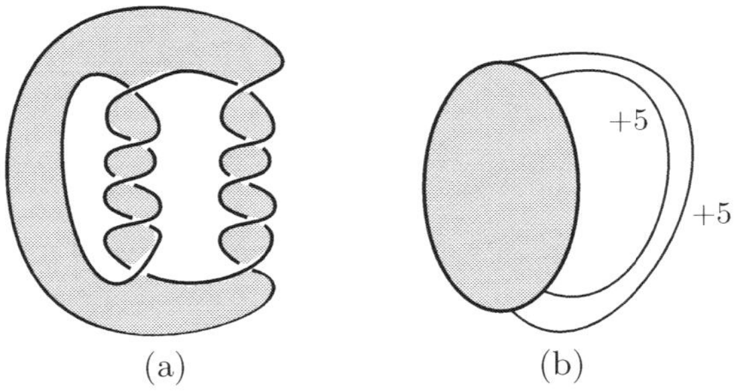

FIGURE 8. (a) A reduced checkerboard colored diagram. (b) After collapsing the bands into thin lines, we get the associated graph Γ. Here, we treat the disk as a fat vertex.

- $S = S_1 \cup S_2$,
- $S_1 \cap S_2$ is a rectangle D which is a rectangular neighbourhood of an interval in S with endpoints in ∂S that is nowhere tangent to ∂S.
- There are three-balls B_1, B_2 such that $S_i \subset B_i (i = 1, 2)$, $B_1 \cup B_2 = S^3, B_1 \cap B_2 = \partial B_1 = \partial B_2$, and $\partial B_1 \cap S = \partial B_2 \cap S = D$.

With the definition of plumbing, we are ready to define arborescent links.

DEFINITION 2.12. Let T be a plane tree with weighted vertices. We construct a link from T by letting each vertex of T correspond to an embedded annulus, where the number of half-twists is the weight of the vertex. If two vertices are connected by an edge, then the two correponding annuli are plumbed together. We define an *arborescent link* associated to T to be the boundary of the surface obtained by plumbing the annuli as specified by T. See Figure 9 for an example of a tree with its corresponding link.

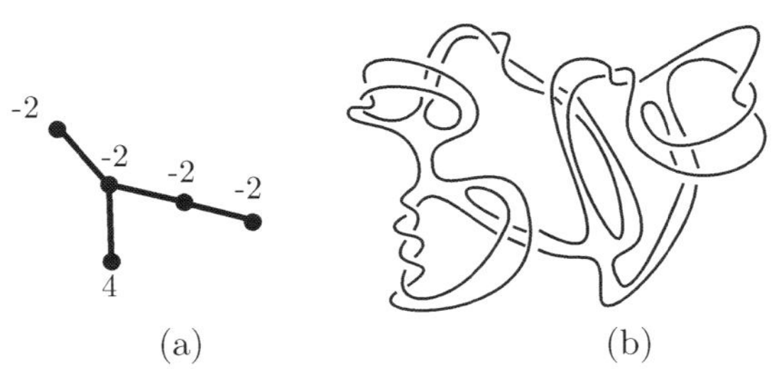

FIGURE 9. (a) The plane tree T containing the instruction on how to plumb annnuli to form an arborescent link. (b) The arborescent link corresponding to T.

2.4. Presenting surfaces in 4-space as unlinks with bands. We begin by recalling the definition of band surgery.

DEFINITION 2.13. Let L be a link and consider a rectangular disk $b : I \times I \to S^3$ embedded in such a way that $L \cap b(I \times I) = b(I \times \partial I)$, where I is the unit interval $[0, 1]$. Then, the link $L' = (L \backslash b(I \times \partial I)) \cup b(\partial I \times I)$, which is called a link obtained from L by the *band surgery* along the band B, where $B = b(I \times I)$.

It's been established that any knotted surface in S^4 can be represented as unlink with some bands attached such that the result of performing surgery along the bands also give an unlink [**16**]. Such a presentation is called a *banded unlink presentation* of a surface.

EXAMPLE 2.14. The leftmost diagram in Figure 10 is an unlink representing a knotted surface in S^4 since the link without the bands is an unlink, and the link we obtain after band surgeries is also an unlink.

FIGURE 10. (Left) A banded-unlink diagram for some surface F. (Middle and Right) The links before and after the band surgeries are both unlinks.

3. Bridge trisections of knotted surfaces in S^4

In [**22**], Meier and Zupan showed that any knotted surface can be decomposed into a union of three *trivial disk systems*, called a *bridge trisection*. Bridge trisections have 4 parameters $(b; c_1, c_2, c_3)$, satisfying $\chi(S) = c_1 + c_2 + c_3 - b$.

DEFINITION 3.1. A c-component D^2-tangle is often simply referred to as a trivial *c-disk system*. A *$(b; c_1, c_2, c_3)$-bridge trisection* of a knotted surface $S \subset S^4$ is a decomposition $(S^4, S) = (X_1, \mathcal{D}_1) \cup (X_2, \mathcal{D}_2) \cup (X_3, \mathcal{D}_3)$ satisfying the following properties:

(1) $(X_i, \mathcal{D}_i)$ is a trivial c_i-disk system.
(2) $(\beta_{ij}, \alpha_{ij}) = (X_i, \mathcal{D}_i) \cap (X_j, \mathcal{D}_j)$ is a trivial b-strand tangle.
(3) $(\Sigma, \mathbf{p}) = (X_1, \mathcal{D}_1) \cap (X_2, \mathcal{D}_2) \cap (X_3, \mathcal{D}_3)$ is a 2-sphere, and $\mathbf{p}$ is a collection of $2b$ points contained in Σ.

We remark that Meier and Zupan showed that a bridge trisection can be encoded by certain types of diagrams.

DEFINITION 3.2. A *tri-plane diagram* P is a triple of trivial tangle diagrams B_1, B_2, and B_3 such that $B_i \cup \overline{B_j}$ is an unlink, where $\overline{B_j}$ denotes the mirror image of B_j.

Since a bridge trisection is another representation of a surface like a banded unlink (Subsection 2.4), there is a way to convert between the two types of diagrams [**22**]. We briefly mention a method of conversion here since we will need it for proofs later.

Suppose that we are given a banded unlink. Our end goal is to get three trivial tangle diagrams. Meier and Zupan showed that there is always a way to isotope the banded unlink so that the bands are horizontally positioned (see [**22**] for more details). This is demonstrated in Figure 11, where in the leftmost image, the bands are not horizontal. At the end of the process (the rightmost image). This is called a banded bridge splitting in [**22**]. Once the banded unlink is in a banded bridge splitting position. The three trivial tangles can be read off: The horizontal line separates an unlink into two tangles B_1 and $\overline{B_2}$. The tangle B_3 is obtained by performing band surgeries on B_1 (see Figure 12). As an exercise, the readers can verify that $B_i \cup \overline{B_j}$ are unlinks for all $i, j \in \{1, 2, 3\}$ (See Figure 13 for one particular pairwise union).

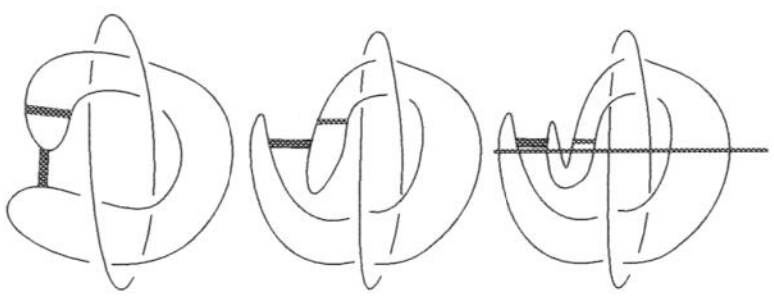

FIGURE 11. From left to right, we isotope the banded unlink so that the bands are in banded bridge splitting.

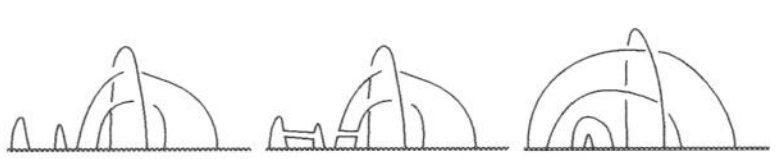

FIGURE 12. A triplane diagram of a knotted surface.

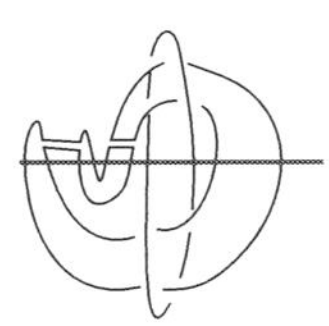

FIGURE 13. Verifying that the union of two trivial tangles in the triplane diagram is an unlink.

The *bridge trisection index* $b(S)$ of a knotted surface S in S^4 is defined to be the minimum number b such that S admits a $(b; c_1, c_2, c_3)$-bridge trisection. The quantity $\min\{c_i\}$ is called the *patch number*, because it is the minimal number of disks, or patches, in one of the disk systems.

PROPOSITION 1.3. *Let S be a knotted surface. Then $\beta(S) = \min_{\mathcal{T}} (\min\{c_1, c_2, c_3\})$, where the minimum is taken over all bridge trisections $\mathcal{T}$ of S.*

PROOF. If S admits a $(b; c_1, c_2, c_3)$-bridge trisection, then Meier-Zupan show in [**22**] that S has an embedding with c_i minima, $b - c_j$ saddles, and c_k maxima, for any choice of distinct i, j, k. Thus S has an embedding with $\min\{c_i\}$ minima, and $\beta(S) \leq \min\{c_i\}$.

For the reverse direction, start with an embedding of S in Morse position with $\beta = \beta(S)$ minima. Note that S can be isotoped so that its critical points are index-ordered, without changing how many critical points of each index exist. Slicing S^4 just above the minima is then an unlink L with β components, and the index 1 critical points give rise to a set of bands b such that surgering L along b results in the unlink just below all the maxima. See for instance [**21**]. As in [**22**], this banded unlink (L, b) can then be isotoped into a banded bridge splitting, again without changing the number of components. The banded bridge splitting induces a $(b; c_1, c_2, c_3)$ bridge trisection, where $c_1 = \beta$. Thus $\beta \geq \min\{c_i\}$. □

COROLLARY 3.3. *Let S be a knotted surface. Then $3\beta(S) - \chi(S) \leq b(S)$.*

PROOF. Combine $\beta(S) \leq \min\{c_i\}$ and the Euler characteristic calculation $\chi(S) = c_1 + c_2 + c_3 - b$. □

As expected, in all the cases that we can prove $\mu(S) = \beta(S)$, we can construct a balanced $(3\beta - \chi(S); \beta)$-bridge trisection.

QUESTION 3.4. *Does there exist a surface S with $b(S) > 3\beta(S) - \chi(S)$?*

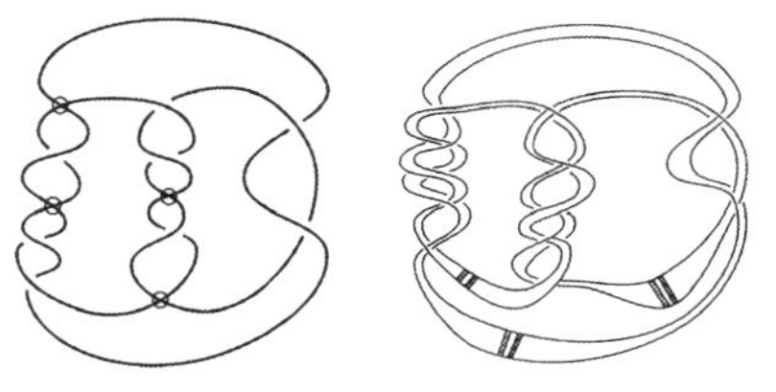

FIGURE 14. From a virtual knot diagram D with r minima (left), one can get a banded link diagram for $Tube(D)$ with $2r$ minima (right).

In the case of twist-spun 2-knots, this is Question 5.2 of [**22**], which we partially answer with Theorem 1.2 in [**19**].

3.1. The Tube map and bridge trisections. The construction of *ribbon bridge trisections* was formulated in [**18**]. In that work, ribbon bridge trisections were defined and utilized by the first author and Meier, Miller, and Zupan to provide the first examples of nonisotopic bridge trisections of isotopic knotted surfaces [**18**].

PROPOSITION 3.5 ([**18**]). *Let S be a ribbon T^2-knot. Suppose that S can be represented by a virtual knot diagram D such that $\beta_{\mathbb{R}^2}(D) = r$, then $b(S) \leq 3r$.*

PROOF. The equatorial cross-section K of $Tube(D)$ has a diagram with $2r$ minima with respect to h as shown on the right of Figure 14 (without the colored bands). The link K together with these bands gives a movie description of ribbon annuli, whose double is $Tube(D)$. A banded unlink diagram of $Tube(D)$ can then be obtained by performing a band move and attaching a dual band (shown in fuchsia) near each yellow band as depicted in the middle of Figure 15. To obtain a dual bridge disk for each yellow band, we can perturb the banded unlink as shown on the right of Figure 15. The process described above produces a banded bridge splitting, and thus a bridge trisection, with $3r$ bridges. □

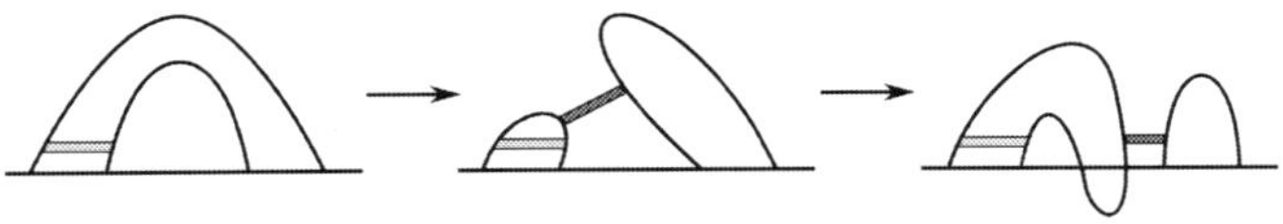

FIGURE 15. To get a bridge trisection, we perform a perturbation to get the rightmost image. This increases the bridge index by 1 for each of the original minima of D, resulting in a bridge trisection index of $3r$. Recall that to convert from banded unlink to a trisection, the authors of [**22**] isotope the banded unlink so that the bands are horizontal and each band has a dual bridge disk.

The following observation follows from the Euler characteristic calculations. We will use it in Corollary 3.5.

COROLLARY 3.6. *Let S be a ribbon T^2-knot. Suppose that S can be represented by a virtual knot diagram D such that $\beta_{\mathbb{R}^2}(D) = r$, then* $\min\{c_i\} \leq r$.

PROOF. The Euler characteristic $\chi(S)$ of S is 0. On the other hand, $\chi(S) = -b+c_1+c_2+c_3$. Equating the two methods gives $3\min\{c_i\} \leq c_1+c_2+c_3 \leq 3r$. □

4. Meridional rank and bridge number of ribbon surfaces

We are now ready for the construction of ribbon tori whose meridional rank equals the bridge number.

4.1. BBKM knot diagrams with virtual crossings. Inspired by ideas from [**1, 2**], we will provide families of ribbon T^2-knot F such that $\mu(F) = \beta(F)$. We begin by defining some local moves on virtual knot diagrams, which will play an important role in our theorems.

DEFINITION 4.1. The *flank-switch move* is shown in Figure 19, and the *flank move* is simply the flank-switch move, but we do not switch the classical crossing. We will be considering the *flank 1-move*, which is a flank move followed by adding an additional crossing as shown in the bottom row of Figure 16. When a classical crossing c in turned into a virtual crossing, we also say that c is *virtualized.*

Observe that if we turn two adjacent crossings into virtual crossings in a twist region, one can remove two newly created virtual crossings by the virtual analog of the Reidemeister II move. Similarly, if one performs two flank-type moves on two adjacent crossings in a twist region, one can use the virtual Reidemeister II move to get rid of two virtual crossings, resulting in two classical crossings in between two virtual crossings. These situations are illustrated in Figure 17. Thus, replacements by a twist region with multiple classical crossings in between two virtual crossings can be realized by our moves.

For the bridge number computations, it is also convenient to notice that the Wirtinger number does not increase after the flank-switch and the flank 1-moves are performed.

LEMMA 4.2. *Suppose that D' is obtained by turning a classical crossing into a virtual crossing, flank 1-moves, and flank-switch moves on a classical knot diagram D , then $\omega(D') \leq \omega(D)$.*

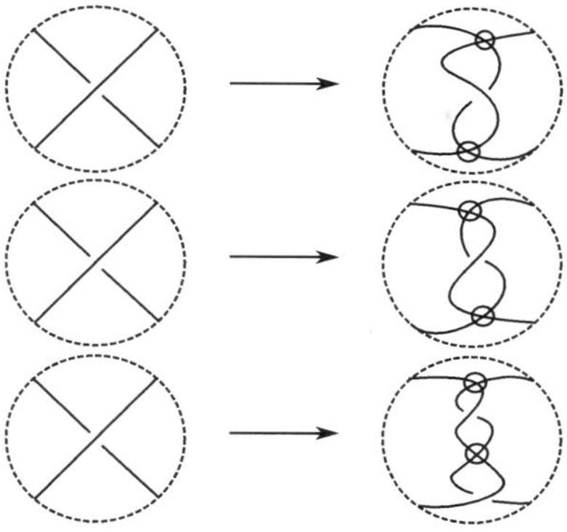

FIGURE 16. (Top) The flank-switch move. (Middle) The flank move. (Bottom) The flank 1-move.

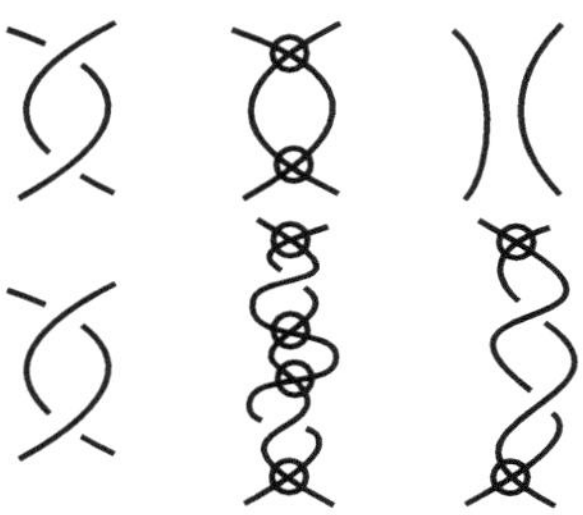

FIGURE 17. (Top) Turning adjacent classical crossings into virtual crossings. (Bottom) Performing two flank-switch moves on adjacent classical crossings.

PROOF. There is a moment during the coloring sequence such that the over-strand at a crossings c of D is colored and the incoming strand is colored. The coloring move can be performed to extend the coloring to the outgoing strand. We break into two cases.

CASE 1. If c is replaced by a virtual crossing, one can simply omit this coloring move at c from the coloring sequence on the virtualized diagram. This shows that the seeds for D persists as seeds for D after some classical crossings have been virtualized (see Figure 18).

CASE 2. If c is replaced by virtual 2-string tangles in the definitions of the flank-switch move and the flank 1-move, then the strands touching the northeast and the northwest endpoints of the virtual tangle are colored blue. The coloring moves can still be performed to extend the coloring to all strands in this virtual 2-string tangle. Figure 19 illustrates this behavior when one flank-switch move is performed. Figure 20 illustrates this behavior when one flank 1-move is performed.

In conclusion, $\omega(D') \leq \omega(D)$ as claimed. □

For the remaining of this section, F will be a ribbon T^2-knot admitting a virtual knot diagram that is obtained by performing moves on a classical BBKM knot diagram following the instructions in Theorem 1.1.

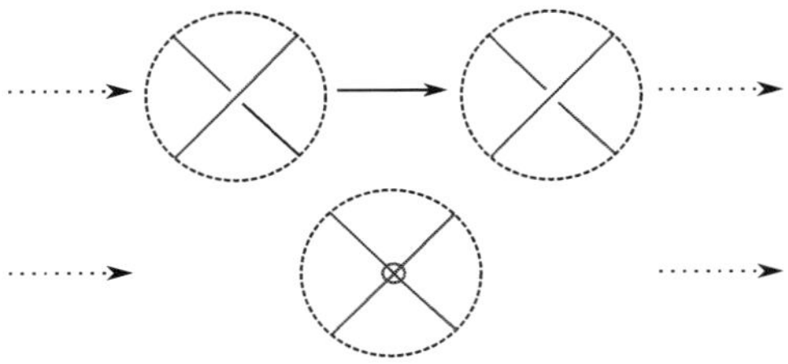

FIGURE 18. (Top) A sequence of diagrams differing only by coloring moves $\cdots \to D_{i-1} \to D_i \to D_{i+1} \to \cdots$. (Bottom) We can use the same seeds for the virtualized diagram, where the sequence will be $\cdots \to D_{i-1} \to D_{i+1} \to \cdots$.

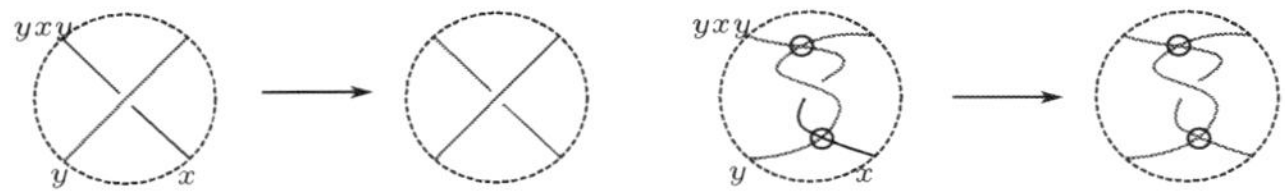

FIGURE 19. Performing a coloring move after a flank-switch move is performed.

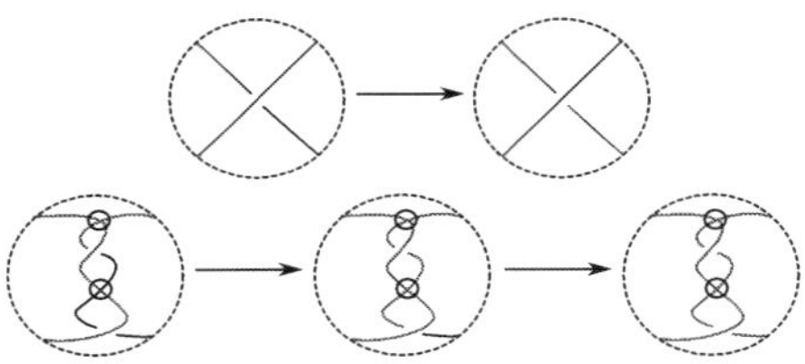

FIGURE 20. After the flank 1-move, we can use the same seeds. The color sequence may be longer, but we are only interested in the number of seeds.

PROOF OF THEOREM 1.1. (1) For the case of single Coxeter generator in a twist region, virtualizing any number of them will preserve the labeling since all strands in the twist region are labeled by the same label. Now, we consider the case of two Coxeter generators in a twist region. Suppose that the two bottom endpoints of a vertical twist region is labeled with two distinct meridians x on the left and y on the right.

CASE 1. Suppose that an odd number of classical crossings are turned into virtual crossings. At the top of the tangle from left to right is either $(xy)^{2k}x$ and $(xy)^{2k+1}x$ or $(xy)^{2k+1}y$ and $(xy)^{2k}y$. Since we want the bottom label and the top label on the same side to be equal, we have that $(xy)^{2k} = 1$ and $(xy)^{2k+2} = 1$. This means that $(xy)^2 = 1$ and therefore we can let the Coxeter weight corresponding to the virtualized twist region be 2.

CASE 2. Suppose that an even number of classical crossings are turned into virtual crossings. At the top of the tangle from left to right is either $(xy)^{2k}x$ and $(xy)^{2k-1}x$ or $(xy)^{2k}y$ and $(xy)^{2k-1}y$. Since we want the bottom label and the top

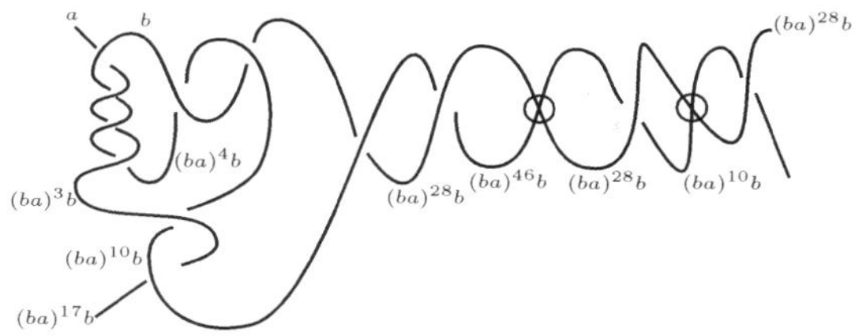

FIGURE 21. Flanking and balancing out in a rational tangle.

label on the same side to be equal, we have that $(xy)^{2k} = 1$ and we can let the Coxeter weight corresponding to the virtualized twist region be $2k$.

(2) The flank-switch move preserves the original labeling as shown in Figure 19.

(3) Note that each time one component of the integral tangle goes under the other component, the labeling of the two sides of the bigon after that crossing have exponents that jump from two sides of the bigon before that crossing by a certain interval r. After one instance of flanking, the exponents of (ab) will drop by r. Adding a new classical crossing will cancel out this drop. Figure 21 demonstrates a specific example $r = 18$ with the jumps in the exponent shown. □

Satoh showed that the Tube map preserves the fundamental group. Thus, the lower bounds from virtual knot diagrams on the meridional rank become lower bounds for the corresponding ribbon T^2-knots (see Subsection 2.1.1 for a discussion on ribbon surfaces and their notations). We can also use the results of [**25**, **27**] to get matching upper bounds. We first elaborate more on how the algorithms in [**25**] works (the proof is essentially the same and we provide it for completeness).

PROPOSITION 4.3 ([**25**]). *Given a diagram D with n overpasses. Then, D is equivalent to D' with n maxima with respect to the standard height function on $\mathbb{R}^2$.*

PROOF. Consider a Gauss diagram G with $\beta_O(L)$ overpasses. Observe that to perform the welded move on Gauss diagrams, one moves two arrow-tails pass each other (see also [**26**]). We first perform a sequence of welded moves so that no two chords c_1, c_2 connecting a fixed overpass O and a fixed underpass U cross each other (see Figure 22). More precisely, we perform welded moves to obtain a Gauss diagram G' such that as we traverse along O counterclockwise, if we encounter the arrow-tail of c_1 before the arrow-tail of c_2, then we must encounter the arrow-head of c_2 before the arrow-head of c_1 as we traverse U counterclockwise. From G', we will produce a virtual link diagram with $\beta_{\mathbb{R}^2}$ maxima. Denote the overpasses and underpasses of G' by $O_1, U_1, O_2, U_2 \ldots, O_{b_1(L)}, U_{b_1(L)}$ in order.

Begin by drawing $\beta_O(L)$ straight vertical segments $\widetilde{O_1}, \ldots, \widetilde{O_{b_1(L)}}$ as shown in Figure 23 as thick black lines. These correspond to $O_1, \ldots, O_{b_1(L)}$. We connect the top of $\widetilde{O_i}$ to the bottom of $\widetilde{O_{i+1}}$ via a monotonically decreasing arc that passes under $\widetilde{O_1}, \ldots, \widetilde{O_{b_1(L)}}$ following the order of the arrow-heads in U_i. Here i is taken cyclically depending on the component of the link. □

COROLLARY 1.2. *Let K be a welded knot obtained from Theorem 1.1, and let $F = Tube(K)$, a ribbon T^2-knot. Then, $\mu(F) = \beta(F)$.*

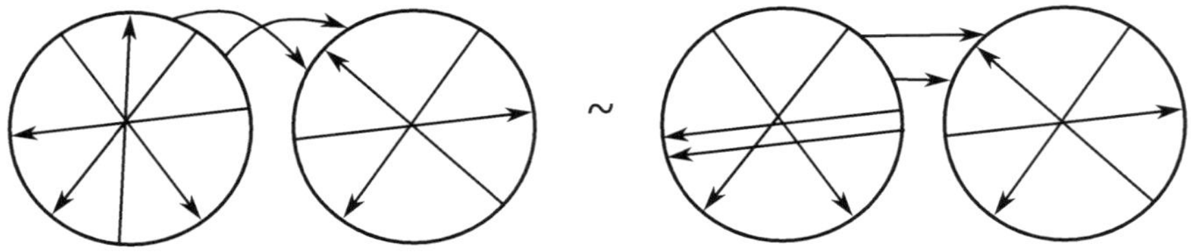

FIGURE 22. Performing welded moves so that the chords connecting A_i to B_j for any fixed i, j do not cross each other.

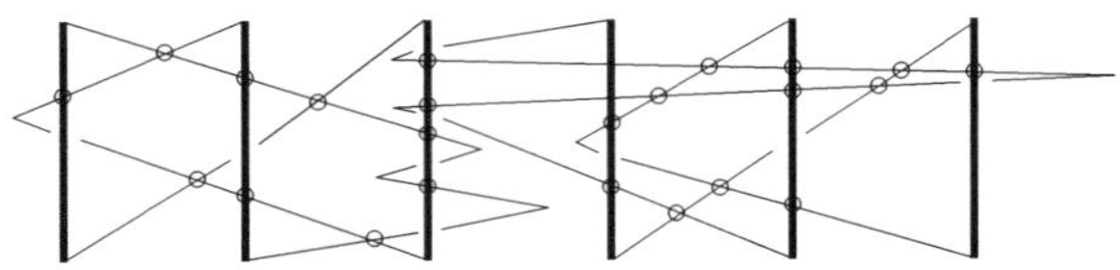

FIGURE 23. A virtual link diagram L with $\beta_O(L)$ maxima constructed from a Gauss diagram G for L, where chords connecting A_i and B_j for fixed i, j do not cross each other.

PROOF. Starting with a classical BBKM knot K with $\mu(K) = n$ represented as a diagram D. Theorem 1.1 shows that performing the flank-switch moves, flank 1-moves, and turning classical crossings into virtual crossings as instructed gives a virtual knot diagram D' such that $\mu(Tube(D')) \geq n$.

For the matching upper bound, Lemma 4.1 says that the same seeds that give rise to $\omega(D) = n$ persist to be seeds for the virtual knot diagram D'. By Theorem 3.3 of [**27**], D' admits a diagram with n overpasses. By Theorem 1.2 of [**25**], there are local moves that one can perform to transform D' to a diagram D'' with n minima with respect to the standard height function on $\mathbb{R}^2$. Furthermore, Tube(D'') is isotopic to Tube(D') after the moves since only welded moves were used and welded moves correspond to an isotopy of the corresponding tori (The proof of this step is reproduced again as Proposition 4.3 for convenience). By Corollary 3.6, these n minima gives rise to n 0-handles for the ribbon surface Tube(D') $= F$. Thus, $\beta(F) \leq n$.

To see an illustration of the correspondence, the readers can consult Figure 9 of [**18**], where each maximum corresponds to an unknot component in a band diagram presentation of a knotted surface. Since each 0-handle corresponds to an unknot component and each 1-handle corresponds to a band, the corollary is proved. □

The following corollary holds.

COROLLARY 1.4. *If F is a ribbon T^2-knot satisfying the hypotheses of Theorem* 1.1, *then* $b(F) = 3n$.

PROOF. Using Proposition 3.5, we get that $b(S) \leq 3n$. If the equality was strict, then we would be able to find a virtual knot diagram representation with fewer than n maxima, which is a contradiction. □

Acknowledgment

We thank the reviewer for helpful comments.

References

[1] Sebastian Baader, Ryan Blair, and Alexandra Kjuchukova, *Coxeter groups and meridional rank of links*, Math. Ann. **379** (2021), no. 3-4, 1533–1551, DOI 10.1007/s00208-020-02124-z. MR4238272

[2] Sebastian Baader, Ryan Blair, Alexandra Kjuchukova, and Filip Misev, *The bridge number of arborescent links with many twigs*, Algebr. Geom. Topol. **23** (2023), no. 1, 75–85, DOI 10.2140/agt.2023.23.75. MR4568000

[3] Sebastian Baader and Alexandra Kjuchukova, *Symmetric quotients of knot groups and a filtration of the Gordian graph*, Math. Proc. Cambridge Philos. Soc. **169** (2020), no. 1, 141–148, DOI 10.1017/s0305004119000136. MR4120787

[4] R. Blair, A. Kjuchukova, R. Velazquez, and P. Villanueva, *Wirtinger systems of generators of knot groups*, Comm. Anal. Geom. **28** (2020), no. 2, 243–262, DOI 10.4310/cag.2020.v28.n2.a2. MR4101339

[5] Ryan Blair, Marion Campisi, Scott A. Taylor, and Maggy Tomova, *Distortion and the bridge distance of knots*, J. Topol. **13** (2020), no. 2, 669–682, DOI 10.1112/topo.12139. MR4092777

[6] Michel Boileau, Yeonhee Jang, and Richard Weidmann, *Meridional rank and bridge number for a class of links*, Pacific J. Math. **292** (2018), no. 1, 61–80, DOI 10.2140/pjm.2018.292.61. MR3708258

[7] Jae-Wook Chung and Xiao-Song Lin, *On the bridge number of knot diagrams with minimal crossings*, Math. Proc. Cambridge Philos. Soc. **137** (2004), no. 3, 617–632, DOI 10.1017/S0305004104007753. MR2103920

[8] Christopher R. Cornwell and David R. Hemminger, *Augmentation rank of satellites with braid pattern*, Comm. Anal. Geom. **24** (2016), no. 5, 939–967, DOI 10.4310/CAG.2016.v24.n5.a2. MR3622310

[9] Y. Diao, C. Ernst, and M. Thistlethwaite, *The linear growth in the lengths of a family of thick knots*, J. Knot Theory Ramifications **12** (2003), no. 5, 709–715, DOI 10.1142/S0218216503002615. MR1999639

[10] H. A. Dye and Louis H. Kauffman, *Minimal surface representations of virtual knots and links*, Algebr. Geom. Topol. **5** (2005), 509–535, DOI 10.2140/agt.2005.5.509. MR2153118

[11] David Gay and Robion Kirby, *Trisecting 4-manifolds*, Geom. Topol. **20** (2016), no. 6, 3097–3132, DOI 10.2140/gt.2016.20.3097. MR3590351

[12] Francisco González-Acuña and Arturo Ramírez, *Epimorphisms of knot groups onto free products*, Topology **42** (2003), no. 6, 1205–1227, DOI 10.1016/S0040-9383(02)00087-3. MR1981354

[13] Mikhail Goussarov, Michael Polyak, and Oleg Viro, *Finite-type invariants of classical and virtual knots*, Topology **39** (2000), no. 5, 1045–1068, DOI 10.1016/S0040-9383(99)00054-3. MR1763963

[14] Allison Henrich and Sam Nelson, *Semiquandles and flat virtual knots*, Pacific J. Math. **248** (2010), no. 1, 155–170, DOI 10.2140/pjm.2010.248.155. MR2734169

[15] Mikami Hirasawa, Naoko Kamada, and Seiichi Kamada, *Bridge presentations of virtual knots*, J. Knot Theory Ramifications **20** (2011), no. 6, 881–893, DOI 10.1142/S0218216511009017. MR2812268

[16] Mark C. Hughes, Seungwon Kim, and Maggie Miller, *Isotopies of surfaces in 4-manifolds via banded unlink diagrams*, Geom. Topol. **24** (2020), no. 3, 1519–1569, DOI 10.2140/gt.2020.24.1519. MR4157558

[17] Atsushi Ichimori and Taizo Kanenobu, *Ribbon torus knots presented by virtual knots with up to four crossings*, J. Knot Theory Ramifications **21** (2012), no. 13, 1240005, 30, DOI 10.1142/S0218216512400056. MR2994592

[18] Jason Joseph, Jeffrey Meier, Maggie Miller, and Alexander Zupan, *Bridge trisections and classical knotted surface theory*, Pacific J. Math. **319** (2022), no. 2, 343–369, DOI 10.2140/pjm.2022.319.343. MR4482720

[19] Jason Joseph and Puttipong Pongtanapaisan, *Meridional rank and bridge number of knotted 2-spheres*, Canad. J. Math. **77** (2025), no. 1, 282–299, DOI 10.4153/S0008414X23000883. MR4861737

[20] Louis H Kauffman, *Virtual knot theory*, Encyclopedia of Knot Theory (2021), 261.

[21] Akio Kawauchi, *A survey of knot theory*, Birkhäuser Verlag, Basel, 1996. Translated and revised from the 1990 Japanese original by the author. MR1417494

[22] Jeffrey Meier and Alexander Zupan, *Bridge trisections of knotted surfaces in S^4*, Trans. Amer. Math. Soc. **369** (2017), no. 10, 7343–7386, DOI 10.1090/tran/6934. MR3683111

[23] J. W. Milnor, *On the total curvature of knots*, Ann. of Math. (2) **52** (1950), 248–257, DOI 10.2307/1969467. MR37509

[24] J. Milnor, *Morse theory*, Annals of Mathematics Studies, No. 51, Princeton University Press, Princeton, NJ, 1963. Based on lecture notes by M. Spivak and R. Wells. MR163331

[25] Yasutaka Nakanishi and Shin Satoh, *Two definitions of the bridge index of a welded knot.* part B, Topology Appl. **196** (2015), no. part B, 846–851, DOI 10.1016/j.topol.2015.05.045. MR3431020

[26] Sam Nelson, *Unknotting virtual knots with Gauss diagram forbidden moves*, J. Knot Theory Ramifications **10** (2001), no. 6, 931–935, DOI 10.1142/S0218216501001244. MR1840276

[27] Puttipong Pongtanapaisan, *Wirtinger numbers for virtual links*, J. Knot Theory Ramifications **28** (2019), no. 14, 1950086, 16, DOI 10.1142/S021821651950086X. MR4077437

[28] Shin Satoh, *Virtual knot presentation of ribbon torus-knots*, J. Knot Theory Ramifications **9** (2000), no. 4, 531–542, DOI 10.1142/S0218216500000293. MR1758871

[29] Martin Scharlemann, *Thin position in the theory of classical knots*, Handbook of knot theory, Elsevier B. V., Amsterdam, 2005, pp. 429–459, DOI 10.1016/B978-044451452-3/50010-1. MR2179267

[30] Daniel S. Silver and Wilbur Whitten, *Knot group epimorphisms*, J. Knot Theory Ramifications **15** (2006), no. 2, 153–166, DOI 10.1142/S0218216506004373. MR2207903

901 Burkemont Avenue, Morganton, North Carolina 28655
Email address: `jason.joseph@ncssm.edu`

901 Palm Walk, Tempe, Arizona 85281
Email address: `ppongtan@asu.edu`

Contemporary Mathematics
Volume **827**, 2025
https://doi.org/10.1090/conm/827/16538

On an invariant for colored classical and singular links

Audrey Baumheckel, Carmen Caprau, and Conor Righetti

Abstract. A colored link, as defined by Francesca Aicardi, is an oriented classical link together with a 'coloration', which is a function defined on the set of link components and whose image is a finite set of 'colors'. An oriented classical link can be regarded as a colored link with its components colored with a sole color. Aicardi constructed an invariant $F(L)$ of colored links L defined via skein relations. When the components of a colored link are colored with the same color or when the colored link is a knot, $F(L)$ is a specialization of the HOMFLY-PT polynomial. Aicardi also showed that $F(L)$ is a stronger invariant than the HOMFLY-PT polynomial when evaluated on colored links whose components have different colors. In this paper, we provide a state-sum model for the invariant $F(L)$ of colored links using a graphical calculus for oriented, colored, 4-valent planar graphs. We also extend $F(L)$ to an invariant of oriented colored singular links.

1. Introduction

Francesca Aicardi [**1, 2**] introduced a special type of oriented links that she called *colored links*. If a function γ is defined on the set C of components of an oriented link L and assigns elements of a finite set N of 'colors', then the link L is called *colored* and the function γ is called a *coloration*. If $|\gamma(C)| = 1$, then the colored link is a classical link; in this case, all components of L are colored with the same color. A coloration γ partitions the set C of components of the oriented link L, hence γ introduces an equivalence relation on C. Two colorations γ and γ' of L are called *equivalent* if there exists a bijection $\gamma(C) \to \gamma'(C)$ that preserves the equivalence classes of C induced by the two colorations.

An invariant of colored links is an invariant of links that takes the same value on ambient isotopic links with equivalent colorations.

Colored links are closely related to tied links. A *tied link* [**3**] is an oriented classical link whose components may be connected by ties. A *tie* is an arc that connects two points on the link, where the two points can belong to different components or the same component of the link. A tie connecting points that belong to the same component of a link can be removed, and such removal does not change the type of tied link. Ties are merely notational devices depicted as springs and are not embedded arcs. The set of ties on a tied link partitions the set C of components

2020 *Mathematics Subject Classification.* Primary 57K12, 57K14.
Key words and phrases. Links, singular links, polynomial invariants for links.
The second author was partially supported by Simons Foundation grant #355640 and NSF grant DMS-2204386.

of a link into equivalence classes. Hence tied links are links whose components are partitioned into classes. Equivalently, one can use colors to indicate the components of the link that are in the same equivalence class. In this sense, tied links are equivalent to colored links. In this paper, we chose to work with colored links for the sake of having less crowded diagrams.

Another type of knotted objects that we work with in this paper are oriented singular links. A *singular link* is an immersion in $\mathbb{R}^3$ of a disjoint union of k circles, where $k \geq 1$, such that the immersion contains finitely many singularities which are transverse double points. The image of the k circles under the immersion are called the *components of the singular link*. A diagram of a singular link is a projection of the singular link in a plane; such a diagram may contain classical crossings, as in classical link diagrams, and singular crossings. A singular crossing in a diagram is depicted as a 4-valent vertex.

It is well known that two singular link diagrams represent *equivalent* (or *ambient isotopic*) singular links if and only if there exists a finite sequence of the extended Reidemeister moves mapping one diagram onto the other (see for example [**7**, **9**]). A proof of this statement can be found in [**4**]. The *extended Reidemeister moves*, depicted in Fig. 1, are local moves on singular link diagrams and include the classical Reidemeister moves $R1$, $R2$ and $R3$, together with the additional $R4$ and $R5$ moves involving singular crossings.

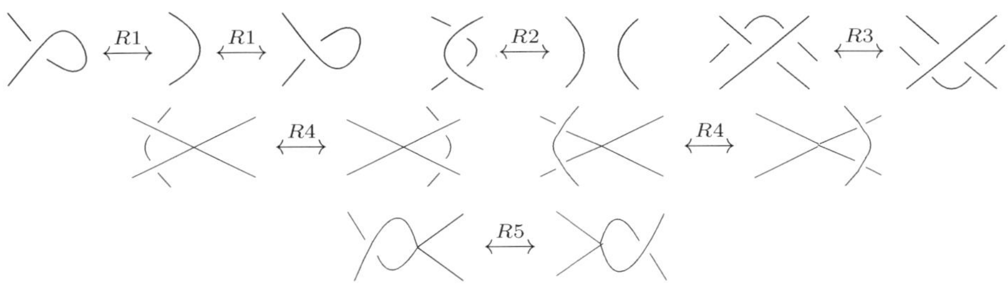

FIGURE 1. Extended Reidemeister moves

We remark that singular links can also be regarded as rigid-vertex embeddings in $\mathbb{R}^3$ of 4-valent graphs. For details, we refer the reader to Kauffman's work in [**7**].

The concept of coloring can be extended to oriented singular links as well. We define the notion of *colored singular link* as an oriented singular link together with a coloration defined on the components of the singular link. An invariant for colored singular links must take the same value on ambient isotopic singular links with equivalent colorations. The diagram on the left in Fig. 2 is an example of a colored (classical) link with three components, while the diagram on the right is an example of a colored singular link with two components.

One way to consider links as colored links or singular links as colored singular links is by coloring all of the link components with the same color. Another way is by coloring all of the link components with different colors.

Aicardi [**2**] defined an invariant $F(L)$ for colored links L. We note that a similar invariant for tied links was defined by Aicardi and Juyumaya in [**3**]. When the link components are colored with a sole color or when the link is a knot, the invariant F is a specialization of the HOMFLY-PT polynomial [**6**, **11**] for oriented classical

FIGURE 2. Examples of colored classical and singular links

links. The reason for studying the polynomial F is that it is a stronger invariant than the HOMFLY-PT polynomial when F is evaluated on colored links whose components have different colors; for details, we refer to Aicardi's work in [**1**].

Given a colored link L, the invariant $F(L)$ is valued in $\mathbb{Q}(x,t,w)$ and is uniquely determined by the following three conditions:

(I) $F\left(\bigcirc\right) = 1$.

(II) Let $D \cup \bigcirc$ be the disjoint union of an unknotted circle with a colored link diagram D, where the colors of D's components are all distinct from the color of the circle. Then

$$F\left(D \cup \bigcirc\right) = \frac{1}{wx} F\,(D)\,.$$

(III) The following skein relation holds,

$$\frac{1}{w}F\left(\diagup\!\!\!\!\diagdown\right) - wF\left(\diagup\!\!\!\!\diagdown\right) = \left(1-\frac{1}{t}\right)F\left()\,(\right) + \left(\frac{1}{w}-\frac{1}{tw}\right)F\left(\diagup\!\!\!\!\diagdown\right),$$

where when forgetting the colors, the four diagrams shown in the skein relation above are parts of link diagrams that are identical almost everywhere, except in a small neighborhood where they differ as shown. The diagrams $\diagup\!\!\!\!\diagdown$ and $\diagup\!\!\!\!\diagdown$ are parts of colored links, where black and red colors indicate any colors, not necessarily distinct. Moreover, the components of the colored links that contain the parts $)\,($, $\diagup\!\!\!\!\diagdown$, and $\diagup\!\!\!\!\diagdown$ are colored with a sole color (the color does not matter, what matters is that it is the same color; that is, the components of the colored links that contain the shown parts belong to the same partition induced by the resulting coloration on the set of the link components.)

The skein relation (III) holds for any two colors of the strands in the two diagrams on the left of the relation. If the two colors are the same, then relation (III) is reduced to the following skein relation:

$$\text{(IV)} \qquad \frac{1}{tw}F\left(\diagup\!\!\!\!\diagdown\right) - wF\left(\diagup\!\!\!\!\diagdown\right) = \left(1-\frac{1}{t}\right)F\left()\,(\right).$$

We refer to the invariant F as the *Aicardi-Juyumaya invariant* for colored links.

Solving for $\frac{1}{w}F\left(\diagup\!\!\!\!\diagdown\right)$ in relation (IV) and substituting it into relation (III), yields the following skein relation:

$$(*)\ \frac{1}{w}F\left(\diagup\!\!\!\!\diagdown\right) - wF\left(\diagup\!\!\!\!\diagdown\right) = \left(t-\frac{1}{t}\right)F\left()\,(\right) + twF\left(\diagup\!\!\!\!\diagdown\right) - \frac{1}{tw}F\left(\diagup\!\!\!\!\diagdown\right).$$

The HOMFLY-PT polynomial P of a classical (single-colored) link is valued in $\mathbb{Z}[\ell, \ell^{-1}, m]$ and is uniquely defined by the condition $P(\bigcirc) = 1$ (unknot) and the skein relation below (see [**2**]):

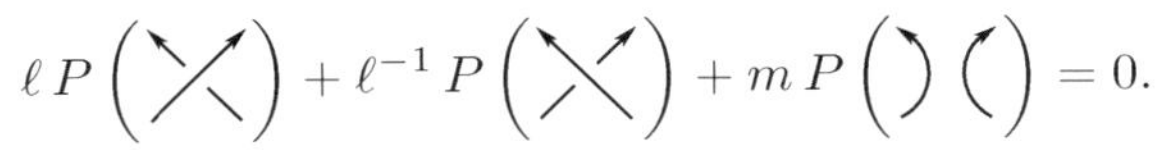

If L is a knot or a link whose components are colored with the same color, then $F(L)$ is determined by relations (I) and (IV), and it coincides with the polynomial $P(L)$ after the following substitutions:

$$\ell = \frac{i}{w\sqrt{t}} \quad \text{and} \quad m = i\left(\frac{1}{\sqrt{t}} - \sqrt{t}\right).$$

The scope of this paper is two-fold. We first extend the invariant F to oriented colored singular links and denote the resulting invariant of colored singular links by $[\cdot]$. We also prove a set of graphical skein relations involving oriented, colored, planar 4-valent graphs; this is done in Section 2. This set of graphical relations provide a recursive way to evaluate oriented, colored, planar graphs with 4-valent vertices. In Section 3, we explain how our graphical skein relations provide a state-sum model for the Aicardi-Juyumaya invariant F for colored links, where the states associated to a diagram of a colored link are oriented, colored, 4-valent planar graphs. This also allows us to define the invariant $[\cdot]$ for colored singular link without relying on the existence of the invariant F for colored links.

2. An extension of the Aicardi-Juyumaya invariant to colored singular links

In this section we extend the Aicardi-Juyumaya invariant to oriented colored singular links. We denote the resulting invariant by $[\cdot]$. We impose first that $[D] = F(D)$, whenever D is a colored (classical) link diagram. Next, we impose the following skein relations for $[\cdot]$:

(1) 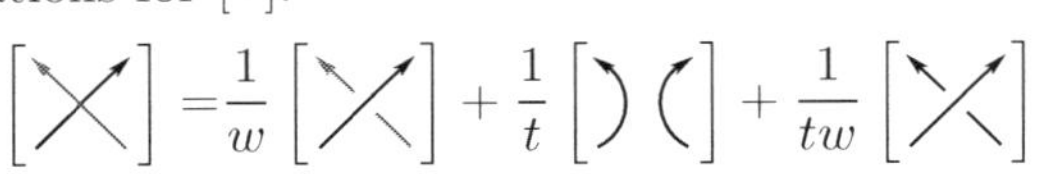

(2) 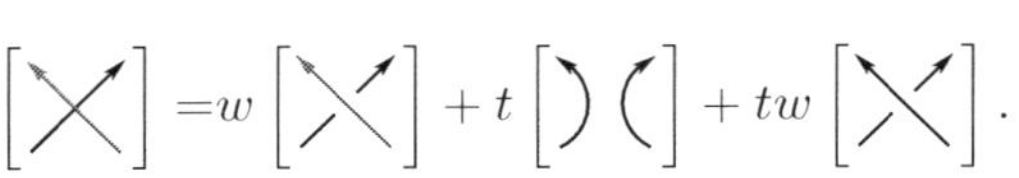

That is, if $\tilde{D}$ is a diagram of a colored singular link, we can use either the skein relation (1) or relation (2) at each of the singular crossings in $\tilde{D}$, to write $[\tilde{D}]$ as a $\mathbb{Q}(x, w, t)$-linear combination of evaluations of colored links. We note that the each diagram in relations (1) and (2) is a part of a larger diagram of a colored singular link whose components have colorations that are globally compatible; in particular, the colorations of the strands shown in a diagram determine the colorations of the link components that the strands are part of. When forgetting the colorations, the diagrams in the two sides of each of these skein relations are parts of larger diagrams that are identical outside of the small neighborhood where the skein relation is applied.

Subtracting relation (2) from relation (1), yields the skein relation

(3) 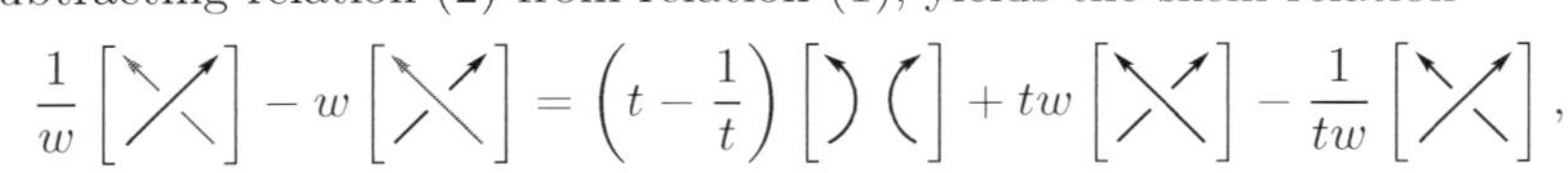

which we know that $[\cdot]$ satisfies, since the invariant F satisfies an equivalent skein relation, as shown in Section 1, and $F(D) = [D]$, for any diagram D of a colored link. Hence, the evaluation $[\tilde{D}]$ of a colored singular link diagram is independent on whether we use the skein relation (1) or (2).

The skein relations (1) or (2) have the following equivalent forms, which will be used later in this section.

(4)
(5)
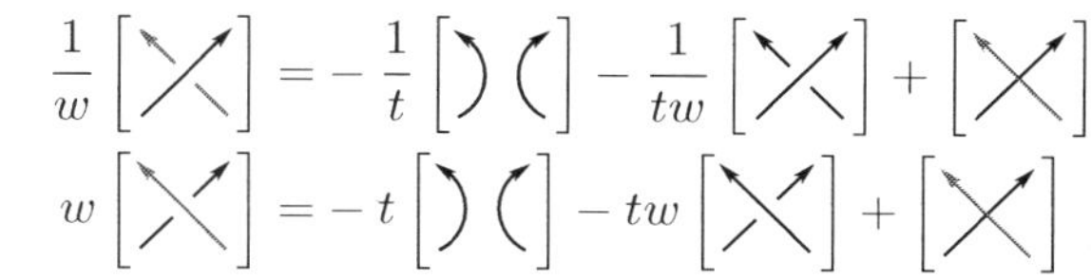

THEOREM 1. *If $\tilde{D}_1$ and $\tilde{D}_2$ are diagrams representing ambient isotopic colored singular links, then $[\tilde{D}_1] = [\tilde{D}_2]$. That is, $[\cdot]$ is an invariant for colored singular links.*

PROOF. We first note that since $[\cdot]$ is an extension of F, then $[\cdot]$ is invariant under Reidemeister moves $R1, R2$ and $R3$. It remains to show that $[\cdot]$ is also invariant under the moves $R4$ and $R5$. Starting with the Reidemeister-type move $R4$, we find that:

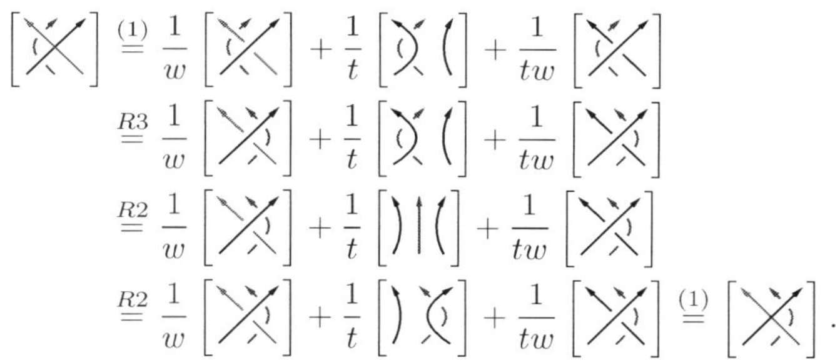

Hence, 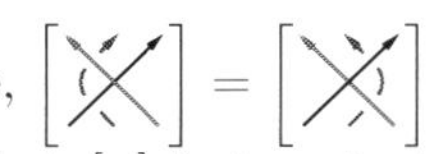. A similar process is used to prove that 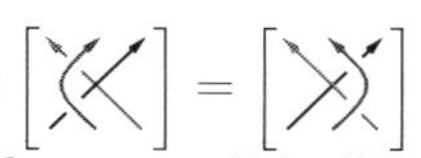 and that $[\cdot]$ is invariant under the other oriented versions of the move $R4$. We prove next the invariance of $[\cdot]$ under the Reidemeister-type move $R5$:

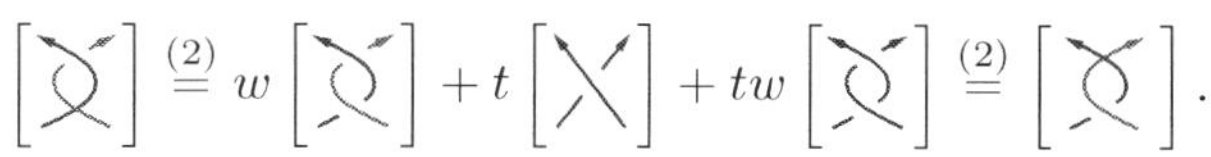

In the first equality above, we applied the skein relation (2) in a small neighborhood of the 4-valent vertex of the diagram on the left-hand side of the move $R5$. In the second step we applied the same skein relation, this time in reverse order, in a small neighborhood near the top part of the three diagrams obtained in the previous step. The proof of invariance under the other oriented versions of the move $R5$ are done in a similar manner. This completes the proof that $[\cdot]$ is an invariant for colored singular links. □

PROPOSITION 2. *Let $L \tilde{\cup} \bigcirc$ be the disjoint union between a colored link L with the standard diagram of the unknot, where the unknot is colored the same as (at least one) of the components of L. Then*

$$[L \tilde{\cup} \bigcirc] = \frac{tw^2 - 1}{w(1-t)}[L] = \left(\frac{tw}{1-t} + \frac{t^{-1}w^{-1}}{1-t^{-1}}\right)[L]. \tag{6}$$

PROOF. We represent $L \tilde{\cup} \bigcirc$ as . From relation (IV), we know that

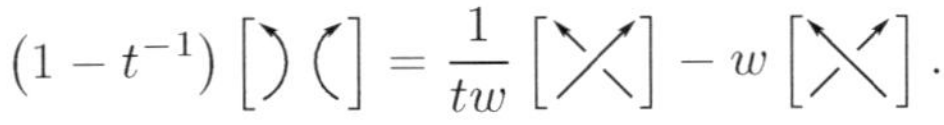

Hence,

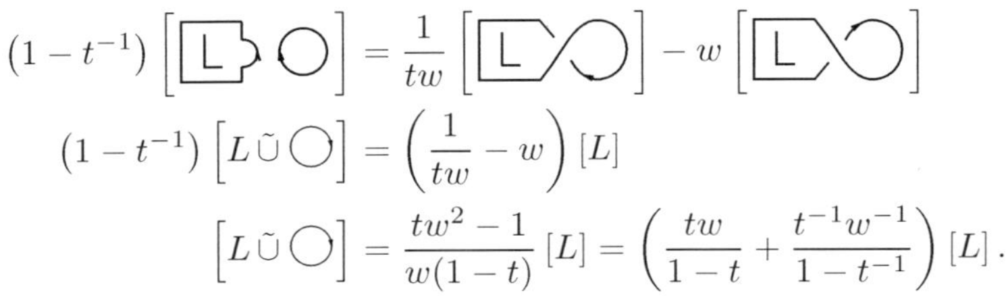

Therefore, identity (6) holds. □

We make use of the skein relations (4) and (5) to prove the following graphical skein relations involving oriented, colored, 4-valent planar graphs. An oriented, colored, 4-valent planar graph can be regarded as a diagram of an oriented colored singular link without classical crossings.

THEOREM 3. *The following graphical skein relations hold for* $[\,\cdot\,]$:

(7) 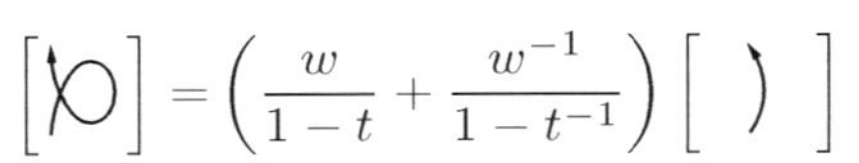

(8) 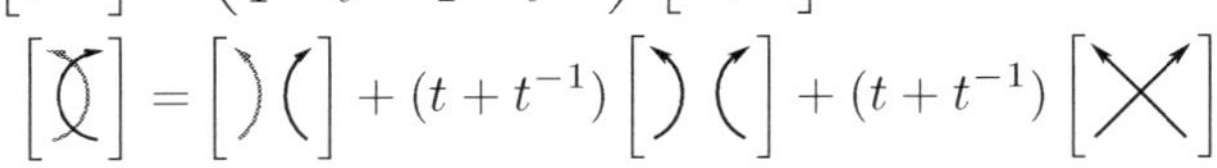

(9) 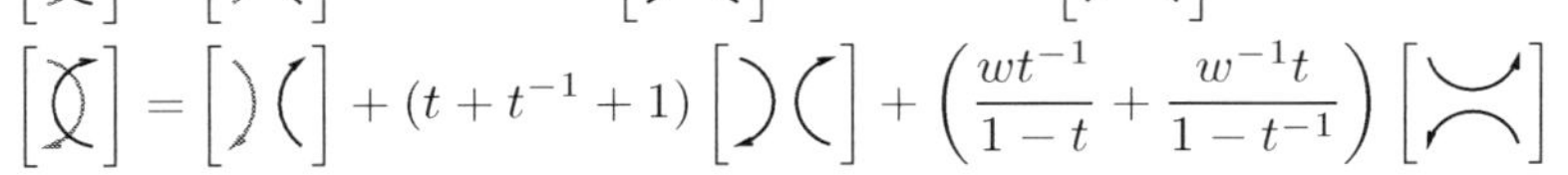

(10) 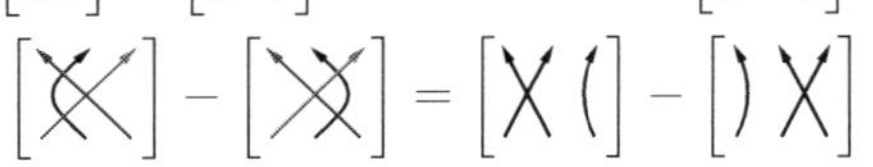

(11) 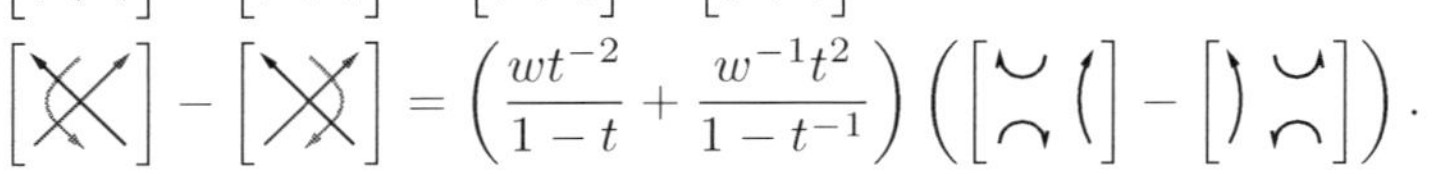

PROOF. We start by proving graphical relation (7). Note that we start by using the skein relation (1), with both strands colored similarly. In the second step we use that $[\,\cdot\,]$ is invariant under the Reidemeister move $R1$.

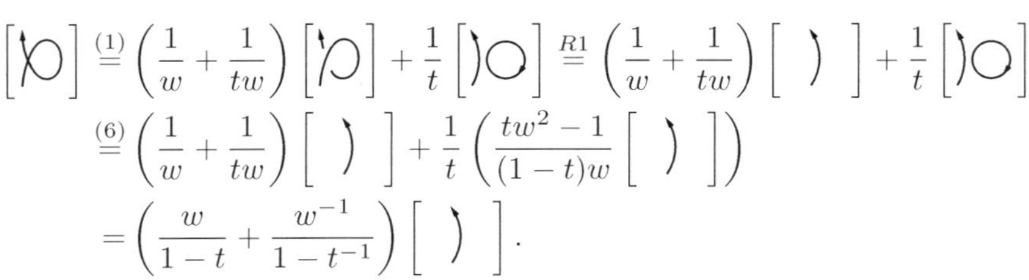

To prove the graphical relation (8) we start from the diagram on the left side of the relation and apply the skein relation (1) in a small neighborhood of the top vertex. In the second step, we apply the skein relation (2) to the resulting first and third diagrams, as we show below.

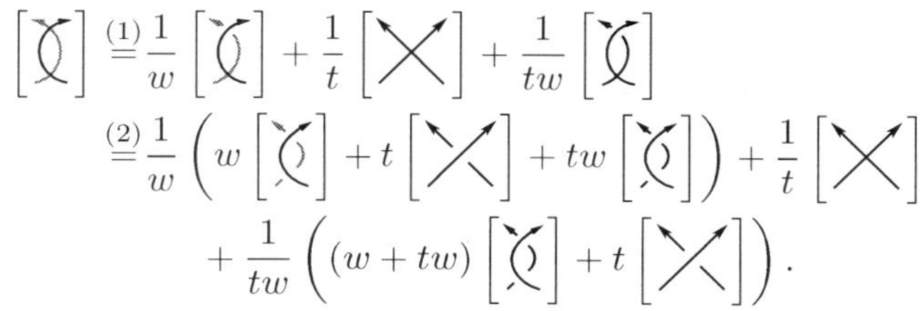

In the next step, we use that $[\cdot]$ is invariant under the Reidemeister move $R2$ and combine like terms. Then we apply the skein relation (4) to the second diagram with a positive crossing and both strands colored with the same color, and obtain the following calculations.

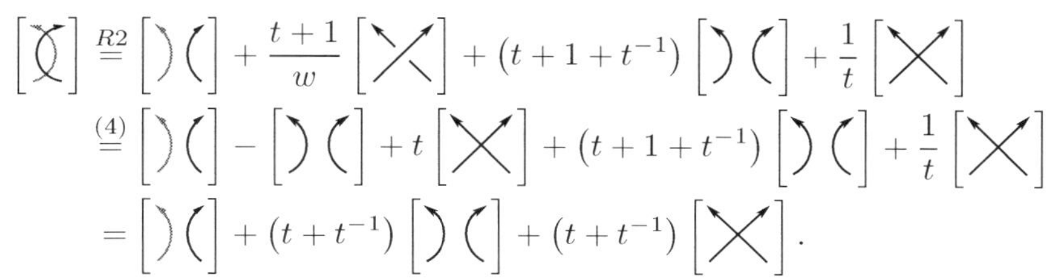

Hence, the graphical relation (8) holds. In the proof of the graphical relation (9), we use a somewhat similar starting point. Note, however, that the strands in the left side diagram of the relation have different orientations than in the relation we just proved.

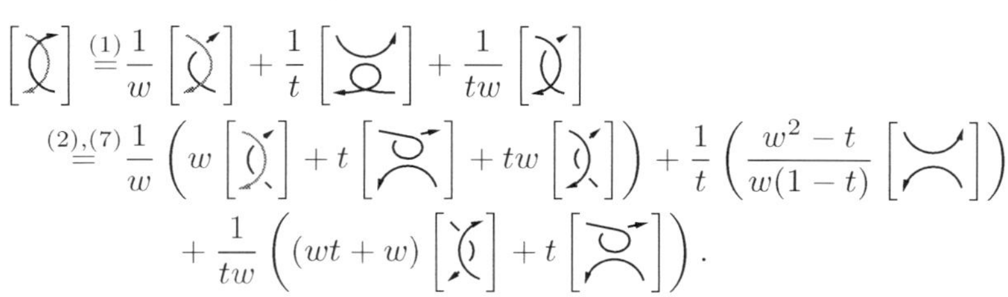

In the second step we also used that $\frac{w^2-t}{w(1-t)} = \frac{w}{1-t} + \frac{w^{-1}}{1-t^{-1}}$. Next we use that $[\cdot]$ is invariant under the moves R1 and R2, to obtain the following:

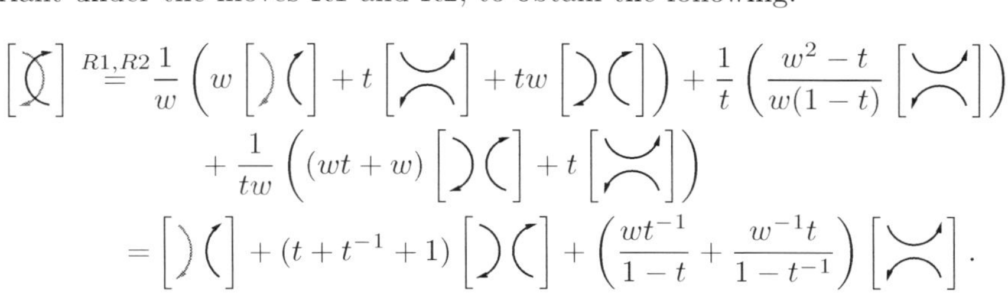

We will now evaluate the two terms of the left side of identity (10). Starting with the first diagram, we resolve first the top left 4-valent vertex using the skein relation (1). After that, in the first and third of the three resulting diagrams, we resolve the bottom left 4-valent vertex by applying the skein relation (2). In the third step we merely simplify expressions.

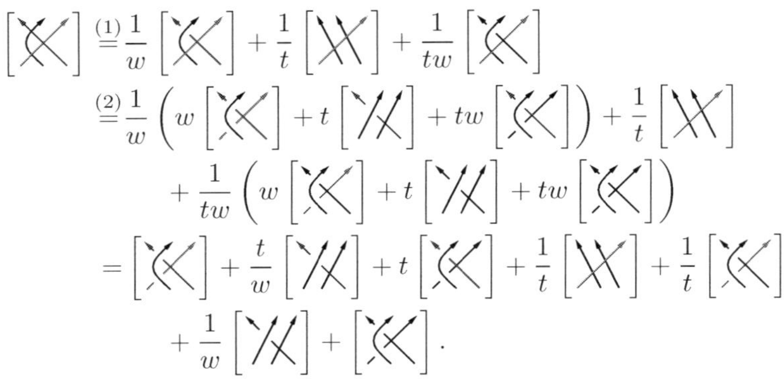

Next we apply skein relation (4) at the positive crossing in the second diagram of the resulting sum above. Finally, in the last step below, we simplify the resulting linear combination of evaluations of colored link diagrams.

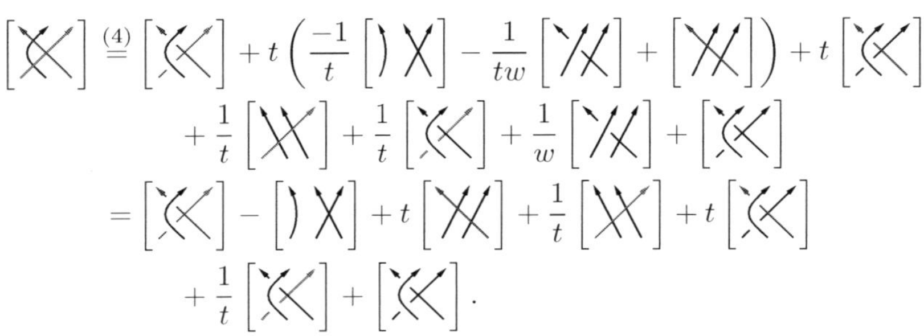

Similar steps are applied to the second diagram appearing on the left side of the identity (10), as we show below.

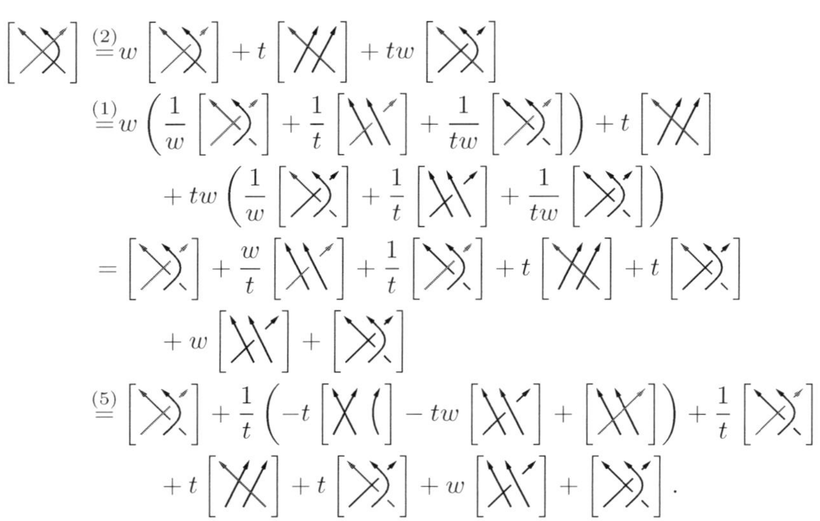

Therefore,

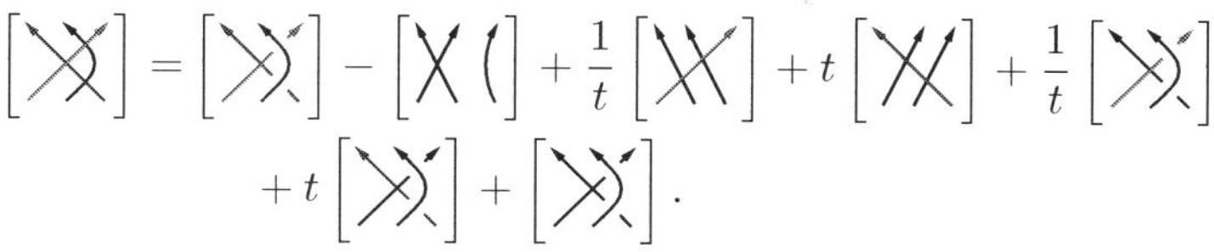

Combining the two results and using that $[\cdot]$ is invariant under the move $R4$, we obtain the following equality, and thus the graphical skein relation (10) holds:

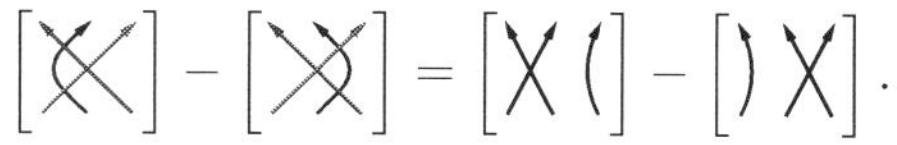

It remains to prove the graphical skein relation (11) and we start by evaluating each of the colored graph diagrams on the left side of the relation. Starting with the first diagram, we obtain the following:

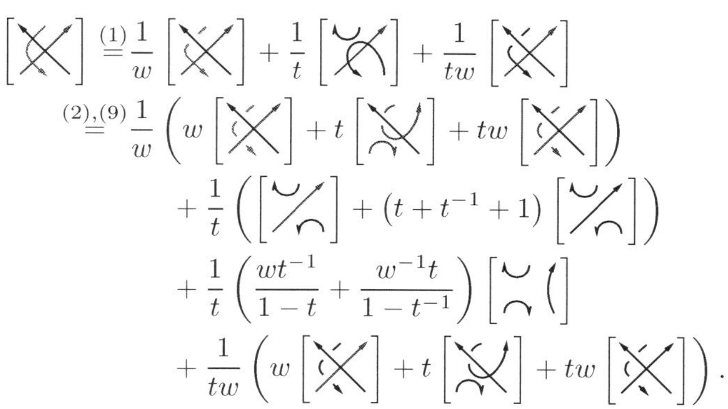

By simplifying the result above, we have:

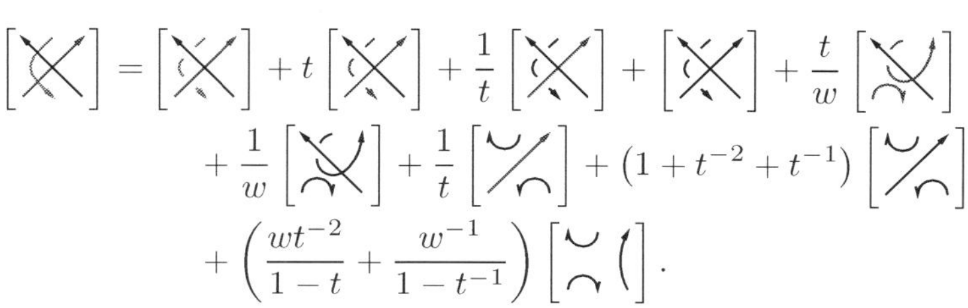

Continuing on with this same relation, we apply skein relation (2) to the fifth and sixth diagrams on the right-hand side of the above equality, followed by applications of the invariance of $[\cdot]$ under the Reidemeister moves $R1$ and $R2$, to obtain the following:

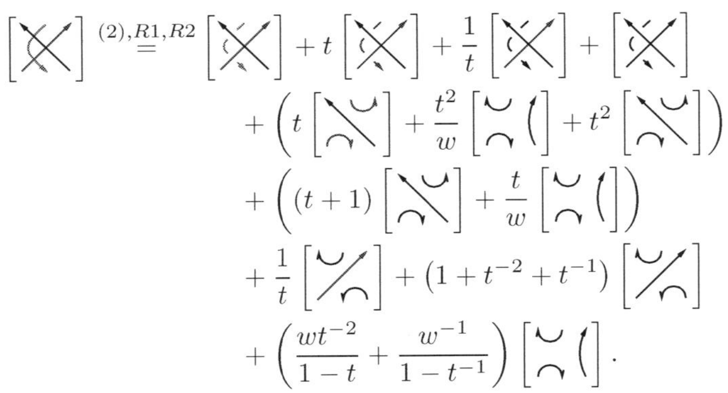

Finally, by simplifying the above equality, we have:

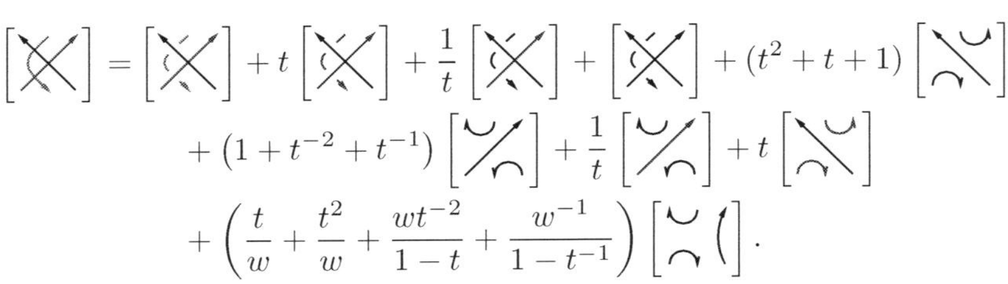

By applying similar steps to the second diagram of the left-hand side of the skein relation (11), we obtain the following equality:

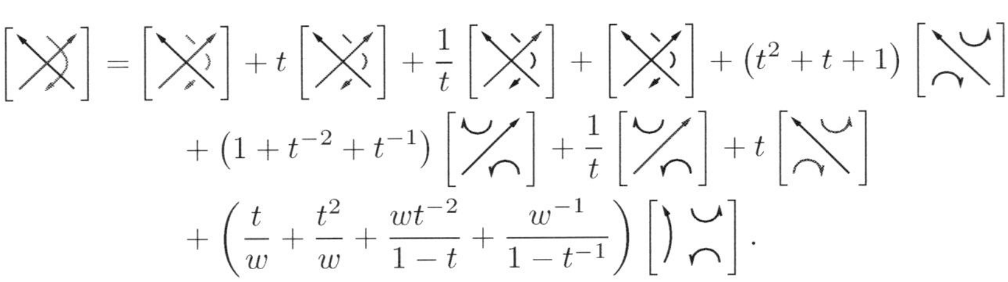

Then, by using that $[\cdot]$ is invariant under the move $R4$, we obtain the desired skein relation:

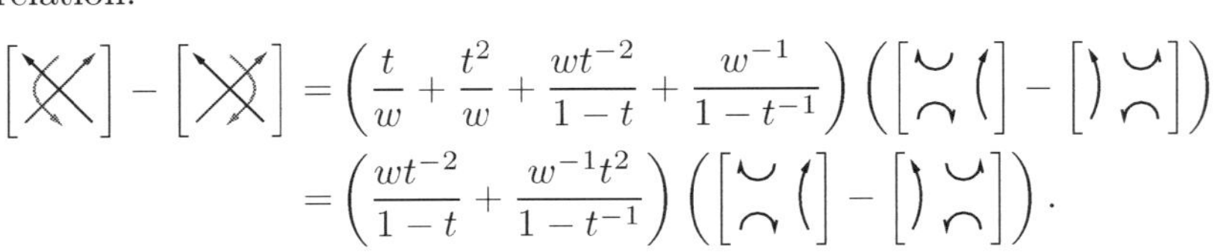

Hence the graphical skein relation (11) holds. □

REMARK 4. We note that if the two strands in the left-hand side diagrams of the skein relations (1) and (2) are colored with the same color, then the following holds:

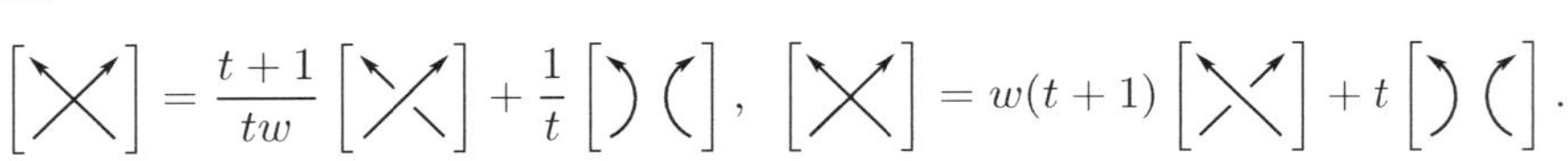

Solving for  and 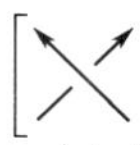 in the above relations and substituting them in (1) and (2), respectively, we obtain the following skein relations:

(12), (13)

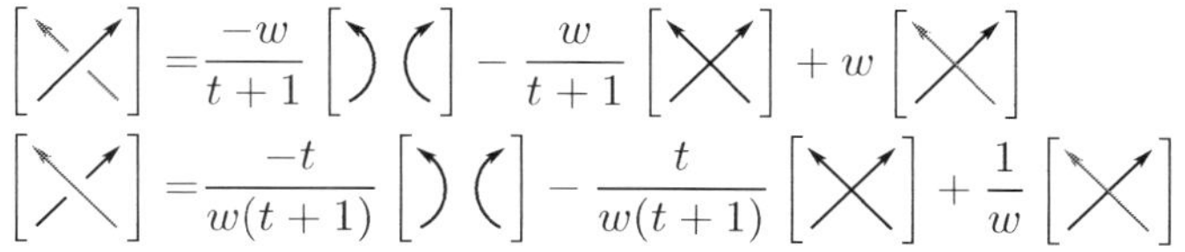

The skein relations (12) and (13) will play a major role in the following section.

3. A state-sum model for the Aicardi-Juyumaya invariant for colored links

The graphical skein relations proved in Theorem 3 provide a recursive way to evaluate oriented, colored, planar graphs with 4-valent vertices, where a 4-valent vertex is diagrammatically represented as a singular crossing. We denote the evaluation of such planar graph G by $[G]$. Specifically, the graphical skein relations of Theorem 3 allow the evaluation $[G]$ to be written as a $\mathbb{Q}(x,t,w)$-formal linear combination of evaluations of colored 4-valent planar graphs with fewer vertices. This also allows us to compute the invariant $[\tilde{D}]$ of a colored singular link without relying on the existence of the invariant F for colored classical links. In particular, the graphical skein relations of Theorem 3 provide a state-sum model for the Aicardi-Juyumaya invariant F for colored (classical) links, where the states associated to a diagram of a colored link are oriented, colored, 4-valent planar graphs, G.

Given any colored link diagram D, we can resolve each of its crossings using the skein relations (12) and (13). After having performed this step at each of the crossings of diagram D, we obtain $[D]$ written as a formal linear combination of evaluations of its states G. Each state G is then evaluated using the graphical relations provided by Theorem 3.

REMARK 5. We remark that this type of approach is not new when working with polynomial invariants in combinatorial knot theory. Kauffman and Vogel [**10**] constructed an invariant for rigid-vertex embeddings of unoriented 4-valent graphs in $\mathbb{R}^3$, which extends the Dubrovnik polynomial to embeddings of graphs. The Dubrovnik polynomial is a version of the 2-variable Kauffman polynomial [**8**] for unoriented links. In [**10**], a graphical calculus for evaluations of planar 4-valent graphs (not colored) was provided. Then Carpentier [**5**] explained how that graphical calculus can be used to provide a state-sum formula for the Dubrovnik polynomial for unoriented links and its extension to rigid-vertex embeddings of 4-valent graphs.

LEMMA 1. *If a connected colored 4-valent planar graph contains at least one vertex and does not contain a loop*  *or a bigon* 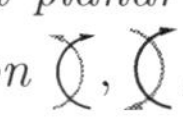*, then using a finite sequence of moves of types* 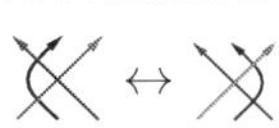 *and* 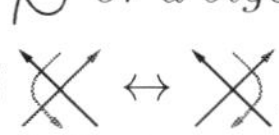*, it is possible to transform the graph into a colored 4-valent planar graph that contains a bigon.*

PROOF. This statement can be proved in a same way as [**5**, Lemma 2]. □

THEOREM 6. *There is a unique rational expression $[G] \in \mathbb{Q}(x, w, t)$ associated to a colored 4-valent planar graph G, such that $[\,\cdot\,]$ satisfies the graphical relations of Theorem 3, as well as the following equalities:*

$$\left[\bigcirc\right] = 1, [O_n^n] = \left(\frac{1}{wx}\right)^{n-1}, \textit{ and } \left[O_n^c \tilde{\cup} \bigcirc\right] = \frac{tw^2 - 1}{(1-t)w}\,[O_n^c], \tag{14}$$

where O_n^c is the standard diagram of the unlink with n components and c colors $(c \leq n)$, and where the circle in the disjoint union $O_n^c \tilde{\cup} \bigcirc$ has the same color as (at least) one of the unknotted circles in O_n^c.

PROOF. The proof is by induction on the number of vertices in a colored 4-valent planar graph and is similar to [**5**, Theorem 3].

Let $[\,\cdot\,]$ and $[\,\cdot\,]^*$ be two rational expressions that satisfy the graphical relations given in Theorem 3 and the additional relations (14). If G has no vertices, then since G is planar, it is a colored unlink containing n unknotted circles. Then, by relations (14), $[G] = [G]^*$.

Suppose that $[\,\cdot\,]$ and $[\,\cdot\,]^*$ yield equal values when evaluated on the same colored 4-valent planar graph with at most k vertices. Let G be a colored 4-valent planar graph with $k+1$ vertices. If G contains a loop or a local configuration of type or , then by applying the skein relation (7), (8) or (9), respectively, the evaluations $[G]$ and $[G]^*$ are written in terms of evaluations of colored 4-valent planar graphs with fewer vertices, and therefore, by the induction hypothesis, $[G] = [G]^*$.

If G does not contain a loop ( ) or a bigon (or), then by Lemma 1, there exists a finite sequence of colored 4-valent planar graphs

$$G = G_0 \to G_1 \to G_2 \to \cdots \to G_s,$$

where for each $0 \leq i \leq s-1$, G_{i+1} is obtained from G_i by applying the move $\leftrightarrow$ or $\leftrightarrow$, and where G_s contains a bigon of type or . Then, by applying the skein relations (10) and (11) followed by the inductive hypothesis, we have that

$$[G_i] - [G_{i+1}] = [G_i]^* - [G_{i+1}]^*, \text{ for all } 0 \leq i \leq s-1.$$

Hence, $[G] - [G_s] = [G]^* - [G_s]^*$. But G_s contains a bigon, and by the previous step in the proof, $[G_s] = [G_s]^*$. It follows that $[G] = [G]^*$.

By the principal of mathematical induction, $[\,\cdot\,]$ and $[\,\cdot\,]^*$ yield the same quantity when evaluated from a graph G with any number of 4-valent vertices, and thus $[\,\cdot\,] = [\,\cdot\,]^*$. □

THEOREM 7. *Let D_1 and D_2 be diagrams representing a colored link. Resolve all crossings in D_1 and D_2 using the skein relations* (12) *and* (13), *to write $[D_1]$ and $[D_2]$ as $\mathbb{Q}(x, w, t)$-linear combinations of evaluations of their states. Then evaluate the corresponding states using the graphical skein relations* (7)–(11), *together with identities* (14). *Then $[D_1] = [D_2]$; that is, the rational expression $[\,\cdot\,] \in \mathbb{Q}(x, w, t)$ is an invariant of colored links. Moreover, $[\,\cdot\,]$ is equal to Aicardi-Juyumaya invariant F.*

PROOF. The proof that $[D_1] = [D_2]$ when the two diagrams differ by a Reidemeister move $R1$, $R2$ or $R3$ is by direct computations, which in some sense are the reverse of the proofs of our graphical skein relations in Theorem 3. □

References

[1] F. Aicardi, *New invariants of links from a skein invariant of colored links*, arXiv:1512.00686, 2015.

[2] Francesca Aicardi, *An invariant of colored links via skein relation*, Arnold Math. J. **2** (2016), no. 2, 159–169, DOI 10.1007/s40598-015-0035-1. MR3504347

[3] Francesca Aicardi and Jesús Juyumaya, *Tied links*, J. Knot Theory Ramifications **25** (2016), no. 9, 1641001, 28, DOI 10.1142/S0218216516410017. MR3544429

[4] Carmen Caprau, *Movie moves for singular link cobordisms in 4-dimensional space*, J. Knot Theory Ramifications **25** (2016), no. 2, 1650013, 23, DOI 10.1142/S0218216516500139. MR3463858

[5] Rui Pedro Carpentier, *From planar graphs to embedded graphs—a new approach to Kauffman and Vogel's polynomial*, J. Knot Theory Ramifications **9** (2000), no. 8, 975–986, DOI 10.1142/S0218216500000578. MR1807602

[6] P. Freyd, D. Yetter, J. Hoste, W. B. R. Lickorish, K. Millett, and A. Ocneanu, *A new polynomial invariant of knots and links*, Bull. Amer. Math. Soc. (N.S.) **12** (1985), no. 2, 239–246, DOI 10.1090/S0273-0979-1985-15361-3. MR776477

[7] Louis H. Kauffman, *Invariants of graphs in three-space*, Trans. Amer. Math. Soc. **311** (1989), no. 2, 697–710, DOI 10.2307/2001147. MR946218

[8] Louis H. Kauffman, *An invariant of regular isotopy*, Trans. Amer. Math. Soc. **318** (1990), no. 2, 417–471, DOI 10.2307/2001315. MR958895

[9] Louis H. Kauffman, *Knots and physics*, 3rd ed., Series on Knots and Everything, vol. 1, World Scientific Publishing Co., Inc., River Edge, NJ, 2001, DOI 10.1142/9789812384836. MR1858113

[10] Louis H. Kauffman and Pierre Vogel, *Link polynomials and a graphical calculus*, J. Knot Theory Ramifications **1** (1992), no. 1, 59–104, DOI 10.1142/S0218216592000069. MR1155094

[11] Józef H. Przytycki and Paweł Traczyk, *Invariants of links of Conway type*, Kobe J. Math. **4** (1988), no. 2, 115–139. MR945888

Department of Mathematics, California State University, Fresno 5245 North Backer Avenue, M/S PB 108, California 93740

Email address: abaumheckel@mail.fresnostate.edu

Department of Mathematics, California State University, Fresno 5245 North Backer Avenue, M/S PB 108, California 93740

Email address: ccaprau@csufresno.edu

Department of Mathematics, California State University, Fresno 5245 North Backer Avenue, M/S PB 108, California 93740

Email address: conorrighetti@mail.fresnostate.edu

Contemporary Mathematics
Volume **827**, 2025
https://doi.org/10.1090/conm/827/16539

A study of Gram determinants in knot theory

Dionne Ibarra and Gabriel Montoya-Vega

ABSTRACT. Historically originated as a sub-field of topology, knot theory is an active area of mathematical investigation that has strong connections with a diverse set of scientific fields such as algebra, biology, and statistical mechanics. A popular and important concept in linear algebra, Gram determinants enjoy a connection with the mathematical theory of knots. In this article, we expose this concept and present several types of Gram determinants in what can be considered as a survey of the current Gram determinants of interest to knot theorists; examples are included to illustrate the definitions. In particular, we pay special attention to a recently defined determinant from a Möbius band and we further study its structure. At the end, some speculation is presented regarding the closed formula for the Gram determinant of type $(Mb)_1$, a problem that arouses serious interest among knot theorists.

CONTENTS

1. Introduction to Gram determinants in knot theory
2. Determinant of type A
3. Generalization of type A
4. Determinant of type B
5. Determinant of type Mb
6. Determinant of type $(Mb)_1$

1. Introduction to Gram determinants in knot theory

The mathematical theory of knots has its historical origins as a sub-area of topology, arguably because of Leibniz's desire that a different type of analysis was needed; he called it *analysis situs* (geometry of position). Although the theory quickly developed into its own research field, it enjoys broad connections to other areas like graph theory, algebra, statistical mechanics, and mathematical biology. In this section we explore how Gram determinants are related to knot theory and we present different types of this determinant. It is important to mention that the initial interest, and most of the current research in Gram determinants related to knot theory, is owed to Edward Witten's belief on the existence of a 3-manifold invariant related to the Jones polynomial. In particular, it was Nicolai Reshetikhin's

2020 *Mathematics Subject Classification.* Primary 57K10; Secondary 57M27.
Key words and phrases. Gram determinants, knot theory, relative Kauffman bracket skein module.

and Vladimir Turaev's construction of such invariant that triggered a number of subsequent works by knot theorists on the relation of these determinants with the mathematical theory of knots.

The notion of a relative skein module is needed in the study of Gram determinants in knot theory. The most popular of such structures is the Kauffman bracket skein module; its structure has shown connections between the module and the geometry and topology of the 3-manifold (see for example [**13**]). The relative Kauffman bracket skein module is of our interest and it is introduced in Definition 1.1.

DEFINITION 1.1. Let M be an oriented 3-manifold and $\{x_i\}_1^{2n}$ be the set of $2n$ framed points on ∂M. Let $I = [-1,1]$, and let $\mathcal{L}^{fr}(2n)$ be the set of all relative framed links (which consists of all framed links in M and all framed arcs, $I \times I$, where $I \times \partial I$ is connected to framed points on the boundary of M) up to ambient isotopy while keeping the boundary fixed in such a way that $L \cap \partial M = \{x_i\}_1^{2n}$. Let R be a commutative ring with unity, $A \in R$ be invertible, and let $S_{2,\infty}^{sub}(2n)$ be the submodule of $R\mathcal{L}^{fr}(2n)$ that is generated by the Kauffman bracket skein relations:

(i) $L_+ - AL_0 - A^{-1}L_\infty$, and
(ii) $L \sqcup \bigcirc + (A^2 + A^{-2})L$,

where $\bigcirc$ denotes the framed unknot and the *skein triple* (L_+, L_0, L_∞) denotes three framed links in M that are identical except in a small 3-ball in M where the difference is shown in Figure 1. Then, the **relative Kauffman bracket skein module** (RKBSM) of M is the quotient:

$$\mathcal{S}_{2,\infty}(M, \{x_i\}_1^{2n}; R, A) = R\mathcal{L}^{fr}(2n)/S_{2,\infty}^{sub}(2n).$$

(A) L_+. (B) L_0. (C) L_∞.

FIGURE 1. The skein triple.

When the manifold is a disc, the free R-module can be equipped with an algebra structure that leads to the classical Temperley-Lieb algebra TL_n (see [**14**], for example). Definition 2.1 introduces a bilinear form on TL_n, and the notion of a Gram matrix.

Physicists N. Temperley and E. Lieb in [**15**] used the Temperley-Lieb algebra to study 2-dimensional Potts models. R. J. Baxter in [**3**] presented the first formal definition of this algebra. Jones independently introduced TL_n in [**9**] while working on von Neumann algebras.

DEFINITION 1.2. Let R be a commutative ring with unity and $d \in R$. Let $n \in \mathbb{N}$ be fixed, then the nth **Temperley-Lieb algebra**, TL_n, is defined to be the unital associative algebra over R with generators $e_1, \ldots, e_{n-1}$, identity element 1_n, and relations:

(1) $e_i e_j e_i = e_i$ for $|i-j| = 1$
(2) $e_i e_j = e_j e_i$ for $|i-j| > 1$
(3) $e_i^2 = de_i$

L. H. Kauffman in [**10**], motivated by utilizing the Kauffman bracket, considered the Temperley-Lieb algebra over $R = \mathbb{Z}[A^{\pm 1}]$, where A is an indeterminate and $d = -A^2 - A^{-2}$. He then constructed a graphical interpretation using tangles.

An n-**tangle** is a rectangular shaped disk with n marked boundary points on the left and n marked boundary points on the right. Kauffman's graphical interpretation of the Temperley-Lieb algebra is obtained from the basis of crossingless tangles. The identity element corresponds to an n-tangle with n parallel arcs in which each ith point on the left is connected to the ith point on the right, and each e_i corresponds to an n-tangle that has two arcs, each connected to the ith and $(i+1)^{th}$ point on the left and right, respectively. An illustration is given in Fig. 2. For simplicity we will label an arc by n to denote n parallel arcs as shown in Fig. 2a.

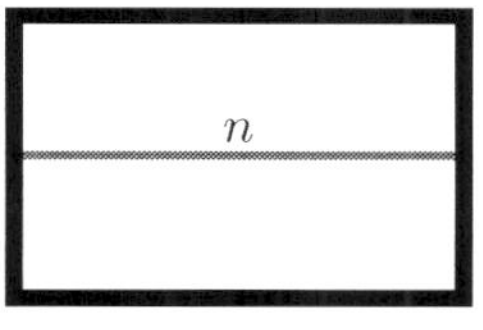

(A) Identity element.

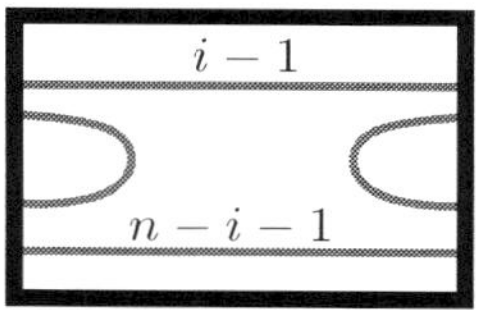

(B) e_i.

FIGURE 2. A graphical interpretation of the generators of TL_n.

DEFINITION 1.3. The n-**tangle algebra** is an R-module with basis elements consisting of n-tangles. The multiplication of two n-tangles is defined by identifying the right side of the first n-tangle to the left side of the second n-tangle, while respecting the boundary points and by letting any resulting trivial curve be denoted by d, see Fig. 3 for an illustrative example. Kauffman's diagrammatic interpretation of the Temperley-Lieb algebra, also known as the **diagrammatic algebra**, is a subalgebra of the n-tangle algebra. It is generated by tangles with no crossings where homotopically trivial curves are denoted by $d \in R$.

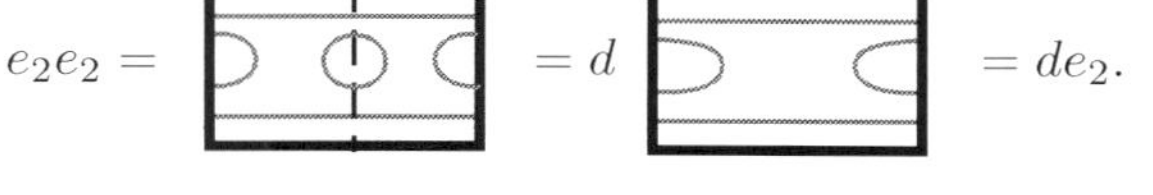

FIGURE 3. An illustration of multiplication.

THEOREM 1.4 ([**10**]). *The diagrammatic algebra is isomorphic to TL_n and can be thought of as a diagrammatic interpretation of it.*

2. Determinant of type A

DEFINITION 2.1. Let $R = \mathbb{Z}[A^{\pm 1}]$, $d = -A^2 - A^{-2}$, and consider the disk D^2 with $2n$ framed points on its boundary. Let $\mathcal{C}_n^A = \{c_1, c_2, \ldots, c_{C_n}\}$ be the set of all diagrams with crossingless connections, up to ambient isotopy, between the $2n$ framed points in D^2. Define a bilinear form $\langle\ ,\ \rangle_A$ in the following way:

$$\langle\ ,\ \rangle_A : \mathcal{S}_{2,\infty}(D^2 \times I, \{x_i\}_1^{2n}) \times \mathcal{S}_{2,\infty}(D^2 \times I, \{x_i\}_1^{2n}) \longrightarrow R.$$

Let $c_i, c_j \in \mathcal{C}_n^A$. Glue c_i with the inversion of c_j along the marked circle, respecting the labels of the framed points. The resulting picture is that of a disk with disjoint null homotopic circles. Thus, we define, $\langle c_i, c_j\rangle_A = d^m$ where m denotes the number of these circles. Figure 4 illustrates an example of the bilinear form when $n = 4$.

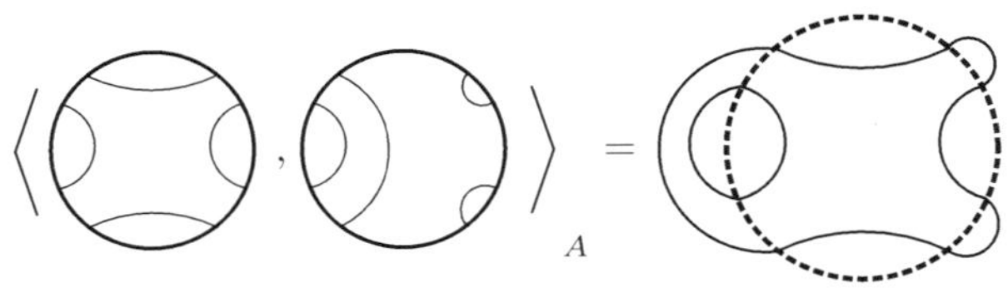

FIGURE 4. The bilinear form on two elements in TL_4. Here, the result is d^2.

The **Gram matrix of type A** is defined as $G_n^A = (\langle a_i, a_j\rangle_A)_{1\leq i,j\leq C_n}$. Its determinant D_n^A is called the **Gram determinant of type A**.

EXAMPLE 2.2. Table 1 shows the Gram matrix of type A for $n = 3$, G_3^A. It can be seen that $D_3^A = (d^2-1)^4 \cdot d^4 \cdot (d^2 - 2d)$.

TABLE 1. The Gram matrix G_3^A.

$\langle\ ,\ \rangle_A$	b_1 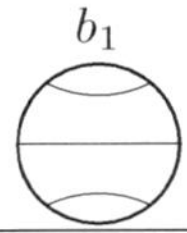	b_2 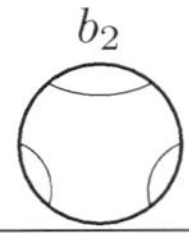	b_3 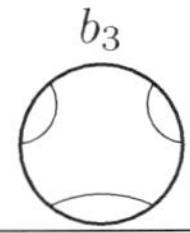	b_4 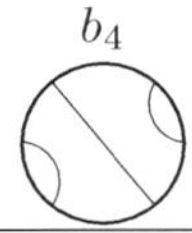	b_5 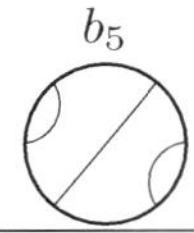
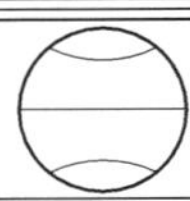	d^3	d^2	d^2	d	d
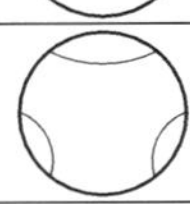	d^2	d^3	d	d^2	d^2
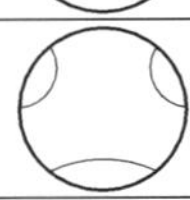	d^2	d	d^3	d^2	d^2
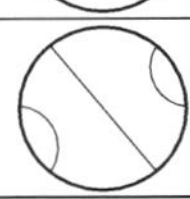	d	d^2	d^2	d^3	d
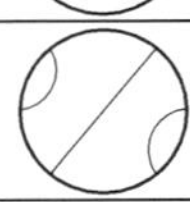	d	d^2	d^2	d	d^3

We now recall the definition of the Chebyshev polynomials of the second kind. Let d be the free variable and denote the polynomial by Δ_n. This polynomial is given by:

$$\begin{aligned}\Delta_0(d) &= 1,\\ \Delta_1(d) &= d,\\ \Delta_n(d) &= d \cdot \Delta_{n-1}(d) - \Delta_{n-2}(d).\end{aligned}$$

Therefore, from Example 2.2 we get $D_3^A = \Delta_1^4 \cdot \Delta_2^4 \cdot \Delta_3$.

A closed formula for the calculation of this determinant was found by Bruce Westbury [**16**] and Philippe Di Francesco [**7**]. One of the tools used by Di Francesco was the Gram-Schmidt orthogonalization together with some combinatorial methods. In [**4**], Xuanting Cai used Jones-Wenzl idempotents to significantly shorten the proof in [**7**]. For a more detailed discussion about this formula and its proof see Lecture 17 in [**1**].

THEOREM 2.3 ([**4, 7, 16**]). *Let $R = \mathbb{Z}[A^{\pm 1}]$. Then,*

$$D_n^A(d) = \prod_{i=1}^{n} \left(\frac{\Delta_i}{\Delta_{i-1}}\right)^{\alpha_i},$$

where $\Delta_i = (-1)^i \dfrac{A^{2i+2} - A^{-2i-2}}{A^2 - A^{-2}}$ *and* $\alpha_i = \dbinom{2n}{n-i} - \dbinom{2n}{n-i-1}$.

3. Generalization of type A

Here we consider a different bilinear form that results in a new type of Gram determinant [**1**]. Similarly as with the determinant of type A, the basis elements are the crossingless connections on $2n$ boundary points on a disk. The construction is presented below.

DEFINITION 3.1. Let the disk D^2, with $2n$ marked points on its boundary, be considered as a rectangle with n points on the top edge and n points on the bottom edge. Define a bilinear form $\langle\ ,\ \rangle_{A^{gen}}$ in the following way:

$$\langle\ ,\ \rangle_{A^{gen}} : \mathcal{S}_{2,\infty}(D^2 \times I, \{x_i\}_1^{2n}) \times \mathcal{S}_{2,\infty}(D^2 \times I, \{x_i\}_1^{2n}) \longrightarrow \mathbb{Z}[d, z].$$

We view the elements of TL_n vertically and multiplying from top to bottom. For $a_i, a_j \in \mathcal{A}_n$, glue a_i with the reflection about the horizontal axis of a_j, which is denoted by $\overline{a_j}$, such that the bottom edge of a_i is identified with the top edge of $\overline{a_j}$. Connect the marked points on the top edge of a_i with those on the bottom edge of $\overline{a_j}$, in the annulus, respecting the ordering of the marked points (see Figure 5). Recall that the described operation is the same as taking the trace of the image of the bilinear form.

The result is an annulus with two types of disjoint circles, homotopically trivial and non - trivial. Thus, we define $\langle a_i, a_j\rangle_{A^{gen}} = d^k z^m$ where k and m denote the number of these circles respectively. We define the **Gram matrix of generalized type *A*** as $G_n^{A^{gen}} = (\langle a_i, a_j\rangle_{A^{gen}})_{1\le i,j\le C_n}$, and denote its determinant by $D_n^{A^{gen}}$.

EXAMPLE 3.2. Consider the basis of Catalan connections on six boundary points. Figure 6 illustrates the calculation of $\langle a_4, a_5\rangle_{A^{gen}}$. The matrix $G_3^{A^{gen}}$ is computed in Table 2, along with its determinant.

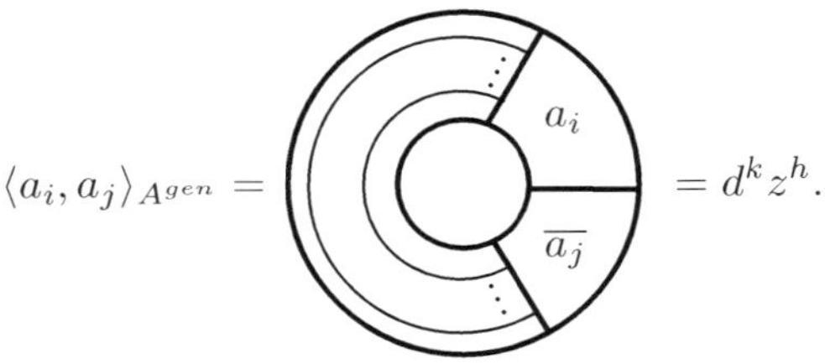

FIGURE 5. The bilinear form of generalized type A.

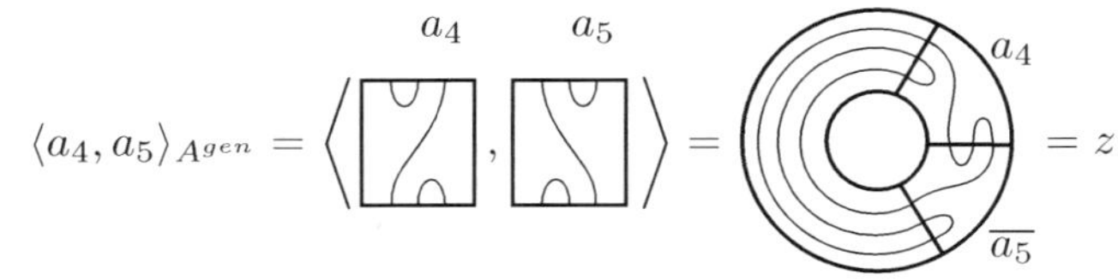

FIGURE 6. The bilinear form of generalized type A.

TABLE 2. The Gram matrix $G_3^{A^{gen}}$.

$\langle\ ,\ \rangle_{A^{gen}}$	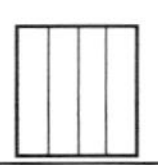	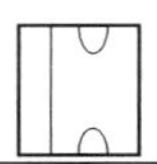	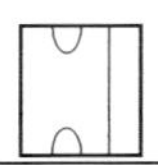		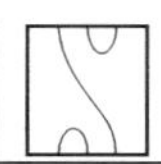
	z^3	dz	dz	z	z
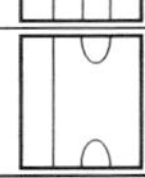	dz	d^2z	z	dz	dz
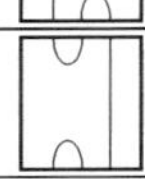	dz	z	d^2z	dz	dz
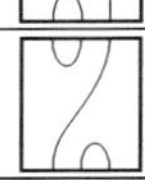	z	dz	dz	d^2z	z
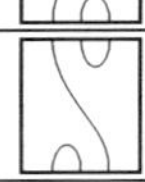	z	dz	dz	z	d^2z

We calculate and factor the determinant:

$$D_3^{A^{gen}} = (d^2-1)^4 z^5 (z^2-2).$$

4. Determinant of type B

It is possible to construct different bilinear forms by changing the ambient surface of the Kauffman bracket skein module. For example, we can explore the notion of a Gram determinant using the annulus and the Möbius band. Paul Martin and Hubert Saleur were the first to consider the type B Gram determinant. Their work focuses on applications to statistical mechanics, and they employ representation theory to prove their determinant in [**11, 12**]. For a historical overview about the development of this type of Gram determinants, the reader is referred to [**8, 14**].

Consider an annulus with $2n$ marked points along the outer boundary; see Figure 7.

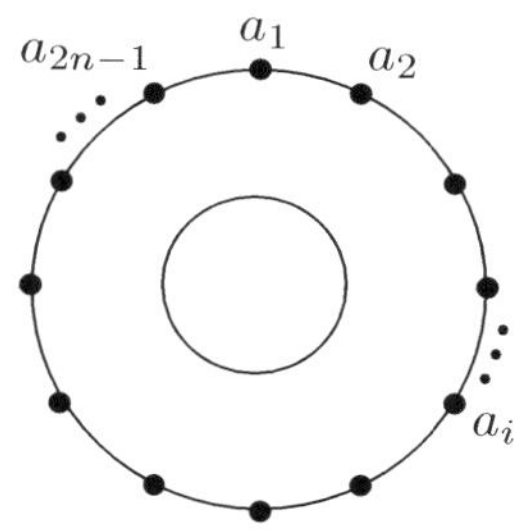

FIGURE 7. An annulus with 2n boundary points.

Similarly as the type A determinant, we can connect these $2n$ points to form crossingless connections in the annulus. Denote by B_n the set of these diagrams. It can be shown that there are $\binom{2n}{n}$ such diagrams. See Figure 8, which illustrates B_2.

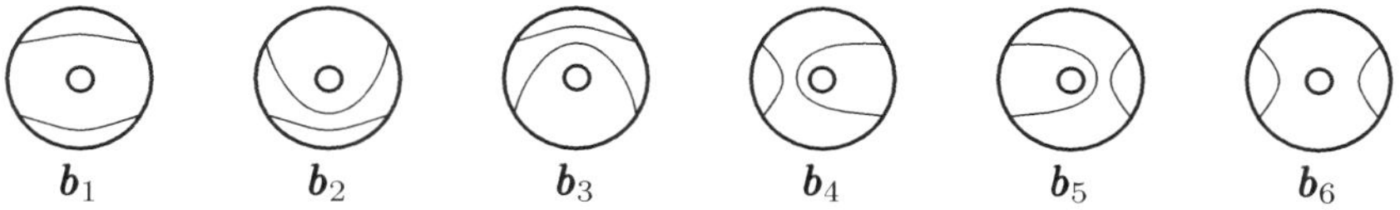

FIGURE 8. The set B_2.

The bilinear form for Type B is defined on these annular diagrams as follows:

DEFINITION 4.1. Let $R = \mathbb{Z}[A^{\pm 1}]$, $d = -A^2 - A^{-2}$, and let A^2 be an annulus with $2n$ marked points on its outer boundary. Let $B_n = \{b_1, b_2, \ldots, b_{\binom{2n}{n}}\}$ be the set of all diagrams of crossingless connection between these $2n$ points. Define the type B bilinear form $\langle\ ,\ \rangle$ in the following way:

$$\langle\ ,\ \rangle : \mathcal{S}_{2,\infty}(A^2 \times I, \{x_i\}_1^{2n}; R, A) \times \mathcal{S}_{2,\infty}(A^2 \times I, \{x_i\}_1^{2n}; R, A) \longrightarrow R[z].$$

Given $b_i, b_j \in B_n$, glue b_i with the inversion of b_j along the marked circle, respecting the labels of the marked points. The resulting picture has disjoint circles, which are either homotopically non-trivial or null homotopic. Then, $\langle b_i, b_j \rangle = z^k d^m$, where k and m denote the number of these circles, respectively.

The Gram matrix of type B is defined as $G_n^B = (\langle b_i, b_j \rangle)_{1 \le i,j \le \binom{2n}{n}}$. Its determinant D_n^B is called the Gram determinant of type B.

EXAMPLE 4.2. The bilinear form $\langle b_5, b_6 \rangle$ is illustrated in Figure 9.

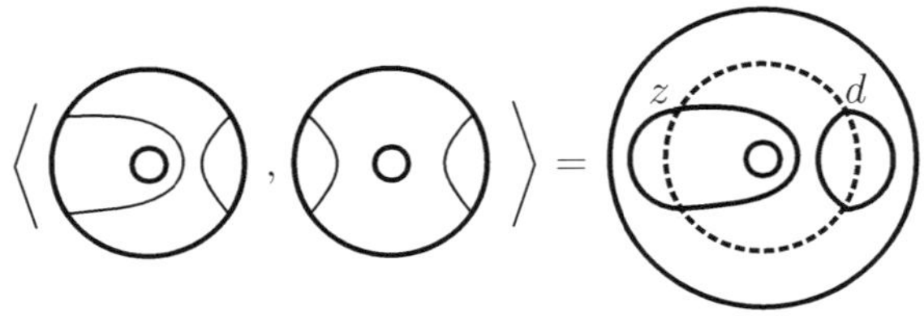

FIGURE 9. The inner product of two elements in B_4. In this case, the result is dz.

TABLE 3. The Gram matrix G_2^B.

$\langle\ ,\ \rangle$	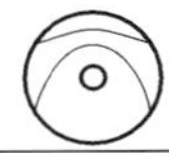	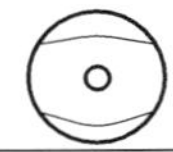		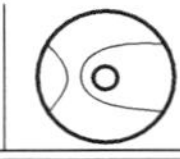	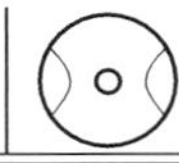	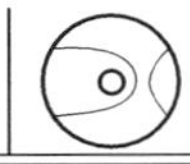
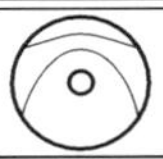	d^2	dz	z^2	z	d	z
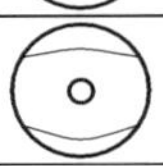	dz	d^2	dz	d	z	d
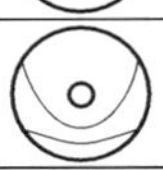	z^2	dz	d^2	z	d	z
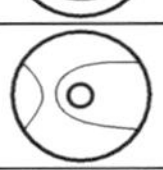	z	d	z	d^2	dz	z^2
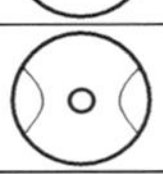	d	z	d	dz	d^2	dz
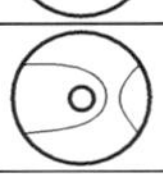	z	d	z	z^2	dz	d^2

EXAMPLE 4.3. Table 3 illustrates the Gram matrix G_2^B. We calculate the Gram determinant for the collection B_2.

We calculate the determinant of this matrix to be:

$$\begin{aligned} D_2^B(d,z) &= -(d-z)^4(d+z)^4(-2+d^2+z)(-2+d^2+z) \\ &= (d^2-z^2)^4((d^2-2)^2-z^2). \end{aligned}$$

Recall that T_n denotes the Chebyshev polynomial of the first kind defined recursively by the equation: $T_{n+1}(d) = dT_n(d) - T_{n-1}(d)$, with the initial conditions $T_0(d) = 2$, and $T_1(d) = d$. The determinant given above can be expressed in terms of Chebyshev polynomials. In general, it is possible to rewrite the Gram determinant of type B in terms of Chebyshev polynomials of the first kind for all n.

The following closed formula was proven by Q. Chen and J. Przytycki and the proof involves the creation of a linear map on the basis $B_{n,0}$ that uses the lollipop method to decorate the inner boundary component with the Jones-Wenzl idempotent. Then, the proof proceeds by showing that the image of the linear map under the basis is a subspace of dimension $\binom{2n}{n} - \binom{2n}{n-k}$ [**6**].

THEOREM 4.4 ([**6**, **12**]). *Let* $R = \mathbb{Z}[A^{\pm 1}]$, *then* $D_n^B = \prod_{i=1}^{n} (T_i(d)^2 - z^2)^{\binom{2n}{n-i}}$.

EXAMPLE 4.5. An example of the bilinear form on two elements in Mb_4 is given below, where ⊛ denotes a crosscap usually denoted by  .

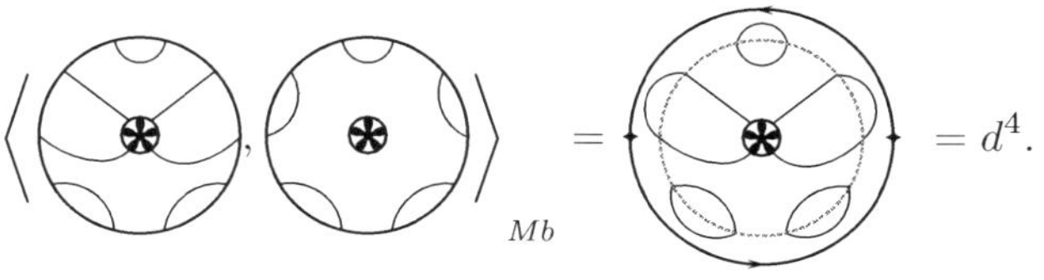

5. Determinant of type Mb

This Gram determinant originates from the study of crossingless connections on a Möbius band. The basis of the relative Kauffman bracket skein module of the twisted I-bundle of the Möbius band is infinite. The Gram matrix of type Mb is created from a finite sub-collection of this basis where we restrict to basis elements with no z and x curves. The bilinear form is defined through the identification of two Möbius bands along their boundaries. This determinant is given in Definition 5.1.

DEFINITION 5.1. Let $Mb_n = \{m_1, \ldots, m_{\sum_{k=0}^{n}\binom{2n}{k}}\}$ be the set of all diagrams of crossingless connections between $2n$ framed marked points on the boundary of the Möbius band $Mb \hat{\times} \{0\}$ in $Mb \hat{\times} I$ with no simple closed curves. Define a bilinear form $\langle\ ,\ \rangle_{Mb}$ on the elements of Mb_n as follows:

$$\langle\ ,\ \rangle_{Mb} : \mathcal{S}_{2,\infty}(Mb \hat{\times} I, \{x_i\}_1^{2n}) \times \mathcal{S}_{2,\infty}(Mb \hat{\times} I, \{x_i\}_1^{2n}) \longrightarrow \mathbb{Z}[d, w, x, y, z].$$

Given $m_i, m_j \in \text{Mb}_n$, identify the boundary component of m_i with that of the inversion of m_j, respecting the labels of the marked points. The result is an element in $Kb \hat{\times} I$ containing only disjoint simple closed curves. The five homotopically distinct simple closed curves in the Klein bottle, including the homotopically trivial curve, are denoted by x, y, z, w, d as illustrated in Figure 10. Then, $\langle m_i, m_j\rangle_{Mb} := d^m x^n y^k z^l w^h$ where m, n, k, l and h denote the number of these curves, respectively.

The Gram matrix of type Mb is defined as $G_n^{Mb} = (\langle m_i, m_j\rangle_{Mb})_{1 \le i,j \le \sum_{k=0}^{n}\binom{2n}{k}}$ and its determinant D_n^{Mb} is called the Gram determinant of type Mb.

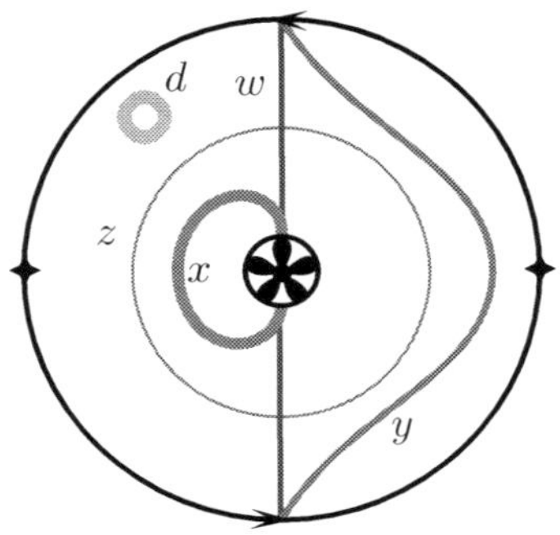

FIGURE 10. Klein bottle and its five homotopically distinct simple closed curve; the curve w intersects both crosscaps, y intersects the outer crosscap, x intersects the inner crosscap, z is a nontrivial curve that does not intersect the crosscaps, and d is the trivial curve.

Qi Chen conjectured the following result for the Gram determinant of type Mb and verified the conjecture for $n \leq 4$. Some work supporting this conjecture can be found in [**1**, **2**, **14**].

CONJECTURE 5.2 (Chen [**5**]). *Let* $R = \mathbb{Z}[A^{\pm 1}, w, x, y, z]$, $d = -A^2 - A^{-2}$, *and* $D_{n,i} = \prod_{k=1+i}^{n} (T_{2k}(d) - 2)^{\binom{2n}{n-k}}$, *where* i *represents the number of curves passing through the crosscap. Then the Gram determinant of type* Mb *for* $n \geq 1$, *denoted by* D_n^{Mb}, *is:*

$$\begin{aligned} D_n^{Mb}(d, w, x, y, z) = &\prod_{k=1}^{n} (T_k(d) + (-1)^k z)^{\binom{2n}{n-k}} \\ &\prod_{\substack{k=1 \\ k \text{ odd}}}^{n} \left((T_k(d) - (-1)^k z)T_k(w) - 2xy\right)^{\binom{2n}{n-k}} \\ &\prod_{\substack{k=1 \\ k \text{ even}}}^{n} \left((T_k(d) - (-1)^k z)T_k(w) - 2(2-z)\right)^{\binom{2n}{n-k}} \\ &\prod_{i=1}^{n-1} D_{n,i}. \end{aligned}$$

PROPOSITION 5.3 ([**2**]). D_n^{Mb} *is divisible by* $(d - z)^{\binom{2n}{n-1}}$.

PROPOSITION 5.4 ([**1**]). D_n^{Mb} *is divisible by* $(w(d + z) - 2xy)^{\binom{2n}{n-1}}$.

PROPOSITION 5.5 ([**14**]). D_n^{Mb} *is divisible by* $((d^2-2-z)(w^2-2)-2(2-z))^{\binom{2n}{n-2}}$.

6. Determinant of type $(Mb)_1$

The Gram matrix of type *Mb* was created by taking a finite sub-collection of the basis of the relative Kauffman bracket skein module of the twisted I-bundle of the Möbius band. However, the number of such elements increase exponentially as n increases. In particular, computing the determinant for $n \geq 5$ has not been achieved due to the size of the matrices. In this section we present a new Gram determinant created in [**8**] by using a sub-collection of Mb_n.

DEFINITION 6.1. Let $(Mb_n)_1 = Mb_{n,0} \cup Mb_{n,1}$ where $Mb_{n,0} = \{m_1, \ldots, m_{\binom{2n}{n}}\}$ is the set of all diagrams of crossingless connections between $2n$ marked points on the boundary of $Mb \mathbin{\hat{\times}} \{0\}$ whose arcs do not intersect the crosscap and $Mb_{n,1} = \{m_1, \ldots, m_{\binom{2n}{n-1}}\}$ is the set of all diagrams of crossingless connections between $2n$ marked points on the boundary of $Mb \mathbin{\hat{\times}} \{0\}$ with exactly one curve intersecting the crosscap. Define a bilinear form $\langle\ ,\ \rangle_{Mb}$ on the elements of $(Mb_n)_1$ by using the same bilinear form as type *Mb*, as follows:

$$\langle\ ,\ \rangle_{Mb} : \mathcal{S}_{2,\infty}(Mb \mathbin{\hat{\times}} I, \{x_i\}_1^{2n}) \times \mathcal{S}_{2,\infty}(Mb \mathbin{\hat{\times}} I, \{x_i\}_1^{2n}) \longrightarrow \mathbb{Z}[d, w, x, y, z].$$

Given $m_i, m_j \in (Mb_n)_1$, identify the boundary component of m_i with that of the inversion of m_j, respecting the labels of the marked points. The result is an element in $Kb \mathbin{\hat{\times}} I$ containing only disjoint simple closed curves. Then $\langle m_i, m_j \rangle_{Mb} := d^m x^n y^k z^l w^h$ where m, n, k, l and h denote the number of these curves, respectively.

The Gram matrix of type $(Mb)_1$ is defined as $G_n^{(Mb)_1} = (\langle m_i, m_j \rangle_{Mb})_{1 \leq i,j \leq |(Mb)_1|}$ and its determinant $D_n^{(Mb)_1}$ is called the Gram determinant of type $(Mb)_1$.

TABLE 4. The Gram matrix $G_2^{(Mb)_1}$.

$\langle\ ,\ \rangle$	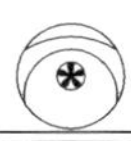	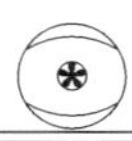	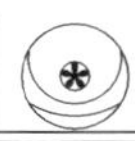	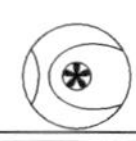	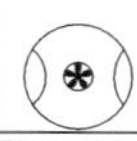	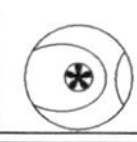	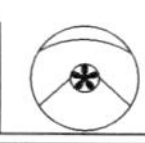	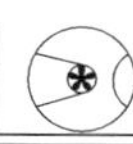	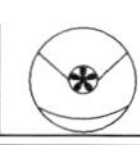	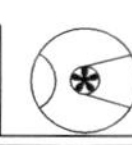
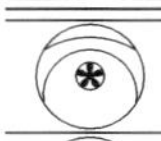	d^2	dz	z^2	z	d	z	dy	y	yz	y
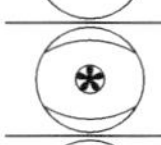	dz	d^2	dz	d	z	d	dy	y	dy	y
	z^2	dz	d^2	z	d	z	yz	y	dy	y
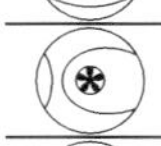	z	d	z	d^2	dz	z^2	y	yz	y	dy
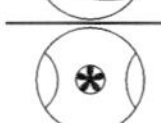	d	z	d	dz	d^2	dz	y	dy	y	dy
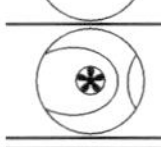	z	d	z	z^2	dz	d^2	y	dy	y	yz
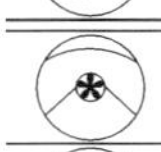	dx	dx	xz	x	x	x	dw	w	xy	w
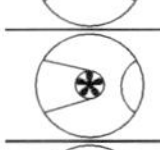	x	x	x	xz	dx	dx	w	dw	w	xy
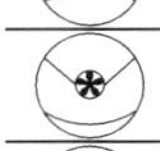	xz	dx	dx	x	x	x	xy	w	dw	w
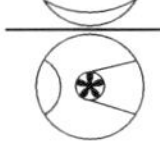	x	x	x	dx	dx	xz	w	xy	w	dw

EXAMPLE 6.2. The first example of $G_n^{(Mb)_1}$ that differs from G_n^{Mb} is when $n = 2$. This matrix is given in Table 4. The set $(Mb_2)_1$ is illustrated in Figure 11.

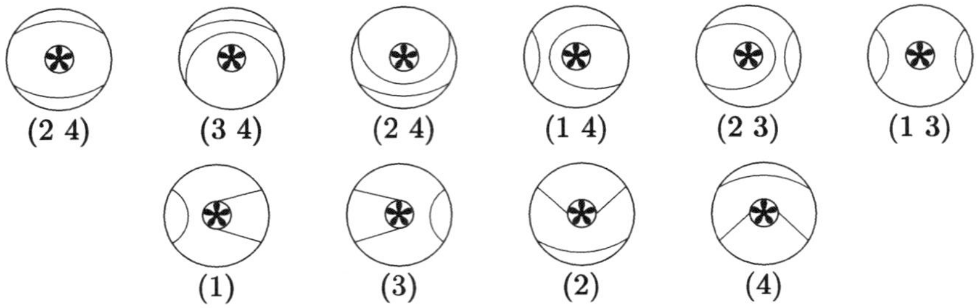

FIGURE 11. The set $(Mb_2)_1$.

$$\begin{aligned} D_2^{(Mb)_1} &= (d^2-4)d^2(d-z)^4(-2+d^2-z)(-2+d^2+z)(-dw+2xy-wz)^4 \\ &= (d-z)^4((d^2-2)+z)((d+z)w-2xy)^4((d^2-2)-z)(d^2(d^2-4)) \\ &= (T_1(d)-z)^4(T_2(d)^2-z^2)((d+z)w-2xy)^4(T_4(d)-2). \end{aligned}$$

The following theorem was proven in [**8**] and it is the main result regarding the structure of the closed formula for the Gram determinant of type $(Mb)_1$.

THEOREM 6.3 ([**8**]).

$$\prod_{k=1}^{n}(T_k(d)+(-1)^k z)^{\binom{2n}{n-k}} \text{ divides } D_n^{(Mb)_1}(d,z,x,y,w).$$

We will extend Theorem 6.3 by presenting a relationship between the determinant of Type $(Mb)_1$ and the determinant of a block matrix constructed from the Gram matrix of Type B and a sub-collection of elements from Type $(Mb)_1$.

DEFINITION 6.4. Consider the set $Mb_{n,1}$ containing the collection of all diagrams of crossingless connections between $2n$ marked points on the boundary of $Mb\hat{\times}\{0\}$ with exactly one curve intersecting the crosscap. Let $G_n^{Mb_{n,1}}$ be the Gram matrix defined on the set $Mb_{n,1}$ using the bilinear form $\langle\ ,\ \rangle_{Mb}$,

$$G_n^{Mb_{n,1}} = (\langle m_i, m_j\rangle_{Mb})_{1\leq i,j\leq \binom{2n}{n-1}},$$

where $m_i, m_j \in Mb_{n,1}$. Denote by $\tilde{G}_n^{Mb_{n,1}}$ the Gram matrix obtained from substituting $y = 0$ and $w = 1$ into $G_n^{Mb_{n,1}}$,

$$\tilde{G}_n^{Mb_{n,1}} = G_n^{Mb_{n,1}}(d,z,x,y=0,w=1).$$

Furthermore, define $\tilde{G}_n^{(Mb)_1}$ to be the block matrix obtained from the direct sum of the Gram determinant of type B and $\tilde{G}_n^{Mb_{n,1}}$,

$$\tilde{G}_n^{(Mb)_1} = G_n^B \oplus \tilde{G}_n^{Mb_{n,1}}.$$

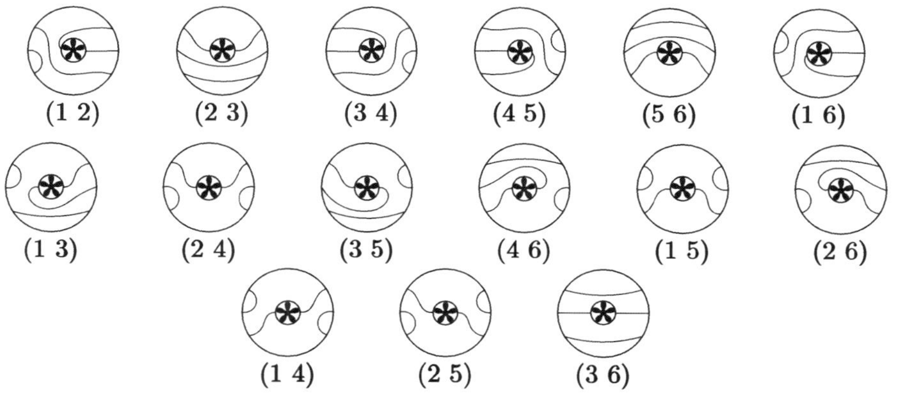

FIGURE 12. The set $Mb_{3,1}$.

EXAMPLE 6.5. The Gram matrix $\tilde{G}_3^{Mb_{3,1}}$, shown in Table 5, is constructed from the elements illustrated in Figure 12.
and its determinant is

$$\det(\tilde{G}_3^{Mb_{3,1}}) = d^{12}(d-1)^2(d+1)^2(d+2)^7(d-2)^7$$
$$= (T_4(d)-2)^{\binom{6}{1}}(T_6(d)-2)^{\binom{6}{0}}.$$

TABLE 5. The Gram matrix $\tilde{G}_3^{Mb_{3,1}}$.

$\langle\, ,\, \rangle$	(12)	(23)	(34)	(45)	(56)	(16)	(13)	(24)	(35)	(46)	(15)	(26)	(14)	(25)	(36)
(12)	d^2	1	1	0	1	1	d	d	0	0	0	d	1	0	1
(23)	1	d^2	1	1	0	1	d	d	d	0	0	0	1	1	0
(34)	1	1	d^2	1	1	0	0	d	d	d	0	0	0	1	1
(45)	0	1	1	d^2	1	1	0	0	d	d	d	0	1	0	1
(56)	1	0	1	1	d^2	1	0	0	0	d	d	d	1	1	0
(16)	1	1	0	1	1	d^2	d	0	0	0	d	d	0	1	1
(13)	d	d	0	0	0	d	d^2	1	0	1	0	1	d	0	d
(24)	d	d	d	0	0	0	1	d^2	1	0	1	0	d	d	0
(35)	0	d	d	d	0	0	0	1	d^2	1	0	1	0	d	d
(46)	0	0	d	d	d	0	1	0	1	d^2	1	0	d	0	d
(15)	0	0	0	d	d	d	0	1	0	1	d^2	1	d	d	0
(26)	d	0	0	0	d	d	1	0	1	0	1	d^2	0	d	d
(14)	1	1	0	1	1	0	d	d	0	d	d	0	d^2	1	1
(25)	0	1	1	0	1	1	0	d	d	0	d	d	1	d^2	1
(36)	1	0	1	1	0	1	d	0	d	d	0	d	1	1	d^2

LEMMA 6.6. *Let $n \geq 2$, then*

$$(d+z)^{\binom{2n}{n-1}} \det(G_n^{(Mb)_1}) = (w(d+z)-2xy)^{\binom{2n}{n-1}} \det(\tilde{G}_n^{(Mb)_1}).$$

PROOF. We will adapt the proof of Proposition 5.4 given in [1] to show that a specific column operation on $\binom{2n}{n-1}$ columns will convert $G_n^{(Mb)_1}$ into a lower triangular matrix at the expense of picking up $\binom{2n}{n-1}$ number of factors of $(d+z)$ in the determinant.

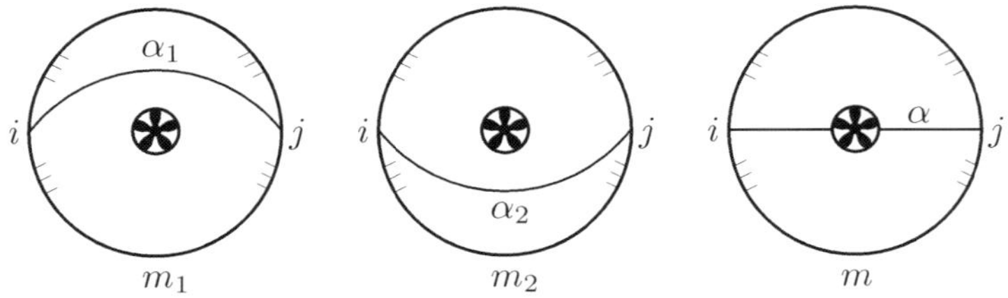

FIGURE 13. The crossingless connection $m \in Mb_{n,1}$ with the curve α connected to the ith and j^{th} marked point and passing through the crosscap. The crossingless connections $m_1, m_2 \in Mb_{n,0}$ are obtained from m by modifying α so that it does not pass through the crosscap.

Let $n \geq 2$. For a crossingless connection m with a curve α that intersects the crosscap, there are only two distinct elements that only differ by α. They are obtained from m by altering α in such a way that it does not intersect the crosscap; as illustrated in Figure 13. By focusing on the matrix obtained from these elements (see Equation (6.1)), we find that the entries all have a common monomial factor, say u, from the fact that the elements only differ by one curve.

(6.1)

$\langle\ ,\ \rangle_{Mb}$	m_1	m_2	m
m_1	ud	uz	uy
m_2	uz	ud	uy
m	ux	ux	uw

The row operation $m \to (d+z)m - y(m_1+m_2)$ will alter the determinant by one factor of $(d+z)$ and yield a column that is divisible by $(w(d+z)-2xy)$.

(6.2)

$\langle\ ,\ \rangle_{Mb}$	m_1	m_2	$(d+z)m - y(m_1+m_2)$
m_1	ud	uz	0
m_2	uz	ud	0
m	ux	ux	$u(w(d+z)-2xy)$

Let $c \in Mb_{n,0}$, then the sub-row $(\langle c, m_1\rangle_{Mb}, \langle c, m_2\rangle_{Mb}, \langle c, m\rangle_{Mb})$ is equal to $u'(d,z,y)$ or $u'(z,d,y)$ where u' is a monomial. This is because m_1, m_2, and m only differ by the arc α, c has no arcs intersecting the crosscap while m_1 and m_2 also have no arcs intersecting the crosscap (yielding either d, z or z, d), and m has only one arc intersecting the crosscap (yielding the variable y). Therefore, after applying the specified column operation on m we obtain $\langle c, (d+z)m - y(m_1+m_2)\rangle_{Mb} = 0$ for all $c \in Mb_{n,0}$.

Let $c' \in Mb_{n,1}$, then the sub-row $(\langle c', m_1\rangle_{Mb}, \langle c', m_2\rangle_{Mb}, \langle c', m\rangle_{Mb})$ is equal to $u'(x,x,w)$, $u'(xd,xz,xy)$ or $u'(xz,xz,xy)$ for some monomial u'. Since c has an arc intersecting the crosscap but m_1 and m_2 do not, then the bilinear form must pick up a curve intersecting the crosscap once; yielding the variable x. Since m_1 and m_2 only differ by one curve then there are two cases to consider.

(1) If the x variable was obtained from the arc α_1 (or α_2, respectively) then $\langle c', m_1\rangle_{Mb} = \langle c', m_2\rangle_{Mb} = u'x$ because there is only one homotopically distinct simple closed curve that intersects the inner crosscap. Furthermore, we also have $\langle c', m_1\rangle_{Mb} = u'w$ where x is changed to w from changing α_1 (or α_2) to α.
(2) If the x variable was not obtained from the arc α_1 (or α_2, respectively) then the pair $(\langle c', m_1\rangle_{Mb}, \langle c', m_2\rangle_{Mb})$ is equal to $u'(dx, zx)$ or $u'(zx, dx)$. Furthermore, $\langle c', m\rangle_{Mb} = u'xy$ because d (or z, respectively) was changed to y by changing α_1 (or α_2) to α and so this change produces a simple closed curve that intersects the outer crosscap.

Therefore, after applying the specified column operation on m if $\langle c', m\rangle_{Mb} = u'xy$ then $\langle c', (d+z)m - y(m_1+m_2)\rangle_{Mb} = 0$ and if $\langle c', m\rangle_{Mb} = u_{c'}w$ then $\langle c', (d+z)m - y(m_1+m_2)\rangle_{Mb} = u_{c'}(w(d+z) - 2xy)$ for all $c' \in Mb_{n,1}$ and some monomial $u_{c'}$ depending on c'.

In summary, after the column operation the determinant picks up a factor of $(d+z)$, the new column C is divisible by $(w(d+z) - 2xy)$, and the entries of the column obtained from elements of $Mb_{n,0}$ are all zero. Furthermore, the column obtained from factoring $(w(d+z) - 2xy)$ out of C is equal to the original column after letting $w = 1$ and $y = 0$.

In this process we focused on a triple obtained from a crossingless connection that intersects the crosscap once. We can obtain similar triples for the $\binom{2n}{n-1}$ distinct crossingless connections in $Mb_{n,1}$. Let M be the matrix obtained by performing a column operation using the triples on the $\binom{2n}{n-1}$ columns formed from crossingless connections that intersect the crosscap once. Then M is a block lower triangular matrix of the form

$$M = \begin{pmatrix} G_n^B & 0 \\ E & D \end{pmatrix},$$

where all $\binom{2n}{n-1}$ columns of D are divisible by $(w(d+z) - 2xy)$ and the matrix obtained from D by factoring out $(w(d+z) - 2xy)$ from every column is equal to $\tilde{G}_n^{Mb_{n,1}}$. Therefore,

$$\begin{aligned}\det(M) &= (w(d+z) - 2xy)^{\binom{2n}{n-1}} \det(G_n^B) \det(\tilde{G}_n^{Mb_{n,1}}) \\ &= (w(d+z) - 2xy)^{\binom{2n}{n-1}} \det(\tilde{G}_n^{(Mb)_1}).\end{aligned}$$

We also have $\det(M) = (d+z)^{\binom{2n}{n-1}} \det(G_n^{(Mb)_1})$ which implies that

$$(d+z)^{\binom{2n}{n-1}} \det(G_n^{(Mb)_1}) = (w(d+z) - 2xy)^{\binom{2n}{n-1}} \det(\tilde{G}_n^{(Mb)_1}). \qquad \square$$

THEOREM 6.7. *For $n \geq 2$,*

$$\det(G_n^{(Mb)_1}) = [((w(d+z) - 2xy))(d-z)]^{\binom{2n}{n-1}} \prod_{k=2}^{n} (T_k(d)^2 - z^2)^{\binom{2n}{n-k}} \det(\tilde{G}_n^{Mb_{n,1}}).$$

PROOF. Recall that, $\tilde{G}_n^{(Mb)_1} = G_n^B \oplus \tilde{G}_n^{Mb_{n,1}}$. By construction and Theorem 4.4,

$$\begin{aligned} \det(\tilde{G}_n^{(Mb)_1}) &= \det(G_n^B)\det(\tilde{G}_n^{Mb_{n,1}}) \\ &= (d+z)^{\binom{2n}{n-1}}(d-z)^{\binom{2n}{n-1}} \prod_{k=2}^{n}(T_k(d)^2 - z^2)^{\binom{2n}{n-k}} \\ &\quad \times \det(\tilde{G}_n^{Mb_{n,1}}). \end{aligned} \tag{6.3}$$

After multiplying Equation (6.3) by $(w(d+z) - 2xy)^{\binom{2n}{n-1}}$ we have

$$\begin{aligned} (w(d+z) - 2xy)^{\binom{2n}{n-1}} \det(\tilde{G}_n^{(Mb)_1}) &= (w(d+z) - 2xy)^{\binom{2n}{n-1}} \\ &\quad \times (d+z)^{\binom{2n}{n-1}}(d-z)^{\binom{2n}{n-1}} \\ &\quad \times \prod_{k=2}^{n}(T_k(d)^2 - z^2)^{\binom{2n}{n-k}} \\ &\quad \times \det(\tilde{G}_n^{Mb_{n,1}}). \end{aligned}$$

Now, by applying Lemma 6.6 to the left hand side of the equation we have

$$\begin{aligned} (d+z)^{\binom{2n}{n-1}} \det(G_n^{(Mb)_1}) &= (w(d+z) - 2xy)^{\binom{2n}{n-1}} \\ &\quad \times (d+z)^{\binom{2n}{n-1}}(d-z)^{\binom{2n}{n-1}} \\ &\quad \times \prod_{k=2}^{n}(T_k(d)^2 - z^2)^{\binom{2n}{n-k}} \det(\tilde{G}_n^{Mb_{n,1}}). \end{aligned}$$

Therefore,

$$\begin{aligned} \det(G_n^{(Mb)_1}) &= (w(d+z) - 2xy)^{\binom{2n}{n-1}}(d-z)^{\binom{2n}{n-1}} \\ &\quad \times \prod_{k=2}^{n}(T_k(d)^2 - z^2)^{\binom{2n}{n-k}} \det(\tilde{G}_n^{Mb_{n,1}}). \\ &= [((w(d+z) - 2xy))(d-z)]^{\binom{2n}{n-1}} \\ &\quad \times \prod_{k=2}^{n}(T_k(d)^2 - z^2)^{\binom{2n}{n-k}} \det(\tilde{G}_n^{Mb_{n,1}}). \end{aligned}$$ □

The following conjecture, originally stated in [**8**], provides a conjectural closed formula for the Gram determinant of type $(Mb)_1$.

CONJECTURE 6.8 ([**8**]). *Let $R = \mathbb{Z}[A^{\pm 1}, w, x, y, z]$. Then, the Gram determinant of type $(Mb)_1$ for $n \geq 1$, is:*

$$D_n^{(Mb)_1} = [(d-z)((d+z)w - 2xy)]^{\binom{2n}{n-1}} \prod_{k=2}^{n}(T_k(d)^2 - z^2)^{\binom{2n}{n-k}} \prod_{k=2}^{n}(T_{2k}(d) - 2)^{\binom{2n}{n-k}},$$

where $T_k(d)$ is the kth Chebyshev polynomial of the first kind and $d = -A^2 - A^{-2}$.

By applying Theorem 6.7, the following conjecture implies Conjecture 6.8. Furthermore, notice that the product in Conjecture 6.9 is conjectured to be a factor of D_n^{Mb}.

CONJECTURE 6.9. *For* $n \geq 2$,

$$\det(\tilde{G}_n^{Mb_{n,1}}) = \prod_{k=2}^{n} (T_{2k}(d) - 2)^{\binom{2n}{n-k}}.$$

Conjecture 6.9 has been verified for $n = 2, 3$ and 4.

Acknowledgments

The first author acknowledges the support by the Australian Research Council grant DP210103136. The second author acknowledges the support of the National Science Foundation through Grant DMS-2212736.

References

[1] Rhea Palak Bakshi, Dionne Ibarra, Sujoy Mukherjee, and Józef H. Przytycki, *A generalization of the Gram determinant of type A*, Topology Appl. **295** (2021), Paper No. 107663, 15, DOI 10.1016/j.topol.2021.107663. MR4241642

[2] R. P. Bakshi, D. Ibarra, S. Mukherjee, and J. H. Przytycki. *A note on the Gram determinant of type Mb*, Contemporary Mathematics, volume 824, pp. 83–89, 2025.

[3] Rodney J. Baxter, *Exactly solved models in statistical mechanics*, Academic Press, Inc. [Harcourt Brace Jovanovich, Publishers], London, 1989. Reprint of the 1982 original. MR998375

[4] Xuanting Cai, *A Gram determinant of Lickorish's bilinear form*, Math. Proc. Cambridge Philos. Soc. **151** (2011), no. 1, 83–94, DOI 10.1017/S0305004111000089. MR2801315

[5] Q. Chen, 2009. Personal communication (email) with J. H. Przytycki.

[6] Qi Chen and Jozef H. Przytycki, *The Gram determinant of the type B Temperley-Lieb algebra*, Adv. in Appl. Math. **43** (2009), no. 2, 156–161, DOI 10.1016/j.aam.2008.12.002. MR2531918

[7] P. Di Francesco, *The meander determinant and its generalizations*, Calogero-Moser-Sutherland models (Montréal, QC, 1997), CRM Ser. Math. Phys., Springer, New York, 2000, pp. 127–144. MR1843567

[8] Dionne Ibarra and Gabriel Montoya-Vega, *A new Gram determinant from the Möbius band*, J. Knot Theory Ramifications **33** (2024), no. 12, Paper No. 2450037, 25, DOI 10.1142/S0218216524500378. MR4822264

[9] V. F. R. Jones, *Index for subfactors*, Invent. Math. **72** (1983), no. 1, 1–25, DOI 10.1007/BF01389127. MR696688

[10] Louis H. Kauffman, *An invariant of regular isotopy*, Trans. Amer. Math. Soc. **318** (1990), no. 2, 417–471, DOI 10.2307/2001315. MR958895

[11] Paul Martin and Hubert Saleur, *On an algebraic approach to higher-dimensional statistical mechanics*, Comm. Math. Phys. **158** (1993), no. 1, 155–190. MR1243720

[12] Paul Martin and Hubert Saleur, *The blob algebra and the periodic Temperley-Lieb algebra*, Lett. Math. Phys. **30** (1994), no. 3, 189–206, DOI 10.1007/BF00805852. MR1267001

[13] Józef H. Przytycki, *Fundamentals of Kauffman bracket skein modules*, Kobe J. Math. **16** (1999), no. 1, 45–66. MR1723531

[14] Józef H. Przytycki, Rhea Palak Bakshi, Dionne Ibarra, Gabriel Montoya-Vega, and Deborah Weeks, *Lectures in knot theory—an exploration of contemporary topics*, Universitext, Springer, Cham, [2024] ©2024, DOI 10.1007/978-3-031-40044-5. MR4738299

[15] H. N. V. Temperley and E. H. Lieb, *Relations between the "percolation" and "colouring" problem and other graph-theoretical problems associated with regular planar lattices: some exact results for the "percolation" problem*, Proc. Roy. Soc. London Ser. A **322** (1971), no. 1549, 251–280, DOI 10.1098/rspa.1971.0067. MR498284

[16] B. W. Westbury, *The representation theory of the Temperley-Lieb algebras*, Math. Z. **219** (1995), no. 4, 539–565, DOI 10.1007/BF02572380. MR1343661

School of Mathematics, Monash University, Australia
Email address: dionne.ibarra@monash.edu

Department of Mathematics, University of Puerto Rico at Río Piedras, San Juan, Puerto Rico
Email address: gabrielmontoyavega@gmail.com

Selected Published Titles in This Series

827 **Carmen Caprau, J. Scott Carter, Neslihan Gügümcü, and Sam Nelson, Editors,** Algebraic Structures in Knot Theory, 2025

825 **Saurabh Verma, María A. Navascués, and Amit Priyadarshi, Editors,** Advances in Dimension Theory, Fractal Functions and Measures, 2025

824 **Hanna E. Makaruk and Robert M. Owczarek, Editors,** Inverse Problems: In Memory of Professor Zbigniew Oziewicz, 2025

823 **Mikhail Bershtein, Anton Dzhamay, and Andrei Okounkov, Editors,** The Versatility of Integrability, 2025

822 **Ahmad Barhoumi, Roozbeh Gharakhloo, and Andrei Martinez-Finkelshtein, Editors,** Recent Developments in Orthogonal Polynomials, 2025

821 **Bogdan D. Suceavă, Adara M. Blaga, Marian Ioan Munteanu, Cezar Oniciuc, Shoo Seto, Gabriel-Eduard Vîlcu, and Lihan Wang, Editors,** Recent Advances in Differential Geometry and Related Areas, 2025

820 **Sudeb Mitra, Rakesh Kumar Parmar, Saminathan Ponnusamy, and Toshiyuki Sugawa, Editors,** Geometric Function Theory and Related Topics, 2025

819 **Howard S. Cohl, Roberto S. Costas-Santos, and Robert S. Maier, Editors,** Applications and q-Extensions of Hypergeometric Functions, 2025

818 **Howard S. Cohl, Roberto S. Costas-Santos, and Robert S. Maier, Editors,** Classical Hypergeometric Functions and Generalizations, 2025

817 **Mikhail Khovanov, Joshua Sussan, and Anton Zeitlin, Editors,** From Representation Theory to Mathematical Physics and Back, 2025

816 **Luca F. Di Cerbo and Laurenţiu G. Maxim, Editors,** Geometry and Topology of Aspherical Manifolds, 2025

815 **Daniel Orr and Joshua Jeishing Wen, Editors,** Macdonald Theory and Beyond, 2025

814 **Susan Montgomery and Siu-Hung Ng, Editors,** Quantum Groups, Hopf Algebras, and Applications, 2025

813 **Thomas Creutzig, Julia Plavnik, and David Ridout, Editors,** Quantum Symmetries, 2025

812 **Ross Geoghegan, Craig R. Guilbault, and Kim Ruane, Editors,** Topology at Infinity of Discrete Groups, 2025

811 **Krishnendu Gongopadhyay and Shashikant A. Katre, Editors,** Geometry, Groups and Mathematical Philosophy, 2025

810 **Robert J. Berman and Yanir A. Rubinstein, Editors,** Convex and Complex: Perspectives on Positivity in Geometry, 2025

809 **Fabrice Baudoin and Luca Rizzi, Editors,** New Trends in Sub-Riemannian Geometry, 2025

808 **Loring W. Tu, Editor,** Group Actions and Equivariant Cohomology, 2024

807 **Galina Filipuk, Editor,** Recent Progress in Special Functions, 2024

806 **Maximilian Engel, Hildeberto Jardón-Kojakhmetov, and Cinzia Soresina, Editors,** Topics in Multiple Time Scale Dynamics, 2024

805 **Anthony Iarrobino, Pedro Macias Marques, Maria Evelina Rossi, and Jean Vallès, Editors,** Deformation of Artinian Algebras and Jordan Type, 2024

804 **Mahir Bilen Can and Jörg Feldvoss, Editors,** A Glimpse into Geometric Representation Theory, 2024

803 **Peter Gothen, Margarida Melo, and Montserrat Teixidor i Bigas, Editors,** Moduli Spaces and Vector Bundles—New Trends, 2024

802 **Ralph M. Kaufmann, Martin Markl, and Alexander A. Voronov, Editors,** Higher Structures in Topology, Geometry, and Physics, 2024

For a complete list of titles in this series, visit the
AMS Bookstore at **www.ams.org/bookstore/conmseries/**.

FSC
www.fsc.org
MIX
Papier | Fördert gute Waldnutzung
FSC® C083411

Kolibri 360 GmbH
Ferdinand-Jühlke-Straße 7
99095 Erfurt, Deutschland
produktsicherheit@kolibri360.de

Algebraic Structures in Knot Theory